American Geophysical Union

ANTARCTIC
RESEARCH
SERIES

Antarctic Research Series Volumes

Volume 79 | ANTARCTIC
RESEARCH
SERIES

Antarctic Peninsula Climate Variability

Historical and Paleoenvironmental Perspectives

Eugene Domack, Amy Leventer, Adam Burnett, Robert Bindschadler,
Peter Convey, and Matthew Kirby
Editors

Ⓢ American Geophysical Union
Washington, D.C.
2003

ANTARCTIC PENINSULA CLIMATE VARIABILITY: HISTORICAL AND PALEOENVIRONMENTAL PERSPECTIVES
Eugene Domack, Amy Leventer, Adam Burnett, Robert Bindschadler, Peter Convey, and Matthew Kirby, Editors

Published under the aegis of the Board of Associate Directors, Antarctic Research Series

Library of Congress Cataloging-in-Publication Data

Antarctic Peninsula climate variability : historical and paleoenvironmental perspectives/
    Eugene Domack . . . [et al.], editors.
        p. cm. — (Antarctic research series ; v. 79)
    Includes bibliographical references.
    ISBN 0-87590-973-6
        1. Antarctic Peninsula (Antarctica)—Climate. 2. Climate changes—Antarctica—Antarctic
Peninsula. 3. Glaciers—Antarctica—Antarctic Peninsula. I. Domack, Eugene, 1956- II. Series.

QC994.9.A59 2003
551.69989—dc22

                        2003060053

ISBN 0-87590-973-6
ISSN 0066-4634

Front cover: View from the bridge wing of the *RVIB Nathaniel B. Palmer* looking southwest across Borkowski Bay towards the Nordensköld Coast and the Drygalski Glacier. Borkowski Bay is the area formerly covered by the Larsen A Ice Shelf. Imaged May 19, 2000. ©2000 Dave Tewksbury

Back cover: False color composite of LANDSAT 7 image acquired February 2000 using bands 2,3 & 5.

Published by
American Geophysical Union
2000 Florida Avenue, N.W.
Washington, D.C. 20009

Printed in the United States of America.

CONTENTS

PREFACE

The Antarctic Peninsula region represents our best natural laboratory to investigate how earth's major climate systems interact and how such systems respond to rapid regional warming. The scale of environmental changes now taking place across the region is large and their pace rapid but the subsystems involved are still small enough to observe and accurately document cause and affect mechanisms. For example, clarification of ice shelf stability via the Larsen Ice Shelf is vital to understanding the entire Antarctic Ice Sheet, its climate evolution, and its response to and control of sea level. By encompassing the broadest range of interdisciplinary studies, this volume provides the global change research and educational communities a framework in which to advance our knowledge of the causes behind regional warming, the dramatic glacial and ecological responses, and the potential uniqueness of the event within the region's paleoclimate record. The volume also serves as a vital resource for public policy and governmental funding agencies as well as a means to educate the large number of ecotourists that visit the region each austral summer.

The content is international in scope with chapters contributed by authors from eleven countries. The organization of the volume follows the order of temporal databases from historical meteorological observations and modeling studies, through expanding time scales of paleoenvironmental archives of ice cores, lake, and marine sediment sequences. Also provided is documentation of the ecological effects of rapid regional warming and the unique capabilities of the Long-Term Ecological Research program, now into its second decade of observations and summary within the western portion of the Antarctic Peninsula. From this temporal perspective we integrate oceanographic and glaciological studies within the region and authors make recommendations for future collaborative research that is urgently needed.

Many papers presented in this volume derive from the work discussed at an international workshop held at Hamilton College in April 2002 (http://academics.hamilton.edu/workshops/antarctica/), but the design and organization of the volume are by necessity of space somewhat more limited than the workshop agenda. As the Antarctic Peninsula working group informally plans to reconvene every other year, this volume will also serve as a guidepost toward these subsequent symposia.

The editors wish to thank the numerous reviewers who handled manuscripts in a timely and constructive manner. We also wish to acknowledge the generous support of the Environmental Studies Program at Hamilton College, the National Science Foundation's Office of Polar Programs (including programs in ocean/climate, geology/geophysics, biology, and glaciology), and Colgate University for support of this endeavor. The volume would not have been possible without the generous and timely contributions of all authors and for the high level of scholarship exhibited within each chapter.

Eugene W. Domack

List of Reviewers

John Anderson
Tony Arnold
Glenn Berger
Dana Bergstrom
Bob Bindschadler
Stefanie Brachfeld
Raymond Bradley
Dave Bromwich
Bill Budd
Adam Burnett
Angelo Camerlenghi
Andrew Carleton
Andrew Clarke
William Connolley
Peter Convey
Sarah Das
Tad Day
David DeMaster
Marianne Douglas
Steve Emslie
Jane Ferrigno
Robert Gilbert
Brenda Hall
Christian Hjort
Terence Hughes
Stan Jacobs
Anne Jennings
Philip Jones

Sharon Kanfoush
Matthew Kirby
Carina Lange
Amy Leventer
Andreas MacKensen
Diane McKnight
Andrew McMinn
Julie Palais
David Peel
Paul Pettre
Jennifer Pike
Leonid Polyak
Marie Poole
Carol Pudsey
Marilyn Raphael
Tony Rathburn
Eric Rignot
Ted Scambos
Reed Scherer
Peter Sedwick
Geoffrey Seltzer
Amelia Shevenell
Stephanie Shipp
Ellen Mosley-Thompson
Ross Virginia
Andrew Watkins
Richard Williams
Jan-Gunnar Winther

ENVIRONMENTAL SETTING OF THE ANTARCTIC PENINSULA

Eugene W. Domack

Department of Geology, Hamilton College Clinton New York

Adam Burnett and Amy Leventer

Department of Geography and Geology, Colgate University, Hamilton New York

" One of the warning signs that a dangerous warming trend is under way in Antarctica will be the breakup of ice shelves on both coasts of the Antarctic Peninsula, starting with the northernmost and extending gradually southward." J. H. Mercer, 1978.

Perhaps nowhere on the surface of the earth have environmental changes taken place with such rapidity and captured the interest of such a diverse community than those observed across the Antarctic Peninsula in the last 10 years. Wholesale decay of ice shelves, long considered to be the harbinger of climate warming, has spurred interest in our attempts to understand the interaction of earth systems on historical to millennial time scales. Because such changes in the cryosphere also impact regional ecosystems the biological community has also become deeply involved in the climate debate. While environmental changes now taking place across the Antarctic Peninsula are historically well documented by a diverse set of meteorological and remote sensing data, considerably less is known concerning the behavior of the atmosphere-ocean-cryosphere system during the past 10,000 years (the interglacial Holocene Epoch). The first purpose of this volume is to integrate our present understanding of current meteorological trends and compare them to records of past environmental change. The hope is that such a comparison will stimulate future studies that are directed at deciphering the key forcing mechanisms operating across the Antarctic Peninsula so that natural signals can be understood within the context of potential anthropogenic changes. The second key objective of the volume is to help further our understanding of how the paleoenvironmental archives are interpreted and what archives need to be acquired. Clearly progress toward both goals must be achieved together. We also hope that the volume will serve as a useful reference for students and senior investigators.

GEOLOGIC AND GEOMORPHOLOGIC BACKGROUND

The Antarctic Peninsula (AP) region encompasses a contrasting physiologic, geologic, and glaciologic terrain that is similar to the southern cordillera of the Andes. The AP consists of a narrow (less than 250 km), elevated landmass (up to 3500 m) that projects from the main continent some 1250-km to the north. This physiography extends Antarctica's glacial carapace into sub polar climates while subjecting it to contrasting oceanographic and meteorological conditions across an east to west gradient (Figures 1 and 2), [see also *BAS*, 2000]. Before the final break up of Gondwanaland, the southern Andean

10.1029/079ARS01

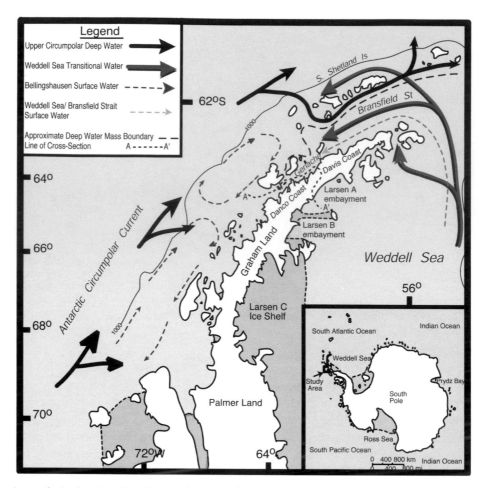

Fig. 1. The Antarctic Peninsula region showing location of major physiographic features and oceanographic circulation patterns [modified after *Shevenell and Kennett*, 2002]. Note frontal boundary between southern limb of Antarctic Circumpolar Current (ACC) and Weddell Sea Transitional Water. Location of cross-section A-A' (Figure 2) is also shown.

arc complex (as it existed along the paleo-Pacific margin) and pre-Jurassic basement rocks were most likely contiguous with the linear trend of the AP. The AP basement rocks are referred to as the Trinity Peninsula Group (and correlates) which consist of intermediate grade metamorphic rocks formed in an accretionary prism [*Barker et al.*, 1991]. The presence of older and higher-grade metamorphic basement rock is documented from only a few localities in the AP [*Barker et al.*, 1991].

Both the AP and the southern Andes are bordered to the east with a back-arc basin stratigraphy of thick Jurassic to Cretaceous marine shales and siltstones that lay behind the, then active, volcanic (subduction related) arc. Deposits continued to accumulate within the basin (The Larsen and/or James Ross Basin) through the Cenozoic. Along the western (Pacific) side of the

Peninsula are thick fore-arc strata that are partially tectonized and consist of volcanogenic and associated sedimentary rocks of Mesozoic to Cenozoic age (Figure 2). These pre-date and post-date final phases of ridge crest subduction along the Pacific margin from Paleogene time onward [*Barker et al.*, 1991; see also *BAS*, 1985].

The Drake Passage was created with the fragmentation and brittle response of the crust to compressive then extensional forces associated with reorganization of seafloor spreading in the Cenozoic. Associated with this event was the uplift and dissection of the Mesozoic arc terrain that has led to the exposure of igneous plutons and related metamorphic/volcanic rocks. These rocks comprise the spine of the Antarctic Peninsula now covered in a permanent ice cap (Figure 2). This uplift episode (variously placed in the Paleogene to Neogene)

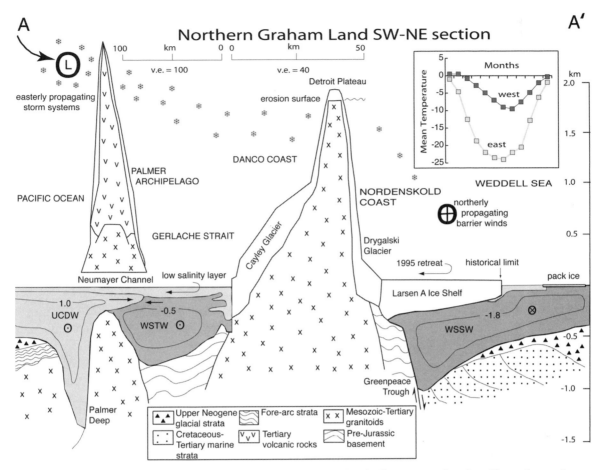

Fig. 2. Schematic cross-section (A-A') across northern Graham Land and adjacent coastal regions illustrating earth system interaction (see Figure 1 for location). Note change in horizontal scale at the western coast of Graham Land. Major water mass characteristics are identified by core temperatures (in-situ) and are identified as Upper Circumpolar Deep Water (UCDW), Weddel Sea Transitional Water (WSTW), and Weddell Sea Shelf Water (WSSW). Geologic data modified after *Elliot* (1997), meteorologic data after *King et al.*, [this volume], and oceanographic settings from personal observation. Note contrasting meterologic and oceanographic conditions on either side of Peninsula as well as recent changes in glacier termini [*Rott et al.*, 1996]. Inset illustrates contrasting mean monthly temperatures between western and eastern side of Peninsula [after *King et al.*, this volume].

likely caused major changes in the accumulation of ice masses across the Peninsula and consequent erosion and depositional patterns across the continental shelf [*Elliot*, 1997]. The event that led to the creation of the Drake Passage also led to the eastward propagation of shearing of both the Southern Andes (left lateral displacement) and the Antarctic Peninsula. Subsequent events have included the creation of volcanic centers associated with the Bransfield Basin (a back arc spreading axis), the James Ross Island volcanic complex, and minor but more incompletely understood volcanism [*Smellie et al.*, 1988; BAS 1985].

The long-term paleoclimate evolution of the Peninsula is equally intriguing as it is marked by some of the old-est glacigenic deposits yet recorded from Antarctica. Eocene glacigenic strata in the South Shetland Islands suggest glaciation as far back as 30 million years ago [*Birkenmajer*, 1991; *Dingle and Lavelle*, 1998]. Hence, initial glaciation took place at the early stages in opening of the Drake Passage and well before the separation of the South Shetland islands from the mainland AP, about 2 million years *ago* [*Barker and Austin*, 1998]. Subsequent climatic episodes are poorly documented except from seismic stratigraphy and drilling across the continental margin that together document major Miocene and Pliocene-Pleistocene expansions of large-scale (ice sheet) glaciation across the broad continental shelf [*Bart and Anderson*, 1995; *Barker et al.*, 1999;

Barker and Camerlenghi, 2002]. Pleistocene glacial-integlacial episodes, so characteristic of the Northern Hemisphere, are less well documented due to the erosional effects of each subsequent glacial event. But the cumulative effect of these glacial expansions has been the dissection of fjords, deep inner shelf basins or troughs, and deposition of a seaward thickening wedge of Neogene-Quaternary diamict across the continental shelf (Figure 2) [*Rebesco et al.*, 1998; *Anderson*, 1999; *Barker et al.*, 1999]. However, glacial/ interglacial cycles are preserved within thick sediment drift deposits of the continental rise which, while not illustrated in Figure 2, are providing significantly new insights into the depositional record of past AP ice sheets [*Lucchi et al.*, 2002]. The present geography is the result of recession of an expanded Peninsula Ice Sheet, post glacial sea level rise, and inferred (but as yet undocumented) isostatic adjustment [*Clapperton and Sugden*, 1982; *Payne et al.*, 1989; *Pudsey et al.*, 1994; *Hjort et al.*, 1997, this volume; *Hall*, this volume].

Swath bathymetry of the surrounding continental shelf is adding a new dimension to our ability to reconstruct patterns of glacial drainage. Soon we will be able to delineate the distribution of all major ice domes during the last glacial maximum [*Wilmott et al.*, this volume; *Gilbert et al.*, this volume; *Canals et al.*, 2002, 2000; *Lowe and Anderson*, 2002, *O'Cofaigh et al.*, 2002]. This will be a vital first step in determining the isostatic load of past ice sheets and the potential differences in ice shelf thickness since deglaciation.

GLACIOLOGIC SETTING

Trans Peninsula contrasts in precipitation and water vapor transport have led to resulting contrasts in glacial character between the eastern and western sides of the Peninsula. These contrasts persist despite the warming trends and decay of ice shelf systems [*Morris and Vaughan*, this volume]. On the western side of the Peninsula large amounts of snow are brought in from the prevailing westerlies and associated cyclonic systems. This leads to high accumulation and lower equilibrium (snow lines) despite markedly warmer summer temperatures. On the Weddell Sea side snowfall is less and (while mean annual temperatures are less) the equilibrium (snow) lines elevations (ELA) are significantly higher. This contrast in accumulation and ELA clearly produces glaciers on the eastern side of the Peninsula that respond (thermally and via melt water) more rapidly to short term climatic perturbations [*Skvarca and DeAngelis*, this vol-

ume; *DeAngelis and Skvarca*, 2003]. While there are contrasts in sediment erodability (such as crystalline rocks on the Peninsula versus volcanic to sedimentary rocks of the James Ross and Seymour Island complex) a greater amount of sediment is clearly produced by tidewater glaciers in the NW Weddell Sea than along the Bellingshausen Sea. The Larsen A and B ice shelves are just one portion of this "low accumulation climate sensitive glacial system". These ice shelves have responded in a most dramatic way because they have low elevation surfaces that are susceptible to short-term increases in surface melt water production [*Scambos et al.*, this volume]. If there were more significant snow-pack the melt water generated during the summer would refreeze, instead of producing an impermeable superimposed ice zone. The outlet and tidewater glaciers of the western side of the Peninsula, especially those north of Marguerite Bay, have very thick snow packs and snow lines that are some tens of meters above sea level. An exception to this includes the glaciers of the South Shetland Islands, which are exposed to significantly more degree days and rainfall than their mainland counterparts [*Domack and Ishman*, 1993].

THE MARINE PERSPECTIVE

While the meteorological contrasts across the western and eastern side of the Peninsula have been emphasized for some time [*Shwerdtfeger*, 1984] a similar, if even more profound, contrast can be seen in the oceanographic setting (Figures 1 and 2). Over three decades of oceanographic observation have demonstrated that the western side of the Antarctic Peninsula is bathed in mid to deep waters of comparative warmth (i.e. > 1.5°C)[*Smith et al.*, this volume]. This warm, circum polar deep water (UCDW) is derived from off-slope upwelling in association with the impingement of the Antarctic Circumpolar Current (ACC). The southern most edge of the ACC abuts the AP shelf as it is driven by westerly winds and is funneled through the Drake Passage and eventually into the Atlantic sector of the Southern Ocean (Figure 3). Against the southern reach of the ACC a colder (<−1.0 °C) and more saline water mass (Weddell Sea Transitional Water, WSTW) encroaches as an East Wind drift out of the northwestern Weddell Sea. The frontal boundary of these two water masses ranges from the southern Bransfield Strait into the southern Gerlache Strait (Figure 2) where there is a noticeable seasonal and inter-annual variability in its position (*Hofmann and Klinck*, 1998). The eastern side of the

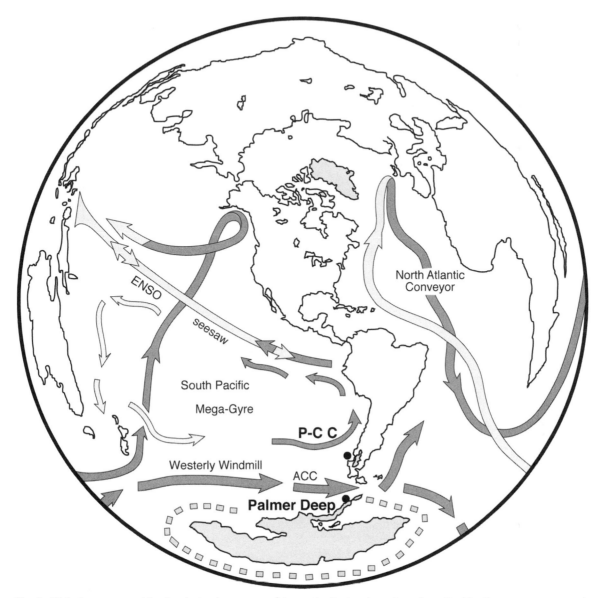

Fig. 3. Global oceanographic circulation in context of Antarctic Peninsula and southern Pacific Ocean processes such as El Niño-Southern Oscillation (ENSO), South Pacific Gyre, the Antarctic Circumpolar Current (ACC), and deep water flow contributed by North Atlantic processes. Location of Peru-Chile Current (P-C C) is also indicated. The northerly limit of sea ice is indicated by a bold dashed line (approximate).

Antarctic Peninsula carries with it the core characteristics of this Modified Weddell Sea water as a western boundary current that flows northward as part of the Weddell Gyre. We informally refer to this water mass as Weddell Sea Shelf Water (WSSW, Figure 2).

While the temperature and salinity characteristics of the shelf and coastal ocean are relatively well known other characteristics such as tides and storm surge are only now becoming clearer [*Padman et al.* 2002; Plate

1]. The spectrum of tidal energy is markedly different on either side of the Peninsula, as suggested by modeling and some actual station records (Plate 1). However we know almost nothing about the role of tidal currents in more general circulation especially in coastal and inner shelf settings where the marine sediment archives are preserved [*Harris et al.*, 1999]. The extent of undermelting of ice shelves is greatly dependent upon the ambient water temperature and the current regime that crosses the

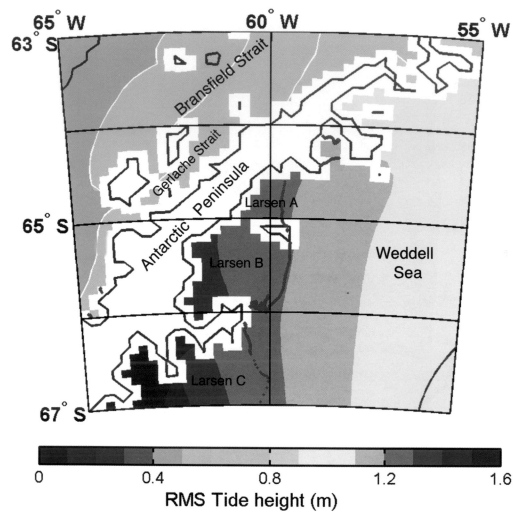

Plate 1: Predicted tidal amplitudes (in meters) surrounding Antarctic Peninsula region as determined by modeling [*Padman et al.*, 2002]. Note maximum tidal energies associated with Larsen Ice Shelf regime. Edge of continental shelf and ice shelves are indicated by bold red line while the coast is indicated by moderate red line.

face and underside of ice shelves [*Jenkins et al.*, 1997; *Rignot and Jacobs*, 2002]. Hence regional contrasts in shelf water temperatures and current energy are important components in ice shelf stability and need to be taken together with the meteorological trends if we are to decipher combined forcing factors for ice shelf collapse.

PALEOCLIMATE ARCHIVES

While ice cores represent some of the most faithful of paleoenvironmental archives we are lacking in long term (complete Holocene) and regionally representative records in the Antarctic Peninsula [*Mosley-Thompson and Thompson*, this volume]. While the variability of precipitation sources and process may preclude a single regionally representative ice core from ever being acquired we still need longer records as a starting place for more complete integration of marine and atmospheric databases. The chemistry of ice cores need further investigation as traditional signals of paleotemperature such as oxygen isotopes provide results that are not entirely consistent with marine sediment signals [*Smith et al.*, 1999; *Mosley-Thompson and Thompson*, this volume]. Terrestrial records such as lake sediments and glacial geomorphology have the same temporal restriction but the reason is very different. Such differences relate to the more recent deglaciation of presently exposed landscapes [*Emslie et al.*, this volume; *Hjort et al.*, this volume; *Hall*, this volume; *Quayle et al.*, this volume] rather than a lack of acquisition of such records.

Marine sediment sequences represent our best high-resolution paleoenvironmental archive to date but chronological issues continue to limit its widespread acceptance. Yet recent studies on inner shelf basins, such as the Palmer Deep (Figure 2), provide a promising record by which forcing mechanisms and other regional records can be compared (Figure 4). The Palmer Deep stratigraphy is remarkable for its relatively firm radiocarbon chronology and in its completeness (~13,000 years) of high-resolution (century scale) events [*Domack et al.*, 2001]. It thus provides an initial framework by which to compare forcing factors for natural variability. One first order factor is solar insolation as controlled by the precessional index over the last 14 ka [*Berger and Loutre*, 1991], (Figure 4). Here the inferred paleoclimatic record is at odds with solar insolation derived from orbital variation alone, despite earlier suggestions to the contrary [*Ingólfsson et al.*, 2001]. Specifically, the Palmer Deep record indicates a progressive cooling and reduced sediment accumulation rate over the last 5 millennia consistent with other regional marine records; [*Lamy et al.*, 2002; *Shevenell et al.*, 1996; *Mosley-Thompson and Thompson*, this volume]. While similar in frequency to the insolation curve (i.e. with an apparent 10 ka half cycle) the cooling trend of the last 5 millennia is clearly out of phase with the nearly 7% increase in solar insolation (at 65° S during December) during this same time (Figure 4). This contrast clearly highlights the requirement of another forcing mechanism for the millennia-scale change in climate that has occurred over the last five thousand years.

One alternative mechanism may be related to solar irradience variation [*Leventer et al.*, 1996] and/or ocean circulation via westerly winds and the ACC [*Domack and Mayewski*, 1999]. In support of the latter hypothesis is the observed coherence of the paleotemperature record of the Peru Chile Current, as observed offshore southern Chile, with the PD record (Figure 4). The remarkable agreement of the Palmer Deep and Peru-Chile Current (PCC) data suggest that the common control may be the southern limb of the South Pacific Gyre, the largest such oceanic gyre in the world's oceans (Figure 3). Both the PCC and ACC are at the eastern end of the gyre and would be expected to reflect similar changes in its strength of circulation and advection of warmer subtropical water. Concurrent southward migration of the Intertropical Convergence Zone within the Atlantic Ocean also produced late Holocene cooling in Central and South America [*Haug et al.*, 2001; *Baker et al.*, 2001]. This event was most likely driven by the tropical Pacific and its response to increased seasonality in the southern hemisphere; presumably under the influence of precessionally-driven solar insolation [*Haug et al.*, 2001].

In contrast, paleoceanographic records from the south Atlantic [*Hodell et al.*, 2003] demonstrate significant divergence in the timing of Late Holocene cooling and middle Holocene warmth. Thus, a decoupling of the teleconnections with the tropical Pacific may be inherent in the system beyond the SE Pacific region. Clearly this is an area that needs further investigation.

Of direct interest to our understanding of ice shelf fluctuation is the inference that the mid Holocene climatic optimum was associated with reduced ice shelf cover along the modern northern limit of ice shelves [*Hjort, et al.*, 2001; *Domack et al.*, 1995; 2001; *Pudsey and Evans*, 2001]. While the style and rate of ice shelf retreat and subsequent reformation in the Holocene are yet to be documented in detail, it is interesting to note that Late Holocene conditions of cooling are associated with an inferred Little Ice Age (LIA, commencing ~700 years

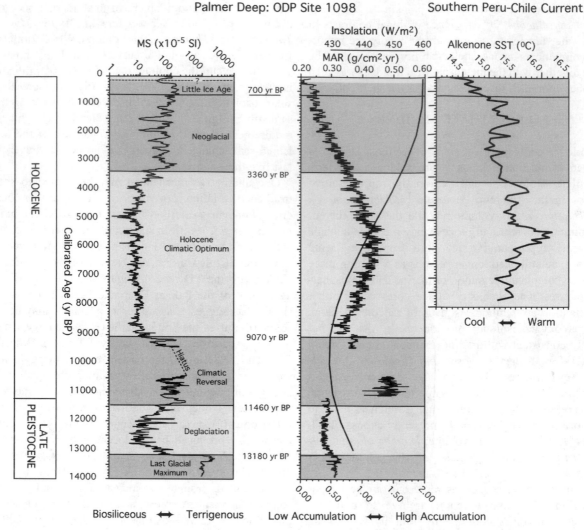

Fig. 4. Paleoclimate reconstruction from Palmer Deep [*Domack et al.*, 2001]. Paleoenvironmental proxies such as Magnetic Susceptibility (MS, in SI units) and Mass Accumulation Rate (MAR) are illustrated against regional solar insolation at 65° S , for December [*Berger and Loutre*, 1991], and paleotemperature reconstruction of southern end of Peru-Chile Current [*Lamy et al.*, 2002]. Note correspondence of both paleo records but opposition of orbitally controlled regional solar insolation. See Figure 3 for location of cores.

BP). This event is correlative to the expansion of ice shelves on both sides of the Peninsula [*Scambos et al.*, this volume; *Gilbert et al.*, this volume; *Brachfeld et al.*, 2003]. However just how far south the ice shelf minima can be recognized remains an important question to address via new field efforts. Preliminary studies suggest that the limit may lie near to the Larsen B system [*Domack et al.*, 2002].

It is further intriguing that paleoceanographic changes are also correlative to Late Holocene events with the withdrawal of UCDW from the western shelf of the AP

during the LIA [*Shevenell and Kennett*, 2002; *Ishman and Sperling*, 2002]. Hence the atmospheric-glaciologic link so strongly supported by historical data sets seems to have had a third component (deep-ocean circulation) associated with it during the late Holocene. That deeper water mass temperatures in the Southern Ocean have indeed warmed during the last 50 years [*Gille et al.*, 2002] would place strong emphasis on the strong coupling between all three systems on up to the present.

While marine and ice core archives are providing useful paleoclimate data the convergence of these studies toward the resolution of multi-decadal events remains to be achieved. It is at this time scale that we desperately need paleo data [*National Academy of Sciences*, 1998]. The recognition of tidal events and seasonal signals in the marine archive is a promising way to achieve greater resolution [*Domack et al.*, this volume; *Yoon* et al., this volume].

METEROLOGIC PERSPECTIVE

Near surface temperature records from the Antarctic Peninsula (AP) have shown considerable warming over recent decades [*King*, 1994; *Comiso*, 2000; *Turner et al.*, 2001; *Vaughan et al.*, 2001]. Most significant is a warming trend of approximately 0.1°C/year during winter at stations along the west coast of the Peninsula [*King et al.*, this volume]. Although not as large, temperature trends are also positive during summer for the western AP and in all seasons on the eastern coast of the Peninsula [*King et al.*, this volume]. *King* [1994] and *Vaughan et al.* [2001] discuss mechanisms that might drive such temperature increases. These mechanisms include: (1) changes in atmospheric circulation and temperature advective patterns across the AP, (2) changing oceanographic processes that create enhanced upwelling of relatively warm circumpolar deep water (CDW) into the western AP region, and (3) changes in surface energy balance, of which the role of sea ice cover is especially important [*Smith et al.*, this volume].

Atmospheric circulation and surface temperatures over the AP are strongly influenced by the position of the Peninsula relative to the circumpolar trough of low pressure (CPT). This relationship can be seen in Figures 1a and 1b of *Simmonds* [this volume] and Figure 2 of *King et al.* [this volume]. The position and strength of the CPT vary as part of a larger-scale zonally symmetric mode of Southern Hemisphere circulation variability in which pressures at all levels vary inversely between the middle and high latitudes. This mode of variability has been given several names, including the high-latitude mode, the Southern Annular Mode (SAM), and the Antarctic Oscillation (AAO). Embedded within the CPT are regions of lower pressure, among which are the Amundsen Sea and Bellingshausen Sea low (ABS low) and the Atlantic low. The ABS low drives predominantly mild northwest wind flow onto the west coast of the AP, which is in contrast to colder, continental flow driven by the Weddell Sea low on the eastern side of the Peninsula (Figure 2). The steep topography of the AP limits the influence of mild northwest flow across the eastern AP

(Figure 2). As a consequence, the near surface temperatures in the west are significantly warmer then those of the east (Figure 2 [*King et al.*, this volume]).

Orographic influence also drives a varied precipitation climatology in the western AP, as transient synoptic disturbances and associated fronts cross the Peninsula [*Turner et al.*, 1995]. Cyclogenesis occurs frequently in the ABS, which contributes to climate variability over a wide range of temporal scales [*Simmonds and Murray*, 1999; *Simmonds*, this volume]. On the intra-annual time scale, the CPT and its core low pressure regions exhibit a semiannual oscillation (SAO) in which the CPT grows stronger and moves poleward during the transition seasons [*van Loon*, 1967; *van Loon and Rogers*, 1984]. During these seasons, transient synoptic disturbances are more intense and most likely to influence the AP climate.

On longer time scales, CPT variability and its associated surface climate impacts have been linked to variations in SAO and AAO. SAO variations occur over a wide range of temporal scales ranging from interannual to interdecadal [*Simmonds and Walland*, 1998; *van den Broeke*, 1998] and several authors have documented a decrease in SAO strength since the late 1970s [*Hurrell and van Loon*, 1994; *Chen and Yen*, 1997]. In a comparison of weak and strong SAO years, *van den Broeke* [2000] found an enhancement of northerly flow, decreased sea ice cover in the ABS, and associated warming over the AP during periods with weaker SAO. As such, warming temperature trends observed in the AP may reflect, in some part, the weakened SAO since the late 1970s. *Marshall* [2002] reexamined this hypothesis and found no significant changes in meridional wind and temperature advection in the west AP region between strong and weak SAO periods. By contrast, *Marshall* [2002] did find significant increases in zonal wind flow with a weakened SAO.

Increased zonal flow is also associated with changes in the AAO, which has trended toward its more positive phase in recent years [*Thompson et al.*, 2000]. *Thompson and Solomon* [2002] attribute warming in the AP to the more positive AAO through a stronger westerly flow, an associated reduction in cold polar outbreaks, and increased warm air advection off the southern Pacific Ocean. These authors suggest that the trend toward increasing AAO be related to decreases in lower stratospheric ozone. However, *King et al.* [this volume] points out that, ozone changes in winter have been minimal and cannot be used to explain winter warming in the AP. The meteorological impacts of high and low AAO conditions are explored by *van den Broeke* and *van Lipzig* [this volume] using a regional climate model for winters (July) during the period 1980-1993. In general, high AAO winters exhibit decreases in surface pressure over Antarctica and an enhancement of westerly flow north of the CPT. More significant is a localized decrease in surface pressure within the ABS low region. This is coupled with an eastward shift in the Atlantic Low, which disrupts cold air advection from the Filchner-Ronne Ice Shelf into the western AP and Weddell Sea basin [*van den Broeke and van Lipzig*, this volume], (see their Figures 4b and 5b). In addition to the flow of mild air into the western AP that accompanies the enhanced westerlies, high AAO conditions may also weaken the Weddell Sea gyre and cause additional warming over the eastern AP and less ice production in the Weddell Sea.

The surface climate and ecology of the western AP also reflects a strong sensitivity to ice cover on the Bellingshausen Sea [*Fraser et al.*, 1992; *Smith et al.*, 1999, this volume]. During winter and spring, when ice cover is at it greatest extent, western AP temperatures exhibit a strong negative correlation with sea ice [*Jacobs and Comiso*, 1997; *King et al.*, this volume]. The directionality of this relationship is somewhat complex given the feedbacks between ice and air temperature and the role that upwelled CDW plays in controlling sea ice in the western AP region [*Jacobs and Comiso*, 1997]. The linkages between AP surface temperature and ice cover, and the ways in which these systems vary interannually, are made more complex by the apparent influence of El Niño-Southern Oscillation (ENSO) related tele-connections. Numerous studies have found associations between ENSO and environmental conditions in the AP and adjacent ocean areas [*Simmonds and Jacka*, 1995; *Yuan and Martinson*, 2000].

Kwok and *Comiso* [2002] explored the relationship between positive, neutral, and negative phases of the Southern Oscillation Index (SOI) and surface meteorological/sea ice variations over the southern oceans. Strong associations were found over the Bellingshausen and Amundsen Seas where negative SOI (warm phase of ENSO) is linked to higher sea level pressure and decreased sea ice extent. The spatial structure of the sea level pressure-ENSO relationship shown by *Kwok* and *Comiso* [2002] exhibits characteristics of the Pacific-South American teleconnection pattern, which has been linked to ENSO [*Simmonds*, this volume], (see their Figure 6).

Rapid warming in the AP raises the question of what role greenhouse gas increases are playing in these temperature increases. *King et al.* [this volume] explores this question using the coupled atmosphere-ocean HadCM3 general circulation model in which an ensemble of four model runs was constructed using observed greenhouse gas and aerosol concentration changes between the 1940s and 1990s. Model results for summer show similar warming to that observed in the instrumental record and implies that summer AP temperature increases may be the result of greenhouse effect. During winter, the model produces a warming trend in eastern AP that is slightly larger than observed. Most important is a significant under-prediction by the model of temperature trends in the western AP during winter. This finding may be a function of the models inability to capture the strength and position of the ABS low and associated sea ice condition, both of which influence AP temperatures. As such, the degree to which greenhouse warming can account for winter temperature changes in the western AP is still unresolved.

REFERENCES

Anderson, J.B., *Antarctic Marine Geology*, Cambridge University Press, 289 pp., 1999.

Baker, P. A., G. O. Seltzer, S. C. Fritz, R. B. Dunbar, M. J., Grove, P. M. Tapia, S. L. Cross, H. D. Rowe, and J. P. Broda, The history of South American tropical precipitation for the past 25,000 years, *Science*, 291, 640-643, 2001.

Barker, P. F., Dalziel, I. W. D., and Story, B. C., Tectonic development of the Scotia Arc Region, in *The Geology of Antarctica*, edited by R. J. Tingey, Oxford Science Publications, *Monograph on Geology and Geophysics*, 27, Clarendon Press, Oxford, 215-248, 1991.

Barker, P. F., A. Camerlenghi, G. D. Acton, and others, *Proc. of the O.D.P. Initial Reports 178* [CD-ROM]. Available from: Ocean Drilling Program, Texas A&M Univ., College Station TX, 77845-9547, USA, 1999.

Barker, P. F., and A. Camerlenghi, Glacial history of the Antarctic Peninsula from Pacific margin sediments, in *Proc. ODP, Sci. Results, 178* [CD ROM], edited by P. F. Barker, A. Camerlenghi, G. D. Acton, and A. T. S. Ramsay, College Station TX (Ocean Drilling Program), 1-40, 2002.

Barker, D. H. N., and J. A. Austin, Jr., Rift propagation, detachment faulting, and associated magmatism in Bransfield Strait, Antarctic Peninsula, *Journal Geophysical Research*, 103, 24017-24043, 1998.

Bart, P. J., and J. B. Anderson, Seismic record of glacial events affecting the Pacific margin of the northwestern Antarctic Peninsula, in *Geology and Seismic Stratigraphy of the Antarctic Margin*. Antarctic Research Series, edited by A.K. Cooper, P. F. Barker, and G. Brancolini, 68, American Geophysical Union, Washington D. C, 75-95, 1995.

Berger, A. and M.F. Loutre, Insolation values for the climate of the last 10 million years. *Quaternary Science Reviews*, 10, 297-317, 1991.

Birkenmajer, K., Tertiary glaciation in the South Shetland Islands, West Antarctica: evaluation of data, in *Geological Evolution of Antarctica*, edited by M. R. A. Thomson, J. A. Crame, and J. W. Thompson, Cambridge University Press, New York, pp. 629-632, 1991.

Brachfeld, S., E. W., Domack, C. Kissel, C. Laj, A. Leventer, S. Ishman, R. Gilbert, A. Camerlengi, and L. B. Eglinton. Holocene history of the Larsen-A Ice Shelf constrained by geomagnetic paleointensity dating. *Geology,* in press.

British Antarctic Survey. *Antarctic Peninsula and Weddell Sea*, 1:3,000,000-scale map. BAS (Misc.) 8, Cambridge, 2000.

British Antarctic Survey. *Tectonic Map of the Scotia arc*, 1:3,000,000 scale map, BAS (Misc.) 3. Cambridge, 1985.

Camerlenghi, A., E. Domack, M. Rebesco, R. Gilbert, S. Ishman, A. Leventer, S. Brachfeld, and A., Drake, Glacial morphology and post-glacial contourites in northern Prince Gustav Channel (NW Weddell Sea, Antarctica), *Marine Geophysical Researches, 22*, 417-443, 2002.

Canals, M., R. Urgeles, and A.M. Calafat, Deep sea-floor evidence of past ice streams off the Antarctic Peninsula, *Geology, 28*, 31-34, 2000.

Canals, M., M.L. Casamor, R. Urgeles, A.M. Calafat, E.W. Domack, J. Baraza, M. Farran, and M. DeBatist, M., Seafloor evidence of a subglacial sedimentary system off the northern Antarctic Peninsula, *Geology, 30*, 603-606, 2002.

Clapperton, C. M., and D. E. Sugden, Late Quaternary glacial history of George VI Sound area, West Antarctica. *Quaternary Research, 18*, 243-267, 1982.

Chen, T.C. and M. C. Yen, Interdecadal variation of the Southern Hemisphere circulation, *Journal of Climate, 10*, 805-812, 1997.

Comiso, J.C., Variability and trends in Antarctic surface temperatures from in situ and satellite infrared measurements, *Journal of Climate, 13*, 1674-1696, 2000.

Convey, P., Antarctic Peninsula climate change: signals from terrestrial biology, this volume.

DeAngelis, H., and P. Skvarca, Glacier surge after ice shelf collapse, *Science, 299*, 1560-1562, 2003.

Dingle, R. V., and M. Lavelle, Antarctic Peninsular cryosphere: early Oligocene (c. 30 Ma) initiation and revised glacial chronology. *J. Geol. Soc.,155*, 433-437, 1998.

Domack, E. W., and S. Ishman, Oceanographic and physiographic controls on modern sedimentation within Antarctic fjords, *Geol. Soc. Amer. Bulletin, 105*, 1175-1189, 1993.

Domack, E. W. and P. A. Mayewski, Bi-polar ocean linkages: evidence from late-Holocene marine and ice core records, *The Holocene, 9*, 247-251, 1999.

Domack, E. W., S.E. Ishman, A.B. Stein, C.E. McClennen, and A.J.T. Jull, Late Holocene advance of the Müller Ice Shelf, Antarctic Peninsula: sedimentologic, geochemical, and palaeontologic evidence, *Antarctic Science, 7*, 159-170, 1995.

Domack, E. W., A. Leventer, R. Dunbar, F. Taylor, S. Brachfeld, C. Sjunneskog, and ODP Leg 178 Scientific Party, Chronology of the Palmer Deep site, Antarctic Peninsula: a Holocene palaeoenvironmental reference for the circum-Antarctic, *The Holocene, 11*, 1-9, 2001

Domack E. W., Duran, D., McMullen, K., Gilbert, R., and A. Leventer. Sediment lithofacies from beneath the Larsen B ice shelf: can we detect ice shelf fluctuation? Eos Trans. AGU, 83 (47), Fall Meet. Suppl., Abstract C52A-04, 2002.

Domack, E. W., A. Leventer, S. Root, J. Ring, E. Williams, D. Carlson, E. Hirshorn, W. Wright, R. Gilbert, and G. Burr, Marine sediment record of natural environmental variability and recent warming in the Antarctic Peninsula, this volume.

Elliot, D., The planer crest of Graham Land, northern Antarctic Peninsula: possible origins and timing of uplift, in *Geology and Seismic Stratigraphy of the Antarctic margin, 2*, edited by P.F. Barker and A.K. Cooper, *Antarctic Research Series, 71*, 51-74, 1997.

Emslie, S. D., P. Ritchie, and D. Lambert, Late-Holocene penguin occupation and diet at King George Island, Antarctic Peninsula, this volume.

Fraser, W.R., W.Z. Trivelpiece, D.G. Ainley, and S.G. Trivelpiece, Increases in Antarctic penguin populations: Reduced competition with whales or a loss of sea ice due to environmental warming?, *Polar Biology, 11*, 525-531, 1992.

Gilbert, R., E. W., Domack, and A. Camerlenghi, Deglacial history of the Greenpeace Trough: ice sheet to ice shelf transition in the northwestern Weddell Sea, this volume.

Gille, S., Warming of the Southern Ocean since the 1950s, *Science, 295*, 1275-1277, 2002.

Hall, B., An overview of the Late Pleistocene glaciation in the South Shetland Islands, this volume.

Harris, P.T., E. Domack, P.L. Manley, Gilbert, R. and A. Leventer, Andvord Drift: A new type of inner shelf, glacial marine deposystem from the Antarctic Peninsula, *Geology, 27*, 683-686, 1999.

Haug, G. H., K. A. Hughen, D. M. Sigman, L. C. Peterson, and U. Röhl, Southward migration of the Intertropical Convergence Zone through the Holocene, *Science, 293*, 1304-1308.

Hjort, C., O. Ingólfsson, M.J. Bentley, and S. Björck, Late Pleistocene and Holocene glacial and climate history of the Antarctic Peninsula region: A brief overview of the land and lake sediment records, this volume.

Hjort, C., M. J. Bentley, and O. Ingólfsson, Holocene and pre-Holocene temporary disappearance of the George VI Ice Shelf, Antarctic Peninsula. *Antarctic Science, 13*, 296-301, 2001.

Hjort, C., O. Ingólfsson, P. Möller, J. M., Lirio, Holocene glacial history and sea-level changes on James Ross Island, Antarctic Peninsula. *Journal of Quaternary Science, 12*, 259-273, 1997.

Hodell, D. A., C.D. Charles, T.P. Guilderson, S.L. Kanfoush, A. Shemesh, and X. Crosta, Abrupt cooling of Antarctic surface waters and sea ice expansion in the South Atlantic sector of the Southern Ocean at 5000 cal yr. B.P., *Quaternary Research, 56*, 191-198, 2002.

Hofman, E. E., and J. M. Klinck, Thermohaline variability of the waters overlying the West Antarctic Peninsula continental shelf, in *Ocean, Ice, and Atmosphere Interactions at the Antarctic continental margin*, edited by S.S. Jacobs and R.F. Weiss, AGU Antarctic Research Series, 75, Washington D.C., 67-82, 1998.

Hurrell, J.W. and H. van Loon, A modulation of the atmospheric annual cycle in the Southern Hemisphere. *Tellus, 46A*, 325-338, 1994.

Ingólfsson, O., and C. Hjort, Glacial history of the Antarctic Peninsula since the Last Glacial Maximum—a synthesis, *Polar Research, 21*, 227-234, 2002.

Ishman, S.E., and M. R. Sperling,, Benthic foraminiferal record of Holocene deep-water evolution in the Palmer Deep, western Antarctic Peninsula. *Geology, 30*, 435–438, 2002.

Jacobs, S.S. and J.C. Comiso, Climate variability in the Amundsen and Bellingshausen Seas, *Journal of Climate, 10*, 697-709, 1997.

Jenkins, A., D. G. Vaughan, S. S. Jacobs, H. H. Hellmer, and J. R. Keys, Glaciologic and oceanographic evidence of high melt rates beneath the Pine Island Glacier, West Antarctica, *Journal of Glaciology, 43*, 114-121, 1997.

King, J.C., Recent climate variability in the vicinity of the Antarctic Peninsula, *International Journal of Climatology, 14*, 357-369, 1994.

King, J.C., J. Turner, G.J. Marshall, W.M. Connolley, and T.A. Lachlan-Cope, Antarctic Peninsula climate variability and its causes as revealed by analysis of instrumental records, this volume.

Kwok, R. and J.C. Comiso, Southern ocean climate and sea ice anomalies associated with the Southern Oscillation, *Journal of Climate, 15*, 487-501, 2002.

Lamy, F., C. Rühlemann, D. Hebbeln, and G. Wefer. High- and low-latitude climate control on the position of the southern Peru-Chile Current during the Holocene, *Paleoceanography, 17*, 1029/2001PA000727, 2002.

Leventer, A., E. W. Domack, S. E. Ishman, S. E. Brachfeld, C. E. McClennen, and P. Manley. Productivity cycles of 200-300 years in the Antarctic Peninsula region: understanding linkages among the sun, atmosphere, oceans, sea ice, and biota *Geol. Soc. Amer. Bulletin, 108*, 1626-1644, 1996.

Lowe, A. L., and J. B. Anderson, Reconstruction of the West Antarctic Ice Sheet in Pine Island Bay during the Last Glacial Maximum and its subsequent retreat history. *Quaternary Science Reviews, 21*, 1879-1897, 2002.

Lucchi, R., M. Rebesco, M. Busetti, A. Caborlotto, E. Colizza, G. Fontan, Sedimentary processes and glacial cycles on the sediment drifts of the Antarctic Peninsula Pacific margin: preliminary results of SEDANO II project, in, *Antarctica at the Close of a Millennium.* edited by J. A. Gamble, D. N. B. Skinner, and S. Henrys, *Royal Society of New Zealand, Bulletin, 35*, 275-280, 2002.

Marshall, G.J., Analysis of recent circulation and thermal advection change in the northern Antarctic Peninsula, *International Journal of Climatology, 22*, 1557-1567, 2002.

Mercer, J. H. West Antarctic ice sheet and CO_2 greenhouse effect: a threat of disaster, *Nature, 271*, 321-325, 1978.

Morris, E. M. and Vaughan, D. J., Spatial and temporal variation of surface temperature on the Antarctic Peninsula and the limit of viability of ice shelves, this volume.

Mosley-Thompson, E. and L. G. Thompson Ice core paleoclimate histories from the Antarctic Peninsula: where do we go from here?, this volume.

National Academy of Sciences. *Decade-to-Century-Scale Climate Variability and Change, A Science Strategy.* National Research Council-National Academy Press, Washington D. C., 142 pp., 1998.

O'Cofaigh, C., C. J. Pudsey, J. A. Dowdeswell, and P. Morris, Evolution of subglacial beforms along a paleo-ice stream, Antarctic Peninsula continental shelf, *Geophysical Research Letters, 29*, 10.1029/2001GL014488, 2002.

O'Cofaigh, C., J. A. Dowdeswell, and C. J. Pudsey, Late Quaternary iceberg rafting along the Antarctic Peninsula continental rise and in the Weddell Sea and Scotia Seas, *Quaternary Research, 56*, 308-321, 2001.

Padman, L., H. A. Fricker, R. Coleman, S. Howard, L. Erofeeva, A new tide model for the Antarctic ice shelves and seas, *Annals of Glaciology, 34*, 247-254, 2002.

Payne, A. J., D. E. Sugden, and C. M. Clapperton, Modelling the growth and decay of the Antarctic Peninsula Ice Sheet. *Quaternary Research, 31*, 119-134, 1989.

Pudsey, C. and J. Evans, First survey of Antarctic sub-ice shelf sediments reveals Mid-Holocene ice shelf retreat, *Geology, 29*, 787-790, 2001.

Pusdey, C. J., P. F. Barker, and R. D. Larter. Ice sheet retreat from the Antarctic Peninsula shelf. *Continental Shelf Research, 14*, 1647-1675, 1994.

Quayle, W. C., P. Convey, L. S. Peck, C. J. Ellis-Evans, H. G. Butler, and H. J. Peat, Ecological responses of maritime Antarctic lakes to regional climate change, this volume.

Rebesco, M, A. Camerlenghi, and A. Zanolla, Bathymetry and morohogenesis of the continental margin west of the Antarctic Peninsula, *Terra Antarctica, 5*, 715-725, 1998.

Rignot, E. S. S. Jacobs, Rapid bottom melting widespread near Antarctic ice sheet grounding lines, *Science, 296*, 2020-2024, 2002.

Rott, H., P. Skvarca, and T. Nagler, Rapid collapse of Northern Larsen Ice Shelf, Antarctica, *Science, 271*, 788-792, 1996.

Scambos, T., C. Hulbe, and M. Fahnestock, Climate-induced ice shelf disintegration in Antarctica, this volume.

Shevenell, A.E. and J. P. Kennett, Antarctic Holocene climate change: a benthic foraminiferal stable isotope record from Palmer Deep, *Paleoceanography, 17*, 10.1029/2000PA000596, 2002.

Shevenell, A., E.W. Domack, and G.M. Kernan, Record of Holocene palaeoclimate change along the Antarctic Peninsula: Evidence from glacial marine sediments, Lallemand Fjord, *Pap. Proc. R. Soc. Tas., 130(2)*, 55-64, 1996.

Shwerdtfeger, W., *Weather and Climate of the Antarctic*, Elsevier, New York, 261 pp., 1984.

Simmonds, I., Regional and large-scale influences on Antarctic Peninsula climate, this volume.

Simmonds, I. and T.H. Jacka, Relationships between the inter-annual variability of Antarctic sea ice and the Southern Oscillation, *Journal of Climate*, *8*, 637-647, 1995.

Simmonds, I. and R.J. Murray, Southern extratropical cyclone behavior in ECMWF analyses during the FROST special observing periods, *Weather and Forecasting*, *14*, 878-891, 1999.

Simmonds, I. and D.J. Walland, Decadal and centennial variability of the southern semiannual oscillation simulated in the GFDL coupled GCM, *Climate Dynamics*, *14*, 45053, 1998.

Skvarca, P. and H. DeAngelis, Impact assessment of regional climate warming on glaciers and ice shelves of the north-eastern Antarctic Peninsula, this volume.

Smellie, J. L., R. J. Pankhurst, M. J. Hole, and J. W. Thomson. Age, distribution and eruptive conditions of Late Cenozoic alkaline volcanism in the Antarctic Peninsula and eastern Ellsworth Land: review, *Brit. Antacti. Surv. Bull, 80*, 21-49, 1988.

Smith, R.C., D. Ainley, K. Baker, E. Domack, S. Emslie, W. Fraser, J.P. Kennett, A. Leventer, E. Mosley-Thompson, S. Stammerjohn and M. Vernet, Marine ecosystem sensitivity to climate change, *Bioscience*, *49*, 393-404, 1999.

Smith, R. C., W. R. Fraser, S. E. Stammerjohn, and M. Vernet, Palmer long-term ecological research on the Antarctic Peninsula, this volume.

Thompson, D. W. J. and S. Soloman, Interpretation of recent Southern Hemisphere climate change, *Science, 296*, 895-899, 2002.

Turner, J., T.A. Lachlan-Cope, J.P. Thomas, and S.R. Colwell, The synoptic origins of precipitation over the Antarctic Peninsula, *Antarctic Science*, *7*, 327-337, 1995

Turner, J., J.C. King, T.A. Lachlan-Cope, and P.D. Jones, Recent temperature trends in the Antarctic, *Nature, 418*, 291-292, 2002.

van den Broeke, M.R., The semi-annual oscillation and Antarctic climate. Part 2: recent changes, *Antarctic Science, 10*, 184-191, 1998.

van den Broeke, M.R., On the interpretation of Antarctic temperature trends, *Journal of Climate*, *13*, 3885-3889, 2000.

van den Broeke, M. R. and N.P.M. van Lipzig, Response of wintertime Antarctic temperatures to the Antarctic oscillation: results of a regional climate model, this volume.

van Loon, H., The half-yearly oscillation in middle and high southern latitudes and the coreless winter, *Journal of Atmospheric Science*, *24*, 472-486, 1967.

van Loon, H. and J. C. Rogers, Interannual variations in the half-yearly cycle of pressure gradients and zonal wind at sea level on the Southern Hemisphere, *Tellus*, *36A*, 76-86, 1984.

Vaughan, D.G., G.J. Marshall, U.M. Connolley, J.C. King, and R. Mulvaney, Devil is in the detail, *Science, 293*, 1777-1779, 2001.

Willmott, V., M. Canals, and J.L. Casamor, Retreat history of the Gerlache-Boyd Ice Stream, northern Antarctic Peninsula: An ultra high resolution seismic reflection study of the deglacial and post-glacial sediment drape, this volume.

Yuan, X. and D.G. Martinson, Antarctic sea ice extent variability and its global connectivity, *Journal of Climate*, *13*, 1697-1717, 2000.

Yoon, H. I., B-K Park, Y. Kim, C. Y. Kang and S. Ho-Kang, Origins and paleoceanographic significance of layered diatom ooze interval from the Bransfield in the northern Antarctic Peninsula around 2500 yrs BP, this volume.

Eugene W. Domack, Geology Department, Hamilton College, 198 College Hill Road, Clinton, New York 13323 (edomack@hamilton.edu).

Adam Burnett, Department of Geography, Colgate University, 13 Oak Drive, Hamilton, New York 13346 (aburnett@mail.colgate.edu)

Amy Leventer, Geology Department, Colgate University, 13 Oak Drive, Hamilton, New York, 13346 (aleventer@mail.colgate.edu)

Meteorological Record and Modeling Results

ANTARCTIC PENINSULA CLIMATE VARIABILITY AND ITS CAUSES AS REVEALED BY ANALYSIS OF INSTRUMENTAL RECORDS

J. C. King, J. Turner, G. J. Marshall, W. M. Connolley, and T. A. Lachlan-Cope

British Antarctic Survey, Natural Environment Research Council, Cambridge, United Kingdom

Climate observations made since the mid twentieth century reveal that the Antarctic Peninsula is a region of extreme climate variability and change. The pattern of change is, however, both seasonally and spatially inhomogeneous. Limited data from the east (Weddell Sea) coast indicate that surface air temperatures here are rising at around 0.03 °C per year in all seasons. On the west (Bellingshausen Sea) coast, summer temperature trends are similar to those prevailing on the east coast but, in winter, warming trends of over 0.1 °C per year are observed, making this the most rapidly warming part of the Southern Hemisphere. Rapid warming is confined to the very lowest levels of the atmosphere and warming of the free troposphere over the Peninsula is not statistically significantly different from the Southern Hemisphere average. Interannual variations in winter temperatures on the west coast are strongly correlated with variations in atmospheric circulation and sea ice extent, suggesting that both atmospheric and ice/ocean processes may be contributing to the long-term warming. However, there is little observational evidence to support long-term atmospheric circulation changes. Coupled atmosphere-ocean general circulation model (AOGCM) experiments, forced with observed greenhouse gas increase, fail to reproduce the observed pattern of warming around the Peninsula. However, current AOGCMs may not be sophisticated enough or of high enough resolution to represent all of the processes that control climate on a regional scale around the Antarctic Peninsula.

1. INTRODUCTION

A fundamental problem facing those attempting to study Antarctic climate variability is the relatively short length of instrumental climate records from the continent. Most Antarctic climate records start in or shortly after the International Geophysical Year (IGY) of 1957 and are thus of barely adequate length for quantifying climate variability and change. The situation is somewhat better in the Antarctic Peninsula sector, where some records extend back to the early 1950s. To the north of the Peninsula lie the South Orkney Islands, where climate records started in the early twentieth century. The locations of the most important records are shown on Figure 1. and details of these records are given in Table 1.

Although none of these records are long by European or North American standards, they do provide important insights into the controls on climate of a region which is undergoing rapid environmental change. The big challenge facing climate researchers is to use these short, quantitative records to improve our understanding of regional climate processes and thus to help to interpret the longer proxy records that are available from sources such as ice and sediment cores.

Copyright 2003 by the American Geophysical Union
10.1029/079ARS02

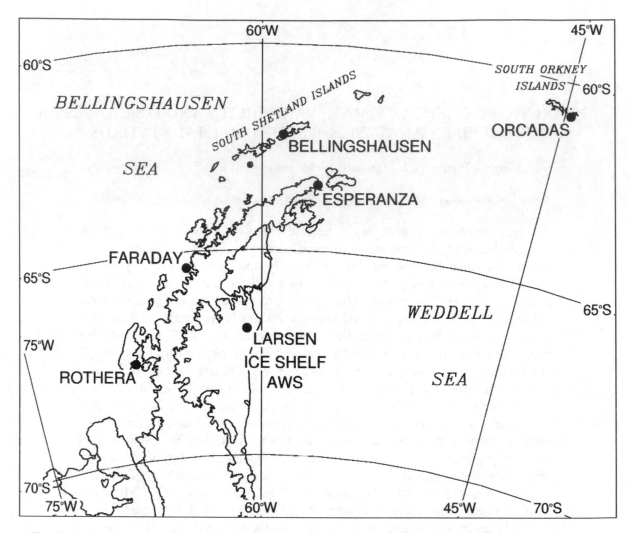

Fig 1. Locations of some of the most important climate records from the Antarctic Peninsula region. Since 1995, the former United Kingdom station "Faraday" has been operated by the Ukraine as Vernadsky station. The combined record from these stations is referred to in this paper as the Faraday record for conciseness.

In this paper, we review what we know about the patterns of climate variability in the Antarctic Peninsula based on analysis of instrumental records. Our focus is thus naturally on the time period from the mid twentieth century to the present-day. In our definition of instrumental records we include not only observations from climate stations but also satellite observations of the atmosphere, ocean and sea ice. We also include the atmospheric analyses and reanalyses produced by the major numerical weather prediction (NWP) centres, that synthesise a wide range of in situ and remotely sensed observations.

2. THE CLIMATE OF THE ANTARCTIC PENINSULA AND THE FACTORS THAT CONTROL IT

2.1 The Climatological Setting

The Antarctic Peninsula crosses the Circumpolar Trough (CPT) of low mean sea level pressure that encircles the Antarctic continent. The central and northern parts of the Peninsula are thus under the influence of the strong westerly winds that prevail north of the trough minimum while, in the very southernmost parts of the

TABLE 1. Details of surface temperature records available from the stations shown in Fig. 1

Station	Latitude (°S)	Longitude (°W)	Elevation (m a.s.l.)	Period of operation	Period analysed (Table 2)
Bellingshausen	62.2	59.0	16	1968–	1968–2000
Esperanza	63.4	57.0	13	1945–8; 1952–	1952–2000
Faraday / Vernadsky	65.2	64.3	11	1947–	1951–2000
Larsen Ice Shelf AWS	66.9	60.9	17	1985–	–
Orcadas	60.8	44.7	6	1903–	1903–2000
Rothera	67.6	68.1	16	1976–	1976–2000

region, coastal easterlies prevail. The CPT, however, is not entirely zonally symmetric and there are climatological low pressure centres within this feature. One of these, the Amundsen-Bellingshausen Sea (ABS) low (Figure 2) has a profound influence on the climate of the Peninsula. The presence of the ABS low means that the winds influencing the west coast of the Peninsula are, on average, from the northwest rather than due west, advecting relatively warm midlatitude air masses toward this region and keeping it mild compared to other regions at a similar latitude.

The mountains of the Peninsula, rising to over 1500 m over much of its length, provide an effective barrier to the low-level northwesterly flow. At low levels on the eastern side of the Peninsula, cold continental air flows northward as a barrier wind, driven by the climatological low pressure centre to the east of the Weddell Sea (Figure 2) [*Schwerdtfeger*,1975; *Parish*, 1983]. The contrasting low-level wind regimes of the two sides of the Peninsula

cause annual mean surface temperatures on the west coast to be some 5-10 °C warmer than those at similar latitudes on the east coast [*Schwerdtfeger*,1975; *Martin and Peel*, 1978].

Differences between east and west coast Peninsula climate are further enhanced by contrasting sea ice regimes on the two sides (Figure 3). Except in the far north, the east (Weddell Sea) coast remains icebound throughout the year in most years. On the western (Bellingshausen Sea) side, sea ice extends to the northern tip of the Peninsula in late winter in most years but melts back to the southern parts of the Bellingshausen Sea by the end of summer. The west coast of the Peninsula differs from all other Antarctic coasts in that it is relatively close to the sea ice edge throughout the winter *and* the prevailing low-level winds blow from the open ocean and across the sea ice to the coast. The implications of this for climate variability and sensitivity are explored further below.

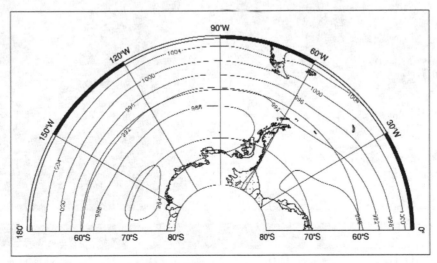

Fig. 2. Annual average mean sea level pressure field (hPa) in the vicinity of the Antarctic Peninsula from the NCEP/NCAR reanalysis for the period 1969-2000

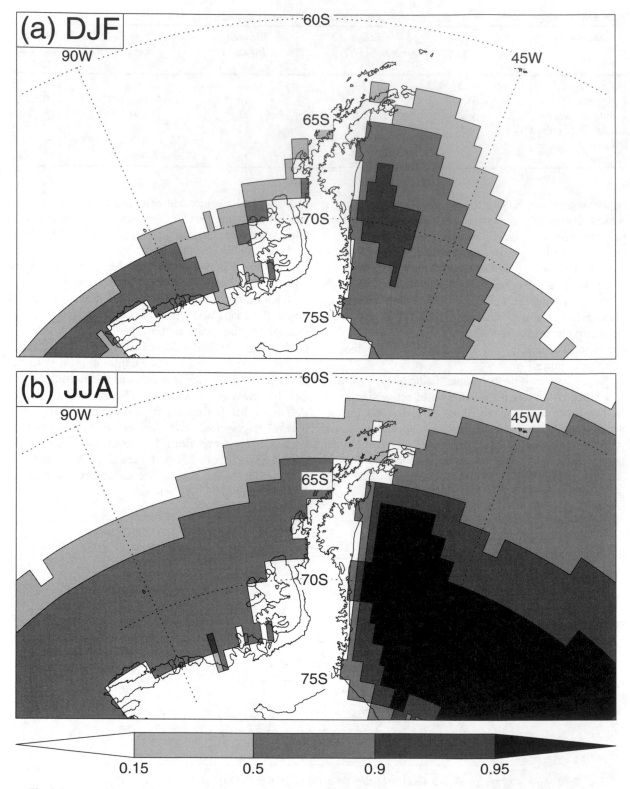

Fig. 3. Mean sea ice concentration (as fraction) 1979-1997 around the Antarctic Peninsula as determined from passive microwave satellite observations in (a) February and (b) September. Data were obtained from the National Oceanographic and Atmospheric Administration Ocean Modeling Branch

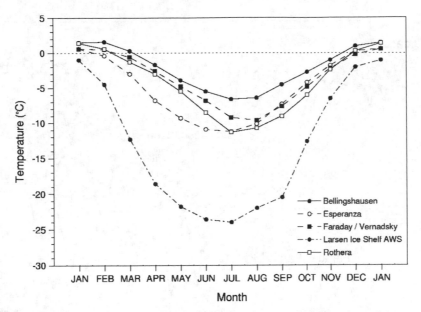

Fig. 4. The annual variation of surface air temperature at selected Antarctic Peninsula stations over the periods shown in Table 1.

2.2 The Mean Climate

2.2.1 Seasonal variation of near-surface temperatures.
Figure 4 shows the annual cycle of near-surface temperatures at selected stations along both coasts of the Peninsula. During the summer (December-February, or DJF) season, temperature gradients across the region are relatively weak. Summer mean temperatures along the west coast range from +1.2°C at Bellingshausen in the north to +0.7°C at Rothera in the south. The east coast is slightly cooler and experiences a stronger meridional temperature gradient. Summer temperatures here range from +0.2°C at Esperanza to –2.5°C at the Larsen Ice Shelf automatic weather station. In winter (June-August, or JJA), both the meridional and zonal temperature gradients are stronger. The winter mean temperature for Bellingshausen is –6.2°C while that for Rothera is –10.1°C. On the east coast the winter is much colder, with a seasonal mean of –23.4°C at the Larsen Ice Shelf AWS.

On the west coast of the Peninsula, and along the northern part of the east coast, the annual cycle of temperature is typical of that seen in much of maritime Antarctica, with a broad summer maximum and the coldest temperatures occurring in July or August. Along the central and southern parts of the east coast, the annual cycle is more reminiscent of that seen over continental Antarctica, with a short, "peaked" summer season and a long "coreless" winter. This confirms the importance of airmasses of continental origin in controlling the climate of the east coast.

2.2.2 Precipitation.
Turner et al. [1995] demonstrated that most precipitation falling on the west coast of the Peninsula came from fronts associated with synoptic-scale weather systems moving across the Bellingshausen Sea. Frontal precipitation is enhanced as air is forced to rise over the steep orography of the Peninsula and analysis of snow accumulation records from stakes, pits and cores indicates that the highest values of accumulation - over 2 m water equivalent per year—are found over the spine of the northern and central Peninsula [*Turner et al.*, 2002]. The mountainous orography of the Peninsula generates large local variations in accumulation.

Precipitation on the west coast peaks during the equinoctial seasons [*Turner et al.*, 1997]. At these times of year the CPT is deepest and is located furthest south so the number and intensity of synoptic-scale cyclones impinging on the Peninsula is at a maximum. Most winter precipitation falls as snow but, during summer, snow and rain occur with approximately equal frequencies.

3. OBSERVED PATTERNS OF TEMPERATURE VARIABILITY AND CHANGE

3.1 Variability of Surface Air Temperature

Time series of annual mean surface air temperature at selected Peninsula stations are shown in Figure 5. All show significant variability on a range of timescales from interannual to interdecadal superimposed on a long-term warming trend. Both variability and trends are

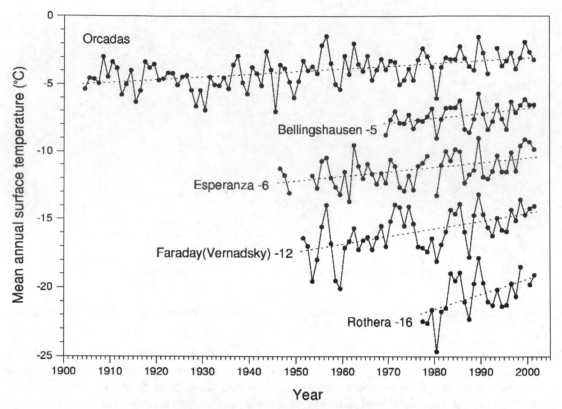

Fig. 5. Time series of annual mean surface air temperature at selected stations in the vicinity of the Antarctic Peninsula. Records other than Orcadas have been offset by the amounts shown for clarity

highest for the more southerly stations on the west coast of the Peninsula. The power spectrum of the annual mean temperature time series for Faraday shows no significant departure from red noise except in the 4-6 year period band. These periods are characteristic of patterns of large-scale climate variability, such as El Niño / Southern Oscillation (ENSO) and the Antarctic Circumpolar Wave (ACW, *White and Peterson* [1998]). This suggests that

such modes of variability may play a role in driving Peninsula temperature variations.

Time series of annual or seasonal mean temperature from stations on the west coast of the Peninsula correlate reasonably well with each other but generally poorly with records from elsewhere in continental Antarctica, indicating that the Peninsula is in a different climatic regime from the rest of the continent

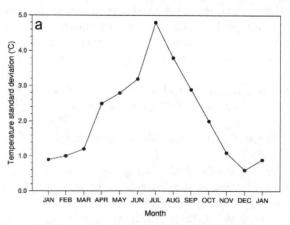

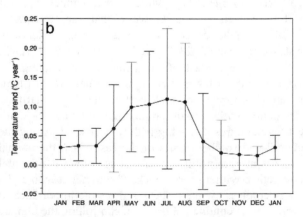

Fig. 6. (a) Interannual standard deviation and (b) linear trend of monthly mean temperatures at Faraday.

TABLE 2. Annual and seasonal temperature trends (in °C per year, with 95% confidence limits) for the records shown in Figure 5. Trends marked a, b and c are significant at below the 1%, 5% and 10% level, respectively. Data used in the analysis can be found at http://www.antarctica.ac.uk/met/gjma/

Station	Annual	Autumn (MAM)	Winter (JJA)	Spring (SON)	Summer (DJF)
Bellingshausen	+0.0366 ± 0.0432[c]	+0.0429 ± 0.0852	+0.0646 ± 0.0929	−0.0036 ± 0.0431	+0.0252 ± 0.0195[b]
Esperanza	+0.0343 ± 0.0264[b]	+0.0414 ± 0.0510	+0.0357 ± 0.0572	+0.0222 ± 0.0374	+0.0438 ± 0.0230[a]
Faraday	+0.0573 ± 0.0412[a]	+0.0649 ± 0.0570[b]	+0.1093 ± 0.0847[b]	+0.0259 ± 0.0423	+0.0246 ± 0.0153[a]
Orcadas	+0.0200 ± 0.0099[a]	+0.0204 ± 0.0149[a]	+0.0258 ± 0.0231[b]	+0.0173 ± 0.0132[b]	+0.0150 ± 0.0056[a]
Rothera	+0.1106 ± 0.1165[c]	+0.0964 ± 0.1366	+0.1661 ± 0.2512	+0.1120 ± 0.1271[c]	+0.0366 ± 0.0430[c]

[*King*, 1994]. Recently, the availability of a high spatial resolution dataset of satellite-derived monthly mean surface temperatures covering the whole of Antarctica [*Comiso*, 2000] has made it possible to define the limits of the region of coherent temperature variations more clearly [*King and Comiso*, 2003]. West coast temperatures correlate strongly with those over the seas just to the west of the Peninsula but rather poorly with temperatures on the east coast. This points to different factors controlling temperature variations on the west and east coasts, at least on the shorter timescales.

Interannual temperature variability and long-term trends both exhibit marked seasonality (Figure 6). Interannual variability is at a minimum during the summer months and rises to a maximum in winter. On the west coast of the Peninsula, warming trends are also much larger in winter than in summer, although the high level of interannual variability during the winter months tends to reduce the statistical significance of the trends during this season. The seasonality of the warming trends varies significantly over the Peninsula region (Table 2). At Bellingshausen in the South Shetland Islands, the summer warming trend is similar in magnitude to that observed further south on the west coast but the winter trend is significantly smaller. On the northern part of the east coast, characterized by Esperanza, the summer warming trend is nearly twice as large as that observed on the west coast. However, there is little seasonal variation in warming rates here and, consequently, winter warming trends are much smaller than those on the west coast.

3.2 Trends in Tropospheric Temperatures

Few data are available for studying the variability of upper-air temperatures over the Peninsula. *Marshall et al.* [2002] constructed a composite tropospheric temperature record for the west coast using radiosonde data from Faraday and Bellingshausen. Analysis of this record (Table 3) shows that temperatures at lower- and mid-tropospheric levels are also increasing. The rate of warming shows little seasonal variation and is of comparable magnitude to the summer surface warming seen in this region. Although the trends in lower- and mid-tropospheric temperatures are somewhat larger than the corresponding southern hemisphere mean trends computed by Angell [2000], the difference is not statistically significant [*Marshall et al.*, 2002].

4. CAUSES OF VARIABILITY AND CHANGE

4.1 Possible Mechanisms

Regional temperature trends may occur for a variety of reasons. Changes in atmospheric circulation may alter the character of airmasses affecting the region. Changing sea surface temperatures and sea ice cover to the west of the Peninsula will modify airmasses reaching the Peninsula through changed surface heat fluxes. Finally, there is the possibility that the warming seen in the Peninsula is related to larger scale changes in the climate system, in particular the surface warming that may be caused by increasing "greenhouse" gas concentrations.

TABLE 3. Annual and seasonal temperature trends (in °C per year, with 95% confidence limits) from the composite west Antarctic Peninsula upper-air temperature dataset for 1956-99 (see *Marshall et al.* 2002 for details). Trends marked a, b and c are significant at below the 1%, 5% and 10% level, respectively.

Pressure level	Annual	Winter (JJA)	Summer (DJF)
500 hPa	+0.0275 ± 0.0217[b]	+0.0292 ± 0.0392	+0.0214 ± 0.0266
700 hPa	+0.0227 ± 0.0225[b]	+0.0155 ± 0.0434	+0.0297 ± 0.0319[c]
850 hPa	+0.0314 ± 0.0242[b]	+0.0235 ± 0.0447	+0.0470 ± 0.0349[b]

In reality, all three of these mechanisms may be playing some role in the changes that we have observed in the Peninsula and they may interact in complex ways. In what follows, we examine the observational evidence for "fingerprints" of each class of mechanisms in an attempt to determine which may be the most important.

4.2 Atmospheric Circulation

The ABS low is a location of high interannual variability in atmospheric circulation. Standard deviations of sea level pressure and lower- and mid-tropospheric geopotential are higher here than anywhere else in the Southern Hemisphere [Connolley, 1997]. As noted above, the ABS low has a profound influence on the climate of the west coast of the Peninsula and variations in its strength and/or location will directly influence Peninsula temperatures through changes in warm air advection. 500 hPa height variations around 65°S, 110°W are strongly anticorrelated with Faraday winter mean temperatures and explain over 40% of the variance in the latter time series. Marshall and King [1998] showed that there were statistically significant circulation anomalies associated with exceptionally warm and cold Peninsula winters by compositing geopotential and wind anomalies for these years. This analysis confirmed that warm winters were associated with a dipole anomaly pattern of low pressure/geopotential in the ABS region and high pressure/geopotential over the Weddell Sea and South Atlantic (Figure 7). Cold winters were associated with a dipole anomaly of opposite sign. There was also evidence for teleconnections to circulation anomalies in the subtropical South Pacific. The anomaly patterns seen in Figure 7 bear considerable resemblance to the Pacific —South American (PSA) mode [Mo and Ghil, 1987; Simmonds, 2003 (this volume)]. This mode is associated with extremes of ENSO [Karoly, 1989] and there is evidence for ENSO contributing to some of the observed variability in the ABS low [Cullather et al., 1996]. The studies listed would suggest that warm (cold) ENSO events should be associated with anomalously high (low) pressure/gepotential in the ABS and hence anomalously cold (warm) conditions in the Peninsula. However, this relationship does not appear to be particularly robust, indicating that other factors may be contributing to circulation variations in the ABS region and that Peninsula temperatures are not controlled by circulation variations alone.

Atmospheric circulation variations thus appear to exert strong control over interannual temperature variability in the Peninsula but do they contribute to the long-term warming trend? This is a difficult question to answer as we do not have reliable atmospheric analyses for this region extending back before about 1968. The National Centers for Environmental Prediction / National Center for Atmospheric Research (NCEP/NCAR) global atmospheric reanalyses extend back to1948 [Kalnay et al., 1996], but the reliability of this product in high southern latitudes is questionable prior to about 1968 [Hines et al., 2000; Marshall, 2002]. Efforts are being made to ensure that the 40 year reanalysis currently being undertaken at the European Centre for Medium Range Weather Forecasting incorporates Antarctic data that were not used in the NCEP/NCAR reanalysis. However, before satellite sounder data became available in the 1970s, the South Pacific sector of the Southern Ocean was almost devoid of observations to constrain analyses. The reliability of reanalyses in this region during the pre-satellite era will thus always be open to question.

Observations from Peninsula stations provide mixed evidence for long-term circulation changes. Turner et al. [1997] showed that the frequency with which precipitation was reported at Faraday increased significantly between 1956 and 1992. This could be interpreted as a trend toward more cyclonic conditions in the ABS region. However, experiments that we have carried out using an atmosphere-only climate model suggest that an upward trend in precipitation could simply be a response to reduced winter sea ice cover in the ABS. Marshall [2002] failed to find any statistically significant trends in tropospheric meridional winds or thermal advection in an analysis of radiosonde data from the Peninsula.

It has been known for some time that the dominant mode of large-scale circulation variability around the Antarctic is a near zonally symmetric oscillation of atmospheric mass between high- and midlatitudes, causing variations in the strength of the circumpolar westerlies (e.g. Rogers and van Loon [1982]). This mode is nowadays referred to as the Southern Hemisphere Annular Mode (SAM) or Antarctic Oscillation (AAO) [Thompson and Wallace, 2000]. When the circumpolar westerlies are strong (weak), the AAO is said to be in a high- (low-) index state. Using a high-resolution regional atmospheric model, van den Broeke and van Lipzig [2003, this volume] show that the high-index state of the AAO is associated with an increased northerly component to the flow over the Peninsula and increased tropospheric temperatures in this region. Thompson and Solomon [2002] have postulated that declining Antarctic stratospheric ozone levels have driven a trend toward the high-index state of the AAO, which may have contributed toward warming in the Peninsula during the austral summer and fall. While this mechanism may be contributing to the small summertime warming, it is unlikely to be playing any role in the much larger winter warming seen on the west coast as no significant ozone depletion has occurred during this season.

4.3 Sea Ice and Oceanic Processes

The west coast of the Peninsula is the only part of coastal Antarctica where there is a strong association between winter temperatures and sea ice extent [*Weatherly et al.*, 1991; *King*, 1994; *Jacobs and Comiso*, 1997]. Anomalously cold winters here are almost invariably associated with anomalously extensive sea ice to the west of the Peninsula and *vice versa*. Variations in sea ice

extent at 70°W explain more than 60% of the variability in July temperatures at Faraday [*King and Harangozo*, 1998]. It is difficult to tell from observations alone whether temperature variations are forcing or responding to sea ice extent variations. Furthermore, ice extent responds to atmospheric circulation variations [*Harangozo*, 1997] and southerly winds that will promote ice growth are also associated with lower than normal temperatures. However, the combination of climatological on-ice winds and a relatively narrow sea ice zone to the west of the Peninsula is likely to favour control of temperatures by ice extent, rather than *vice versa*. Figure 8 shows modeled winter temperatures at Faraday in a run of the Hadley Centre HadAM3 atmosphere-only global climate model forced with observed sea ice extents and sea surface temperatures. The model closely reproduces observed temperature variations, supporting

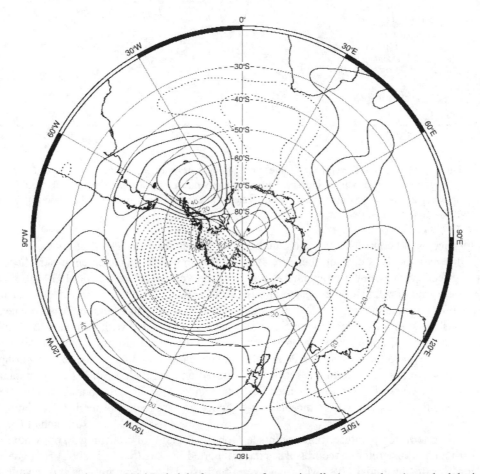

Fig. 7. Winter (June-August) mean 500 hPa height for a group of exceptionally (greater than 1 standard deviation from the long-term mean) warm winters at Faraday (1971, 1983, 1989, 1998, 2000.) minus that for a group of exceptionally cold winters (1969, 1976, 1977,1978, 1980, 1987). Contours are at intervals of 10 geopotential metres, with negative values shown dashed. Data are from the NCEP/NCAR reanalysis.

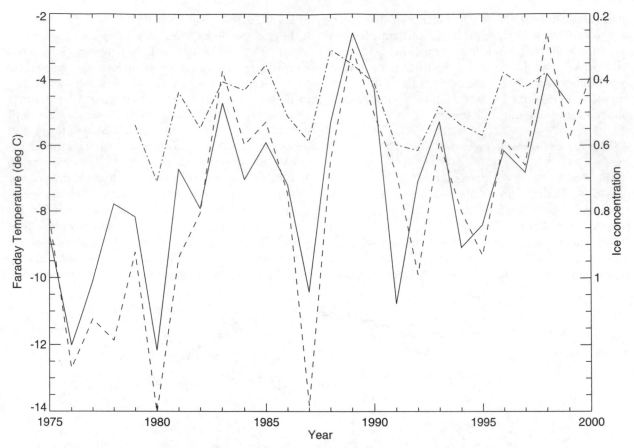

Fig. 8. Modeled winter temperatures at Faraday (solid line) in an atmosphere-only GCM experiment forced with observed sea ice extent and sea surface temperature variations. Observed Faraday temperatures for the same period (dashed line) are shown for comparison. Also shown are observed winter mean sea ice concentrations (dot-dash line) for the region bounded by 60°S and 70°S, 60°W and 80°W—note the inverted scale.

the hypothesis that ice extent controls temperature in this region. Even if this hypothesis is correct, there are still likely to be strong feedbacks as lower temperatures will promote further ice growth while higher temperatures will inhibit growth or promote melting. Statistically significant correlations between ice extent and temperature are only found in winter and spring, the seasons when there is extensive ice cover to the west of the Peninsula [*King*, 1994]. During summer and autumn, sea ice is restricted to the southern parts of the Bellingshausen Sea and thus exerts little influence over west Peninsula temperatures.

Association of long-term temperature trends with sea ice changes is difficult because reliable records of sea ice extent only date back to the mid-1970s, when passive microwave radiometers were first flown on polar orbiting satellites. Given the high level of interannual variability in ice extent in the Peninsula sector, these records are barely long enough to establish trends with a high

level of confidence. The ABS sector is the only Antarctic region in which sea ice extent has declined in all seasons of the year over the period 1979-1998 [*Zwally et al.*, 2002]. However, the greatest decline in this sector has occurred during the summer season, when ice extent is not well correlated with Peninsula temperatures. Trends during the winter are much smaller but temperatures are much more sensitive to ice extent during this season. Based on the relationship between interannual temperature and ice extent variations, *King* [1994] estimated that an ice edge retreat of order 1 degree latitude in the ABS over 40 years would be sufficient to account for the observed rate of warming. The winter season trend reported by *Zwally et al* [2002] is about half this value but has a large uncertainty as it is calculated from a 20-year record. From analysis of a very limited number of ship observations, *King and Harangozo* [1998] suggest that sea ice in the ABS sector during the late 1950s (a very cold period) was more extensive than at any time during the period for which we have satellite observations. The long-term trends may thus be larger than those based on the recent record.

Sea ice extent in the ABS will vary in response to changes in both atmospheric [*Harangozo*, 1997] and oceanic forcing. As discussed above, evidence for long-term atmospheric circulation changes is mixed and very few data are available on long-term oceanic change in this region. The ABS sector is unusual in that it is the only part of the Antarctic continental shelf that is subject to intrusions of relatively warm Circumpolar Deep Water (CDW) [*Deacon*, 1984, p.108]. If this water mass could be brought into the upper part of the water column, it would act as a source of heat that would inhibit sea ice formation or promote melting. Changes in the rate of CDW intrusion could thus result in long-term sea ice extent changes in this sector. The mechanisms behind CDW intrusion are still not fully understood but are probably associated with the proximity of the Antarctic Circumpolar Current (ACC) to the shelf break in this sector. Changes in the position or strength of the ACC resulting from variations in the strength of the circumpolar westerlies, or changing ocean temperatures in the vicinity of the ACC, are thus a likely source of variability. Ocean sediment records from the continental shelf to the west of the Peninsula show evidence for oscillation between CDW-dominated and shelf water-dominated regimes in this sector during the late Holocene [*Shevenell and Kennett*, 2002].

4.4 Regional Response to "Greenhouse" Warming

The response of the global climate system to changing atmospheric concentrations of "greenhouse" gases has been studied using coupled atmosphere-ocean general circulation models (AOGCMs). Such experiments demonstrate that increasing "greenhouse" gas concentrations over the past century have contributed significantly to observed global warming. The models also reproduce the observed broad-scale geographic variation in warming quite well, with the greatest warming seen in the polar regions where feedbacks involving sea ice and snow cover are most powerful. However, models are still limited in their ability to reproduce regional detail.

We have examined the results of a series of runs of the Hadley Centre AOGCM, version HadCM3. An ensemble of four runs was carried out, each forced with identical observed "greenhouse" gas and anthropogenic aerosol changes and differing only in the initial conditions used. Figure 9 shows modeled surface air temperature changes

between the 1940s and 1990s averaged over the four ensemble members. During the summer (December-February) season the modeled rate of warming in the Peninsula is similar in magnitude to that observed at Peninsula stations. The summertime warming observed in the Peninsula is thus not inconsistent with that to be expected as a result of increasing greenhouse gas concentrations. During the winter, modeled warmings around most of Antarctica are larger than in summer, probably as a result of sea ice–temperature feedbacks. All four ensemble members show a consistent pattern of rapid warming in the Weddell Sea sector, with modeled warming rates on the east coast of the Peninsula somewhat greater than those observed. Modeled warming rates in this sector of up to 8 °C per century are consistent with the observed rate of warming over the Antarctic sea ice zone over 1979-1998 reported by Comiso [2000]. In contrast with the observations, modeled winter warming rates on the west coast are smaller than those on the east coast . The modeled winter warming rate on the west coast (2-3 °C per 100 years) is much smaller than the observed warming rates for this region shown in Table 2. However, the modeled changes in the region to the west of the Peninsula vary greatly between the ensemble members. Taken at face value, these results might caution against attributing the observed strong winter warming in this sector to "greenhouse" warming enhanced by regional feedback processes. It is not clear, however, whether the HadCM3 model represents regional climate processes sufficiently well to make it a useful tool for studying regional change in the Antarctic. The strength and location of the ABS low are poorly reproduced in the model. Consequently, mean sea ice extent in the ABS and mean Peninsula temperatures are not well represented either. The ocean component of HadCM3 has a resolution of 1.25° latitude by 1.25° longitude and this may be insufficient to resolve the rather subtle changes in ice extent that, as demonstrated above, have a large impact on Peninsula winter temperatures.

5. CONCLUSIONS

Instrumental climate records collected in the Antarctic Peninsula region from the mid 20th century show that this is a region of extreme climate variability that has undergone significant warming. The spatial and seasonal pattern of the warming is complex; by studying it we are able to gain some insight into the processes that may be driving climate change in this region.

During the summer season, surface air temperatures throughout the Peninsula region have been rising at around 0.02 °C per year over the past 50 years, with a

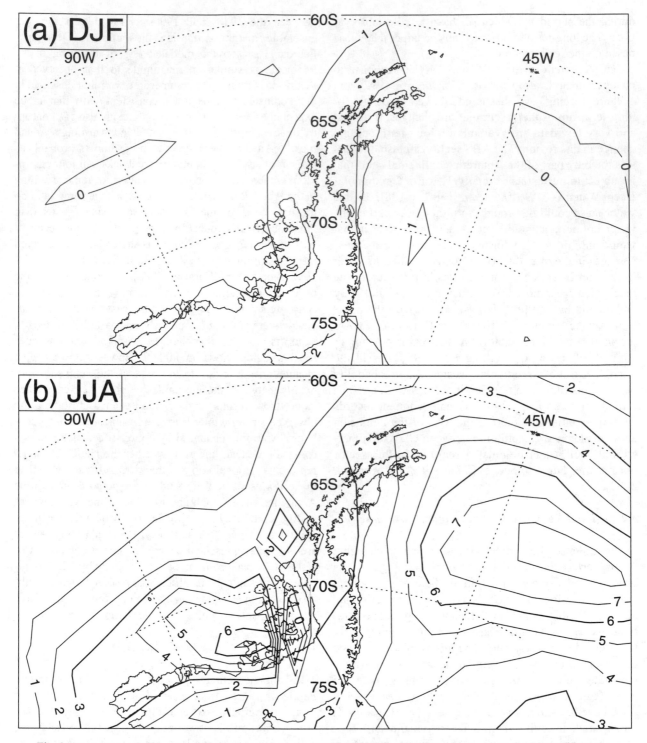

Fig. 9. Average rates of change of surface air temperature 1940-2000 in the Peninsula sector from four ensemble members of a run of the HadCM3 AOGCM forced with observed greenhouse gas and anthropogenic aerosol changes. (a) Summer (DJF), (b) Winter (JJA). Warming/cooling rates are shown in °C per 100 years.

similar rate of warming seen in lower- and mid-tropospheric temperatures. This is not significantly different from the mean rate of warming across the Southern Hemisphere over the same period and is not inconsistent with the regional warming rates predicted by AOGCMs forced with observed "greenhouse" gas increases. The summer warming in this region is, therefore, in some senses, not exceptional. What is exceptional is the regional environmental change seen in response to this warming. Because the Peninsula extends far northward, the 0°C summer isotherm passes through the region. Small temperature rises drive this isotherm increasingly further south, dramatically increasing melt rates in its vicinity thus leading to the rapid deglaciation discussed elsewhere in this volume.

In winter, the pattern of warming is more complex. The limited number of observations available from the east coast of the Peninsula indicate that warming rates here are no greater in winter than in summer. However, on the west coast, winter warming rates are 3-5 times greater than those prevailing during the summer, making the region the most rapidly warming (even in an annual mean sense) part of the Southern Hemisphere. The high warming rates are, however, confined to the very lowest part of the atmosphere. Winter warming rates in the free troposphere are similar to those observed in summer. The surface amplification of the warming signal in winter, and the absence of amplification in summer (when all but the very southern part of the ABS is ice free) points toward the importance of feedbacks involving atmosphere-ice-ocean coupling in controlling the winter warming. The strong anticorrelation between Peninsula west coast winter temperatures and ABS sea ice extent on an interannual timescale is further evidence that these processes are at work. On the east coast, conditions are less favorable for such feedbacks. This coast is icebound throughout most winters and is mostly under the influence of cold southerly winds.

Even if the rapid winter warming on the west coast can be attributed to a decline in sea ice cover, we still need to understand what is causing the latter. Winter ice cover will respond to changes in both atmospheric and oceanic circulation. Very few data are available for studying long-term oceanic change in this region and, as discussed in section 4.2, evidence for changed atmospheric circulation is mixed and any changes that have occurred must have been rather subtle. Experiments with global climate models do not lend immediate support to the idea that the observed changes result from a local amplification of "greenhouse" warming. However, current global models are probably not detailed enough to capture all of the complex interactions between atmosphere, ice and oceans that drive climate variability in the Peninsula region. Development of a regional coupled atmosphere-ice-ocean model is a high priority if we are to understand the causes of variability and change in this complex and fascinating region.

Acknowledgments. We thank colleagues at the Hadley Centre for providing the results from runs of their models. We are indebted to the many observers who, over the years, have recorded the climate at Antarctic Peninsula stations and also to those who had the foresight to archive and collate such observations.

REFERENCES

Angell, J.K., Global, hemispheric and zonal temperature deviations derived from radiosonde records., in *Trends Online: A compendium of data on global change*, Carbon Dioxide Information Analysis Center: Oak Ridge National Laboratory, Oak Ridge, TN, 2000.

Comiso, J.C., Variability and trends in Antarctic surface temperatures from in situ and satellite infrared measurements, *J. Climate*, *13*, 1674-1696, 2000.

Connolley, W.M., Variability in annual mean circulation in southern high latitudes, *Climate Dynamics*, *13*, 745-756, 1997.

Cullather, R.I., D.H. Bromwich, and M.L. VanWoert, Interannual variations in Antarctic precipitation related to El Nino southern oscillation, *J. Geophys. Res.*, *101*, 19109-19118, 1996.

Deacon, G., *The Antarctic Circumpolar Ocean*, Cambridge University Press, 180pp., 1984.

Harangozo, S.A., Atmospheric meridional circulation impacts on contrasting winter sea ice extent in two years in the Pacific sector of the Southern Ocean, *Tellus*, *49A*, 388-400, 1997.

Hines, K.M., D.H. Bromwich, and G.J. Marshall, Artificial surface pressure trends in the NCEP-NCAR reanalysis over the southern ocean and Antarctica, *J. Climate*, *13*, 3940-3952, 2000.

Jacobs, S.S., and J.C. Comiso, Climate variability in the Amundsen and Bellingshausen Seas, *J. Climate*, *10*, 697-709, 1997.

Kalnay, E., and 21 others, The NCEP/NCAR 40-year reanalysis project, *Bull. Amer. Meteorol. Soc.*, *77*, 437-472, 1996.

Karoly, D.J., Southern Hemisphere circulation features associated with El Niño - Southern Oscillation events, *J. Climate*, *2*, 1239-1252, 1989.

King, J.C., Recent climate variability in the vicinity of the Antarctic Peninsula, *Int. J. of Climatol.*, *14*, 357-369, 1994.

King, J.C. and J.C. Comiso, The spatial coherence of interannual temperature variations in the Antarctic Peninsula. *Geophys. Res. Lett.*, 30(2), 1040, doi: 10.1029/2002GL015580, 2003.

King, J.C., and S.A. Harangozo, Climate change in the western Antarctic Peninsula since 1945: observations and possible causes, *Ann. Glaciol., 27,* 571-575, 1998.

Marshall, G.J., Analysis of recent circulation and thermal advection change in the northern Antarctic Peninsula. *Int. J. Climatol.*, *22*, 1557-1567, 2002.

Marshall, G.J., and J.C. King, Southern Hemisphere circulation anomalies associated with extreme Antarctic Peninsula winter temperatures, *Geophys. Res. Lett.*, *25*, 2437-2440, 1998.

Marshall, G.J., V. Lagun and T.A. Lachlan-Cope, Changes in Antarctic Peninsula tropospheric temperatures from 1956-99; a synthesis of observations and reanalysis data. *Int. J. Climatol., 22,* 291-310, 2002.

Martin, P.J., and D.A. Peel, The spatial distribution of 10m temperatures in the Antarctic Peninsula, *J. Glaciol.*, *20*, 311-317, 1978.

Mo, K.C., and M. Ghil, Statistics and dynamics of persistent anomalies, *J. Atmos. Sci.*, *44*, 877-901, 1987.

Parish, T.R., The influence of the Antarctic Peninsula on the wind field over the western Weddell Sea, *J. Geophys. Res.*, *88*, 2684-2692, 1983.

Rogers, J.C., and H. van Loon, Spatial variability of sea-level pressure and 500 mb height anomalies over the Southern Hemisphere, *Mon. Weather Rev.*, *110* , 1375-1392, 1982.

Schwerdtfeger, W., The effect of the Antarctic Peninsula on the temperature regime of the Weddell Sea, *Mon. Weather Rev.*, *103*, 45-51, 1975.

Shevenell, A.E. and J. P. Kennett, Antarctic Holocene climate change: A benthic foraminiferal stable isotope record from Palmer Deep, *Paleoceanography*, *17(2)*, 9-1–9-12, 2002

Simmonds, I, Regional and large-scale influences on Antarctic Peninsula climate. (this volume)

Thompson, D.W.J., and J.M. Wallace, Annular modes in the extratropical circulation. Part I: Month- to-month variability, *J. Climate*, *13*, 1000-1016, 2000.

Thompson, D.W.J., and S. Solomon, Interpretation of recent Southern Hemisphere climate change, *Science*, *296* (5569), 895-899, 2002.

Turner, J., T.A. Lachlan-Cope, J.P. Thomas, and S.R. Colwell, The synoptic origins of precipitation over the Antarctic Peninsula, *Antarctic Science*, *7*, 327-337, 1995.

Turner, J., S.R. Colwell, and S. Harangozo, Variability of precipitation over the coastal western Antarctic Peninsula from synoptic observations, *J. Geophys. Res.*, *102*, 13999-14007, 1997.

Turner, J., T. A. Lachlan-Cope, G. J. Marshall, E. M. Morris, R. Mulvaney, and W. Winter, Spatial variability of Antarctic Peninsula net surface mass balance, *J. Geophys. Res.*, *107(D13)*, 4-1–4-18, 2002

van den Broeke, M.R. and N.P.M. van Lipzig, Response of wintertime Antarctic temperatures to the Antarctic Oscillation: results of a regional climate model. (this volume)

Weatherly, J.W., J.E. Walsh, and H.J. Zwally, Antarctic sea ice variations and seasonal air temperature relationships, *J. Geophys. Res.*, *96*, 15119-15130, 1991.

White, W.B., and R.G. Peterson, An Antarctic circumpolar wave in surface pressure, wind, temperature and sea-ice extent, *Nature*, *380* (6576), 699-702, 1996.

Zwally, H.J., J.C. Comiso, C.L. Parkinson, D.J. Cavalieri, and P. Gloersen, Variability of Antarctic sea ice 1979-1998, *J. Geophys. Res.*, *107* (C5), 9-1–9-19, 2002.

John C. King, J. Turner, G. J. Marshall, W. M. Connolley, and T. A. Lachlan-Cope, British Antarctic Survey, Natural Environment Research Council, High Cross-Madingley Rd., Cambridge, UK CB30ET

ANTARCTIC PENINSULA CLIMATE VARIABILITY
ANTARCTIC RESEARCH SERIES VOLUME 79, PAGES 31-42

REGIONAL AND LARGE-SCALE INFLUENCES ON ANTARCTIC PENINSULA CLIMATE

Ian Simmonds

School of Earth Sciences, University of Melbourne, Victoria, Australia 3010

The geography and latitude of the Antarctic Peninsula (AP) differ significantly from those of most of Antarctica, and hence it is not surprising that its weather and climate exhibit special characteristics. We present here modern aspects of the climate of the AP derived from a recently-released meteorological 're-analysis' data set which arguably presents the best meteorological representation of this part of the world. The Peninsula lies within the circumpolar pressure trough throughout the year and it is situated at the eastern end of a region of very frequent, intense cyclones. A trajectory analysis shows that, through the agency of these storms, the Peninsula is subject to a broad range of transport types and air masses. Attention is paid to the complex interaction between the atmospheric circulation and sea ice. We also discuss many of the large-scale influences on AP weather and climate. Included in these are the El Niño-Southern Oscillation, the semiannual oscillation, vacillations of the 'Antarctic mode', and the Antarctic circumpolar wave.

1. INTRODUCTION

The Antarctic Peninsula (AP) has a number of special characteristics which makes its weather and climate rather different from those of the rest of Antarctica. It is the most northerly part of the continent, and hence is the most subject to midlatitude influences. The Peninsula is long, narrow, and has relatively high elevations which exceed 2 km over most of its length. The AP is a major climatic divide with maritime conditions on the west side and a continental environment on the east side. Its north-south orientation influences in a variety of ways longitudinally-propagating weather systems. In addition, the AP forms the southern boundary of the Drake Passage and hence is subject to the many consequences of that constriction of the Southern Ocean. One of these is close proximity to the Antarctic Circumpolar Current which propagates oceanic signals from the Pacific to the Atlantic oceans. The location of the Peninsula also dictates that its weather and climate is associated with global scale processes (teleconnections etc.).

The AP, particularly on its western side, is amongst the most accessible regions of Antarctica, and is one of the few parts of the continent to have a continuous climate record dating back to before the International Geophysical Year [*King,* 1994]. As such it represents a very valuable archive of changes which have occurred over the region. Indeed the rate at which the western side has warmed over recent decades [*King,* 1994; *Doran et al.,* 2002] is as great as any other region in the world.

2. SOME ASPECTS OF THE CLIMATOLOGY OF THE ANTARCTIC PENINSULA REGION

Historically, a variety of inadequacies have made the task of obtaining reliable meteorological analyses in the Antarctic region very difficult [*Hutchinson et al.,* 1999; *Hines et al.,* 2000; *Simmonds et al.,* 2003]. In recent times the use of 'four-dimensional data assimilation' and 'reanalysis' has made available sets of historical analyses in which one can have considerable confidence. One such set is described by *Kalnay et al.* [1996] and *Kistler*

10.1029/079ARS03

et al. [2001], and *Simmonds and Keay* [2000a, b] have reported on aspects of high southern latitude circulation as revealed in these analyses.

Recently this reanalysis data set was improved and updated by the National Centers for Environmental Prediction and the Department of Energy [*Kanamitsu et al.*, 2000, 2002; *Kistler et al.*, 2001], and we refer to it as the NCEP-2 set. The NCEP-2 set has been found to be a significant upgrade of the original product, and we use it here to provide what could be argued as the most trustworthy picture of regional weather and climate in the AP region. The analyses we have used were available every 6 hours and cover the period 1 January 1979 to 29 February 2000.

The average summer (December to February) distribution of mean sea level pressure (MSLP) in the AP region derived from the compilation is presented in Figure 1(a). (Note that the plot, as for others in this paper, is not drawn with the stereographic projection usually used, but rather with the orthonomic projection centered at 60°W, 65°S. This projection produces an image of the planet which would be obtained when viewed from an infinite distance. For our purposes it hence has the advantage of presenting the greatest resolution at the central point, as opposed to the lowest resolution there in the stereographic case [*Royer et al.*, 1990].) The northern part of the region is host to strong westerlies, while the circumpolar trough (CPT) encircles the continent and displays extrema off West Antarctica and Dronning Maud Land. It will be noticed that the CPT cuts the AP at its halfway point, near Marguerite Bay. The winter (June to August) pattern (Figure 1(b)) shows a rather similar structure, but also some interesting differences. For example, the mean westerlies in the eastern Pacific are weaker in winter, and that the mean pressures are somewhat higher in the vicinity of the AP in that season. The climatological low pressure system to the west of the Peninsula moves significantly westward in the winter season.

Of central importance to the weather and climate of the AP region is the frequency with which extratropical cyclones are spread across the domain. We have diagnosed these frequencies using the Melbourne University cyclone tracking scheme [*Simmonds and Murray*, 1999; *Simmonds et al.*, 1999] and present in Figure 2 the derived cyclone frequencies for summer and winter. There is a tendency for the higher cyclone frequencies to be found near the regions of the deeper parts of the CPT. In both seasons significant numbers of cyclones are found in the broad region to the west of the AP. These systems are known to be associated with the vast majority of pre-

cipitation reports on the western side of the Peninsula [*Turner et al.*, 1995]. Potential vorticity considerations suggest that cyclone numbers should decrease dramatically over elevated topography, and hence there are fewer systems diagnosed over the AP. Continental conditions obtain in the Weddell Sea and it is little surprise that the cyclone count there is also relatively modest.

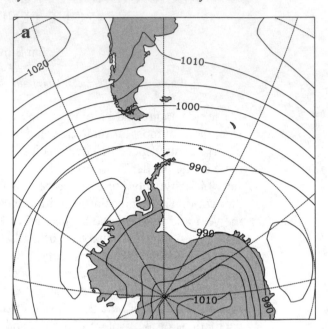

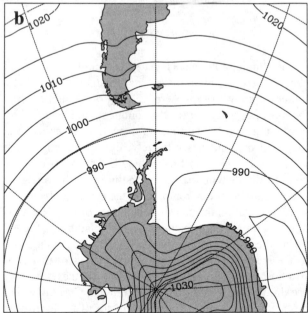

Fig. 1. Average (a) summer (December – February) and (b) winter (June – August) distributions of mean sea level pressure in the Antarctic Peninsula region. The contour interval is 5 hPa.

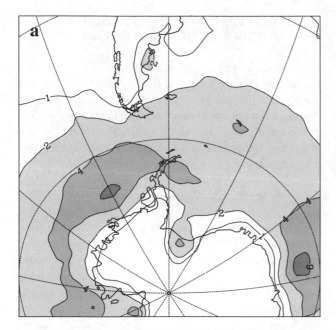

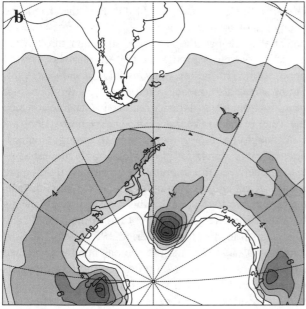

Fig. 2. Cyclone system density (the mean number of cyclones per analysis found in a 1000 (deg. lat.)^2 area) in (a) summer and (b) winter. The contour interval is 2, with an additional isoline at 1.

The presence of large numbers of cyclones to the west of the Peninsula means that the region is one of very high temporal variability on a wide range of timescales from synoptic and interdecadal [*Simmonds and Murray*, 1999]. A consequence of this is that there is a rich vari-

ety of air trajectories which reach the AP, and that variety may not be readily appreciated from consideration only of the structure of the mean MSLP distributions (Figure 1). This diversity obviously has considerable implications for the transport of dust, pollens etc. and hence the interpretation of the presence of (possibly exotic) microfossils in the palynorecord in any Southern Hemisphere region must pay due regard to these atmospheric considerations. *Burckle and Delaney* [1999] have commented that it is likely that wind systems have transported microfossil material from large portions of the southern hemisphere to the Antarctic polar ice cap. (In a different context, *Pole* [2001] and others have interpreted plant fossil records to signify the air transport of palynotaxa from Australia to New Zealand.) The need to be cognizant of the role of the atmospheric circulation is particularly evident in the AP region which we have seen to be one (in the modern record, at least) of complex mean circulation and high levels of synoptic activity, which induce a large assortment of atmospheric particle paths to the area. To present an aspect of this we have calculated the origin points of all 850 hPa (about 1.5 km above the surface) four-day trajectories which reach 304.5°E, 63.1°S (near Esperanza on the northern tip of the Peninsula) using the accurate technique described in *Perrin and Simmonds* [1995]. The wide variety of origin points can best be conveyed by presenting their spatial frequency distribution. Figure 3(a) shows that greatest frequency of summer departure points is found some 40° to the west of Esperanza, as one might have expected. However, it will be appreciated that the distribution function is quite broad; significant numbers of trajectories start from more than 90° upstream of the AP, and a considerable proportion originate from east of the AP. Note that the frequency of source points over Tierra del Fuego (and the implications for aeolian transport of terrestrial materials) is as much as half of the maximum referred to above. Indeed significant numbers of starting points are found as far north as 45°S over Chile and Argentina. The high frequencies from southern South America are partially explained by the strong northerlies to be found in front of the very frequent mobile cyclones as they approach the Drake Passage. A similar, if a little more diffuse, plot for winter is displayed in Figure 3(b). The greater spread reflects more active and varied circulations in that season. The proportion of beginning points found over South America is smaller than in summer but still noteworthy. (The local maximum over Dronning Maud Land may have limited meaning given that the mean surface pressures in this region are considerably

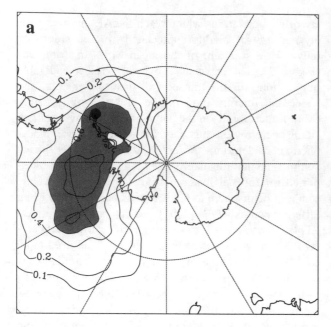

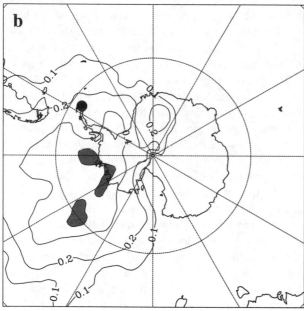

Fig. 3. Frequency distribution of origin points of all (a) summer and (b) winter 4-day 850 hPa trajectories which terminate at 63.1°S, 304.5°E (indicated by a solid dot). The contour interval is 0.2 origin points per 1000 (deg. lat.)² per trajectory, and an extra contour has been added at 0.1.

less than 850 hPa.) It is clear that the process of identifying the remote location of transported materials in the AP must make due reference to the nature of atmospheric variability through the relevant epoch.

3. INTERACTIONS OF ATMOSPHERIC CIRCULATION AND SEA ICE

A number of studies, including those of *King* [1994], *Harangozo* [1997], *King and Harangozo* [1998] and *van den Broeke* [2000], have emphasized the complexity of the interaction of the atmospheric circulation and sea ice in the environs of the AP. We shall firstly explore the manner in which extratropical cyclones and sea ice distribution interact.

3.1 Cyclone – Sea Ice Interactions

There are many good reasons to believe that high latitude cyclones and sea ice (and its concentration) interact in many and complex ways [*Yuan et al.*, 1999]. It would be expected that areas of sea ice with low concentrations would result in large fluxes of latent and sensible heat to the atmosphere, and these in turn would encourage the development of incipient atmospheric disturbances [*Watkins and Simmonds*, 1995]. On the other hand, the strong winds associated with cyclonic systems have the potential to change the sea ice distribution through the advection as well as through their effects on stresses on the upper surface of the ice [*Watkins and Simmonds*, 1998]. Strong winds also impact in intricate ways on the rate of heat loss from leads in the ice, and hence the rate at which the water in them refreezes [*Dare and Atkinson*, 1999].

Although these arguments appear strong *Godfred-Spenning and Simmonds* [1996] found only modest associations between the interannual variations of Antarctic sea ice extent and cyclone numbers. To explore this matter further, we here consider the statistical relationships between Antarctic sea ice extent (from the SMMR and SSM/I instruments) for the period 1979–1996 [*Watkins and Simmonds*, 1999] and the cyclone numbers obtained by *Simmonds and Keay* [2000b] when shorter timescales are considered. For each year of these data sets we deal with the residuals after the 'seasonal cycle' (the first two harmonics) is removed. For each winter season six non-overlapping 2-week averages are then taken. The correlations of these mean residuals of sea ice extent and cyclone numbers (to the north and to the south of the mean ice edge) in the same longitudinal sector are determined and contoured in the manner of *Godfred-Spenning and Simmonds* [1996]. The statistical significance (95% confidence level) of the correlations was established by Monte Carlo simulations.

Figure 4(a) shows that in the longitudinal sectors of interest scattered significant negative correlations are

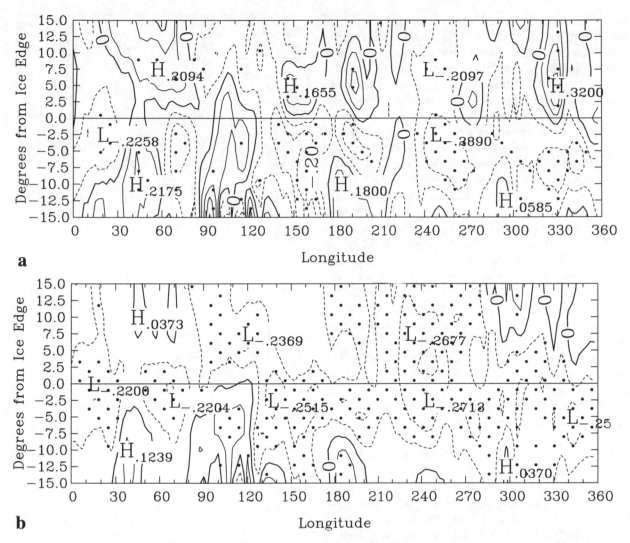

a

b

Fig. 4. (a) Mean correlation between longitudinally co-located winter 2-week averaged ice extent (averaged over 30° longitude wide sectors) and cyclone density (25° longitude sectors) (seasonal cycle removed) for 1979-96. Longitude is presented on the abscissa, while the ordinate indicates the position of 5°-wide latitudinal 'strips' relative to the mean sea ice edge. (Hence, the 'zero latitude' ribbon occupies the same 5° latitudinal zone as the mean sea ice edge.) Stippling denotes regions of the plot over which the mean correlations differ significantly (95% confidence level) from zero (Monte Carlo test). (b) As for part (a), but for the 3-day averaged data and for the period 1988-96.

found between the ice extent and cyclone numbers to the south of the mean ice edge. That is, more ice in the Amundsen, Bellingshausen and Weddell sectors is statistically associated with fewer cyclones to the south of the mean ice edge. To determine the sensitivity of the correlation pattern to the averaging period used the analysis is repeated for 3-day averages (in this case only considering the period 1988–96 because the (earlier) SSMR data were collected only every second day). Figure 4(b) shows the nature of the correlation plot then changes

considerably. On this 'synoptic' timescale the regions identified above remain as those of significant correlations, but it is seen that there are now strong associations with cyclone numbers at, and to the north of, the ice edge, in the broad longitudinal region centered on the AP.

At first sight it may appear unexpected that the nature of cyclone numbers and sea ice extent depends so strongly on the timescale considered. However, we remind the reader that the arguments presented above implicitly apply to short or synoptic scales, and the associations one would have expected in the broad AP region indeed show up in the relevant correlations (Figure 4(b)). The correlations between the two-week averaged data are much weaker, and are weaker still when the interannual variations are considered (Figure 1(c) of *Godfred-Spenning and Simmonds* [1996]). To understand this dependence, one can certainly imagine that, for example, if the ice edge suddenly retreated for whatever reason, large heat fluxes to the atmosphere would follow almost immediately. However, if the retreat persisted the local atmosphere would become warmer and moister, which would in turn be responsible for a reduction of the fluxes. Hence on timescales longer than the synoptic scale the atmosphere comes into equilibrium with the characteristics of the underlying surface and the above arguments become less compelling. Put another way, the temporal covariances of basic near-surface climate parameters is very strong in the subantarctic region [*Simmonds and Dix,* 1989] and hence the nature of the apparent associations between them will depend on the timescale being considered. In particular, deductions of cyclone behavior from (proxies of) AP sector ice cover in past epochs should be arrived at with caution.

3.2 Relationships Between the Interannual Anomalies of Sea Ice Extent and Geopotential Height

The above considerations point to the complexities involved when one is considering the interaction of the sea ice and cyclone behavior in the AP region. One of these is the change of behavior when synoptic or seasonal excursions cause the CPT to cross the sea ice edge. *Gordon and Taylor* [1975] and *Enomoto and Ohmura* [1990] have discussed how when the CPT lies to the south of the ice edge in a given sector (and hence the ice edge is subjected to westerlies) the Ekman circulation is responsible for carrying sea ice to the north and it in turn is replenished by ice formed at higher latitudes. By contrast, when the pressure trough lies to the north of the ice edge the Ekman transport is to the south and hence discourages expansion of the sea ice edge. The operation of this mechanism is partly responsible for the seasonally asymmetric behavior of sea ice extent *viz.* slow growth and rapid retreat [*Gloersen et al.,* 1993; *Simmonds and Jacka,* 1995; *Watkins and Simmonds,* 1999].

With these points in mind it is of interest to diagnose the extent to which the interannual variations of sea ice extent and atmospheric height fields are related. To quantify this we have determined the mean spring (September to November) height fields at the 12 pressure levels (those indicated on the ordinate of Figure 5) of the

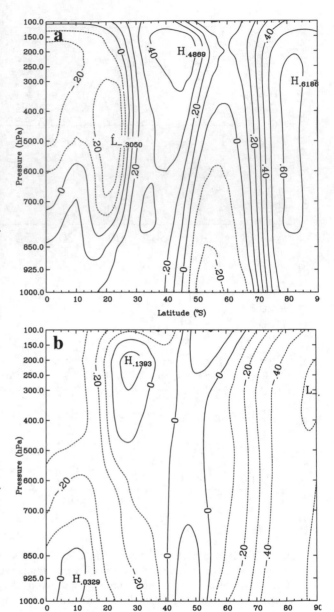

Fig. 5. Correlations of the interannual variabilities of sector-averaged mean spring (September to November) sea ice extent and geopotential height for the period 1979-96 (all variables have been detrended). The sector averaging was for (a) 240-300°E and (b) 300°E-0°E.

NCEP-2 record for each of the years 1979–96. The longitudinal mean of these heights in the Amundsen/Bellingshausen Seas (ABS) sector (240–300°E) have been determined, and the correlations of the (detrended) interannual variations of these heights and of spring sea ice extent averaged over the same sector calculated. The resulting correlations are plotted as a function of latitude and pressure level in Figure 5(a).

The structure of the plot allows us to explore aspects of the complex processes operating. The small correlation coefficients at lower levels near the mean ice edge in the sector (about 65°S) indicates that there is little association with local mean pressure and sea ice variations. However, the correlations show decreases to the north, meaning that interannual increases in ice extent are associated with northward migration of the CPT in the sector. The vertical profile of the correlations in the broad vicinity of the ice edge also reveals some very interesting behavior. If sea ice and local temperature profile variations were linked in a reasonably simple manner one would expect spring seasons with more sea ice to be those with colder temperatures in the lower troposphere. This would mean (via the hydrostatic equation) that geopotential height anomalies would become progressively more negative with height. In fact, the Figure exhibits the opposite of this (correlations increase arithmetically with height) strongly suggesting that the local considerations are far from explaining the co-variations. This seems to be so even close to the surface. Consistent with these considerations *Harangozo* [1997] has remarked that in the South Pacific sector ice advance leading to above-normal ice extent takes place only when a lowering of air temperatures is accompanied by equatorward flow. In a similar vein *Smith and Stammerjohn* [2001] comment that 'much of the variability in both air temperature and sea ice in the WAP region has been shown to be influenced by contrasting maritime (warm, moist) and continental (cold, dry) climate regimes'. *Renwick* [2002] has also emphasized the role meridional flow anomalies in influencing Southern Hemisphere (SH) sea ice extent. *Harangozo* [2000] found that the El Niño-Southern Oscillation (ENSO) only indirectly influences west AP winter temperatures via alterations in the local ice extent which in turn is sensitive to the local, ENSO teleconnected, meridional circulation variations. The apparent relationships between ABS surface air temperature and sea ice extent were shown by *Kwok and Comiso* [2002] to be associated, in part, with the Southern Oscillation.

The semiannual oscillation (SAO) is one possible candidate for larger scale forcing of anomalous behavior in the Antarctic. *Harangozo* [1997] shows in his sample that changes of warm air advection, which in turn influence sea ice extent, emanate from adjustments of the SAO in the South Pacific. *Simmonds and Jones* [1998] have shown that the SAO exhibits significant interannual and interdecadal variability and *van den Broeke* [2000]

has recently demonstrated how SAO variability and that of wintertime ABS sea ice extent are coupled. The numerical experiments of *Walland and Simmonds* [1998] have also quantified the complex interactions between the SAO, sea ice and cloud.

Many studies have indicated that the ENSO phenomenon significantly impacts on seasonal conditions in the ABS. *Simmonds and Jacka* [1995] showed the Southern Oscillation Index to be positively correlated with Pacific sea ice in spring (their Figure 2(c)), and hence there is a strong suggestion that over the period of our record a significant proportion of the variations of ABS sea ice and geopotential heights is forced by a third agent, the ENSO. *Xie et al.* [1994] have also found strong links between ENSO indices and sea ice extent anomalies Weddell Sea, as have *Yuan and Martinson* [2000] for the Weddell and ABS. In addition, *Bromwich et al.* [2000] have confirmed strong correlations between the moisture budget over the broader West Antarctic sector and the Southern Oscillation Index.

A similar plot obtained for the sea ice and geopotential height averaged over Weddell sector (300°E-0°E) is presented in Figure 5(b). The magnitude of the correlations are smaller in this sector, and the latitudinal structure of the correlations at low levels indicates that positive anomalies of sea ice extent are associated with a *southward* shift of the trough. In the Weddell Sea a physical argument could be made that when the trough is further south the enhanced advective and Ekman transports from the 'ice factory' in the southern parts of the Weddell result in a northerly excursion of the sea ice edge. As in the ABS there is a tendency for the arithmetic values of the correlations to increase with height, but in the Weddell sector this tendency is very weak and confined to the lower layers of the atmosphere.

4. TELECONNECTIONS AND REMOTE INFLUENCES ON THE ANTARCTIC PENINSULA

It is now well established that the atmosphere can support low-frequency teleconnection patterns, in which the influence of local forcings can be felt at remote locations. The fact that tropical processes, and the ENSO in particular, can affect the Antarctic (and specifically the AP) is now recognized [*Harangozo*, 2000].

The identification and quantification of the most important modes of low-frequency variability in the SH have been until recently made difficult by the relatively poor quality of the atmospheric analyses, particularly at upper levels where the teleconnection patterns show themselves most clearly. The NCEP-2 reanalysis set offers the opportunity to identify these modes with arguably one of the best sets of SH analyses available. To identify these dominant modes we have undertaken a Principal Component Analysis (PCA) of the SH 200 hPa stream function. (We have chosen to analyse stream function, rather than geopotential height as is often done, to facilitate the identification of the tropical components (should they exist) of teleconnection patterns.) The six-hourly stream function fields were deseasonalized by removing both the linear trend and the first four annual harmonics. The data were then low-pass filtered to retain only fluctuations having periods greater than 50 days.

The second PCA mode (which explains 7.5% of the variance) is displayed in Figure 6. The pattern shows a node in the tropical Pacific in the vicinity of the dateline, then nodes of alternating sign to the east of New Zealand, near the ABS, and to the north of the Weddell Sea. This pattern has become known as the 'Pacific-South American' pattern, and its presence has been shown to be closely associated with ENSO for dynamical reasons which are now fairly well understood [*Mo and White*,

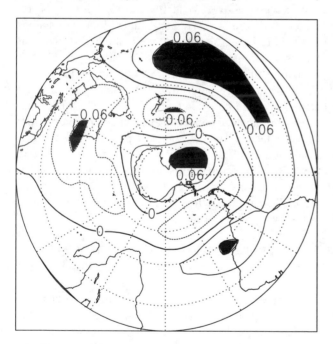

Fig. 6. Second principal component of low-frequency Southern Hemisphere 200 hPa streamfunction variability. (Analyzed data were deseasonalized by removing both the linear trend and the first four annual harmonics, then low-pass filtered to retain only fluctuations having periods greater than 50 days.) The contour interval is 0.03, and regions over which the magnitude of the function exceeds 0.06 are shaded. The zero contour is bolded and negative contours are dashed.

1985; *Karoly*, 1989; *Chen et al.*, 1996; *Karoly et al.*, 1996; *Renwick*, 1998; *Harangozo*, 2000]. This suggests that ENSO has an influence on conditions in the ABS and, to a lesser extent, the Weddell Sea, and provides a dynamical framework in which the ENSO connections discussed in the last section can be interpreted. *Garreaud and Battisti* [1999] have shown that during El Niño events similar teleconnection patterns in MSLP can be found across the Pacific, with anomalously high pressures found in the ABS region and somewhat lower pressures in the Weddell. A rather remarkable finding in their work was that a very similar pattern was deduced when they considered only the multidecadal band pass filtered data. *Karoly et al.* [1996] had also come to a similar conclusion using a limited amount of station data over the SH.

5. CLIMATIC INFLUENCE ON PENINSULA ICE SHELVES

Among the most spectacular examples of weather and climate interactions with the Peninsula cryosphere has been the disintegration over the last decade of a number of ice shelves, including parts of the Larsen ice shelf. Recently the northern section of the Larsen B shelf shattered and separated from the continent. In excess of 3,000 km^2 of the shelf disintegrated over a 35-day period beginning on 31 January 2002.

Scambos et al. [2000] reviewed surface and satellite data during periods of rapid retreat of AP ice shelves. They found that characteristics associated with the retreats included deeply embayed ice fronts, extensive calving of small elongated icebergs, the presence of melt ponds on the shelf surface in the vicinity of the break-ups, and 'punctuated events'. Their analysis revealed that most breakup events occurred during melt seasons. They suggested that melt ponds are 'a robust harbinger' of ice shelf retreat, and that infiltration of meltwater into crevasses is the main mechanism by which ice shelves weaken and draw back. This process had been suggested by *MacAyeal et al.* [1998] in connection with the Hemmen Ice Rise, near the northeast corner of the Filchner-Ronne Ice Shelf. They found that the mélange of sea ice and blown snow that fills many shelf-front rifts acts as a structurally cohesive unit, but that that unit may be more vulnerable to warming and more likely to fail during large storm events than the thicker shelf ice. Hence local (summer) climate imposes the conditions and thresholds for weakening of AP ice shelves. In particular, *Torinesi et al.* [2003] have pointed to the importance of the number and length of melt periods occurring each year. Similar conclusions had been drawn by a number of authors, including *Rott et al.* [1996] and *Skvarca et al.* [1999]. We also point out that *Rignot and Jacobs* [2002] have recently concluded that bottom melt rates of Antarctic glaciers are much greater than had been previously supposed, and that rates are highly sensitive to water temperature and warm water flowing into ice shelf cavities. Their work focused on large outflow glaciers and conditions at their grounding lines. Even though, by contrast, the Larsen sheets were sustained primarily by snow accumulation, it is not unreasonable to conjecture that bottom melting would have played a role in the Larsen breakups.

The punctuated nature of these disintegrations suggests that other influences may be important determinants of the precise timing of the start of retreat. A number of authors have commented that retreat or disintegration have started during storms and, as remarked above, the rift-filled mélange may be more likely to fail during large storm events [*MacAyeal et al.*, 1998]. In late January 1995 the northernmost part of the Larsen Ice Shelf disintegrated, with the loss of about 4200 km² of the shelf [*Rott et al.*, 1996; *Doake et al.*, 1998]. The initiation of this disintegration was coincident with a storm in the region. The synoptic map at 00 UTC on 25 January (Figure 7(a)) shows a cyclone with a central pressure deeper than 965 hPa situated just to the west of Alexander Island, and associated strong northwesterlies influencing the northern half of the AP. Twenty-four hours later (Figure 7(b)) the system has moved to the southern part of the Weddell Sea and is still associated with strong westerly (i.e., 'offshore') flow in the region of the Larsen shelf. The central pressure of this system was more than 10 hPa lower than the average for systems in these regions [*Jones and Simmonds*, 1993]. It could be argued that this particular storm was important for the precise timing of the disintegration.

We note finally that *Turner et al.* [2002] have recently explored the high latitude atmospheric circulation of the late spring and early summer of 2001/02. They found it to be highly anomalous, with October to December pressures at South Georgia being the highest on record, and those at Halley to be the lowest. The distribution of (normalized) pressure anomalies during that period (their Figure 3) indicate very strong northwesterly anomalies over the northern AP (and to the north of the Weddell Sea). This mean pattern established in the period leading to the disintegration of the northern part of the Larsen B shelf beginning in late January 2002 has many features in common with the synoptic maps just prior to the 1995 disintegration (Figure 7).

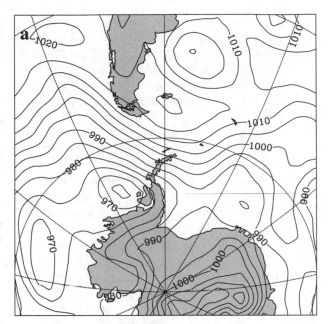

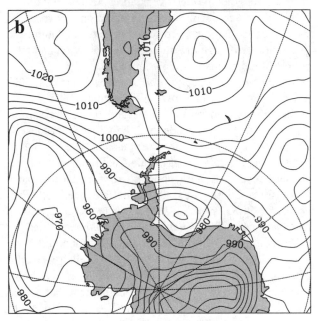

Fig. 7. Synoptic maps at 00 UTC on (a) 25 and (b) 26 January 1995. The contour interval is 5 hPa.

6. CONCLUDING REMARKS

We have seen that in many respects the environment of the AP is very atypical of the Antarctic. Being an elongated but narrow geographical feature it is subject to strong marine influences. Its weather and climate respond to a wide variety of local and regional forcings, including feedbacks associated with sea ice, the presence of an unceasing progression of cyclonic systems across its domain, and the close proximity of the Antarctic Circumpolar Current as it makes its way through the Drake Passage. It is also subject to large scale influences via ENSO and the Pacific-South American teleconnection pattern, vacillations in the semiannual oscillation, and excursions in the position of the circumpolar trough. Each of these factors lead to a rich matrix of weather and climate variability in the region.

For many years the 'high-latitude mode' (also referred to as the 'annular mode' or 'Antarctic mode') has been identified as an important component of large scale SH atmospheric variability [*Kidson*, 1975; *Mo and White*, 1985; *Karoly*, 1990; *Ghil and Mo*, 1991; *Mechoso et al.*, 1991]. This mode is a seesaw pattern between middle and high southern latitudes in geopotential height at most levels in the troposphere, and is apparent on a wide range of timescales. The oscillation of the mode is associated with a strengthening and weakening of the midlatitude westerlies. Such changes have implications for many of the topics we have been discussing here, including the distribution of mean pressure, cyclone activity, and teleconnection patterns from low latitudes.

Thompson et al. [2000] and *Hartmann et al.* [2000] have identified a trend toward high index polarity (i.e., higher pressure in mid latitudes and lower pressure at high latitudes) of this mode over the last few decades. At present there is considerable debate as to whether, and in what ways, the trend is associated with changes in, and consequences of, anthropogenically-produced traces gases (Greenhouse gases, stratospheric ozone, etc.) [*Fyfe et al.*, 1999; *Krivolutsky*, 1999; *Hartmann et al.*, 2000; *Kushner et al.*, 2001; *Thompson and Solomon*, 2002].

We comment that another potentially important large scale influence on AP climate is the so-called Antarctic circumpolar wave (ACW) [*White and Peterson*, 1996; *Peterson and White*, 1998], the maintenance of which may be associated with ENSO. The question as to whether this is a robust feature of the Southern Ocean is still unresolved. During the period considered by *White and Peterson* (*viz.* 1985-94) the wave appeared to exhibit an orderly eastward motion and was consistent with the temporal variations of cyclone frequency [*White and Simmonds*, 2003]. However, outside this period the presence of the ACW is by no means obvious in MSLP or sea ice [*Bonekamp et al.*, 1999; *Gloersen and Huang*, 1999; *Simmonds*, 2003]. A number of studies have suggested that the ACW may undergo low-frequency modulations or be made up of multiple components [*Baines and Cai*, 2000; *Carril and Navarra*, 2001; *Venegas*, 2003]. An understanding of the role that this feature may play in AP climate must await further observational and theoretical insights into its character.

Acknowledgments. The author expresses his gratitude to Kevin Keay, Pandora Hope and Harun Rashid for their assistance in the preparation of this manuscript. Funding from the Antarctic Science Advisory Committee and from the Australian Research Council was instrumental in allowing some of this research to be undertaken. An invitation to a meeting organised by Eugene Domack, under a National Science Foundation grant, was a key motivation for writing this paper.

REFERENCES

Baines, P. G., and W. J. Cai, Analysis of an interactive instability mechanism for the Antarctic Circumpolar Wave, *J. Climate*, 13, 1831-1844, 2000.

Bonekamp, H., A. Sterl and G. J. Komen, Interannual variability in the Southern Ocean from an ocean model forced by European Centre for Medium-Range Weather Forecasts reanalysis fluxes, *J. Geophys. Res.*, 104, 13317-13331, 1999.

Bromwich, D. H., A. N. Rogers, P. Kallberg, R. I. Cullather, J. W. C. White and K. J. Kreutz, ECMWF analyses and reanalyses depiction of ENSO signal in Antarctic precipitation, *J. Climate*, 13, 1406-1420, 2000.

Burckle, L. H., and J. S. Delaney, Terrestrial microfossils in Antarctic ordinary chondrites, *Meteoritics & Planetary Science*, 34, 475-478, 1999.

Carril, A. F., and A. Navarra, Low-frequency variability of the Antarctic Circumpolar Wave, *Geophys. Res. Lett.*, 28, 4623-4626, 2001.

Chen, B., S. R. Smith and D. H. Bromwich, Evolution of the tropospheric split jet over the South Pacific ocean during the 1986–89 ENSO cycle, *Mon. Wea. Rev.*, 124, 1711-1731, 1996.

Dare, R. A., and B. W. Atkinson, Numerical modeling of atmospheric response to polynyas in the Southern Ocean sea ice zone, *J. Geophys. Res.*, 104, 16691-16708, 1999.

Doake, C. S. M., H. F. J. Corr, H. Rott, P. Skvarca and N. W. Young, Breakup and conditions for stability of the northern Larsen Ice Shelf, Antarctica, *Nature*, 391, 778-780, 1998.

Doran, P. T. et al., Antarctic climate cooling and terrestrial ecosystem response, *Nature*, 415, 517-520, 2002.

Enomoto, H., and A. Ohmura, The influences of atmospheric half-yearly cycle on the sea ice extent in the Antarctic, *J. Geophys. Res.*, 95, 9497-9511, 1990.

Fyfe, J. C., G. J. Boer and G. M. Flato, The Arctic and Antarctic oscillations and their projected changes under global warming, *Geophys. Res. Lett.*, 26, 1601-1604, 1999.

Garreaud, R. D., and D. S. Battisti, Interannual (ENSO) and interdecadal (ENSO-like) variability in the Southern Hemisphere tropospheric circulation, *J. Climate,* 12, 2113-2123, 1999.

Ghil, M., and K. Mo, Intraseasonal oscillations in the global atmosphere. Part II: Southern Hemisphere, *J. Atmos. Sci.,* 48, 780-790, 1991.

Gloersen, P., W. J. Campbell, D. J. Cavalieri, J. C. Comiso, C. L. Parkinson and H. J. Zwally, Arctic and Antarctic Sea Ice, 1978–1987: Satellite Passive-Microwave Observations and Analysis. NASA Scientific and Technical Information Program, NASA SP-511, 290 pp, 1993.

Gloersen, P., and N. E. Huang, In search of an elusive Antarctic circumpolar wave in sea ice extents: 1978–1996, *Polar Research,* 18, 167-173, 1999.

Godfred-Spenning, C. R., and I. Simmonds, An analysis of Antarctic sea-ice and extratropical cyclone associations, Int. *J. Climatol.,* 16, 1315-1332, 1996.

Gordon, A. L., and H. W. Taylor, Seasonal change of Antarctic sea ice cover, *Science,* 187, 346-347, 1975.

Harangozo, S. A., Atmospheric meridional circulation impacts on contrasting winter sea ice extent in two years in the Pacific sector of the Southern Ocean, *Tellus,* 49A, 388-400, 1997.

Harangozo, S. A., A search for ENSO teleconnections in the west Antarctic Peninsula climate in Austral winter, *Int. J. Climatol.,* 20, 663-679, 2000.

Hartmann, D. L., J. M. Wallace, V. Limpasuvan, D. W. J. Thompson and J. R. Holton, Can ozone depletion and global warming interact to produce rapid climate change?, Proceedings of the National Academy of Sciences of the United States of America, 97, 1412-1417, 2000.

Hines, K. M., D. H. Bromwich and G. J. Marshall, Artificial surface pressure trends in the NCEP-NCAR reanalysis over the Southern Ocean and Antarctica, *J. Climate,* 13, 3940-3952, 2000.

Hutchinson, H. A. et al., On the reanalysis of Southern Hemisphere charts for the FROST project, *Wea. Forecasting,* 14, 909-919, 1999.

Jones, D. A., and I. Simmonds, A climatology of Southern Hemisphere extratropical cyclones, *Climate Dyn.,* 9, 131-145, 1993.

Kalnay, E. et al., The NCEP/NCAR 40-year reanalysis project, *Bull. Amer. Meteor. Soc.,* 77, 437-471, 1996.

Kanamitsu, M., W. Ebisuzaki, J. Woolen, J. Potter and M. Fiorino, Overview of NCEP/DOE Reanalysis-2, in Second International WCRP Conference on Reanalysis, WCRP-109, WMO/TD-No. 985, Wokefield Park, Reading, UK, 23-27 August 1999, pp. 1-4, 2000.

Kanamitsu, M., W. Ebisuzaki, J. Woollen, S.-K. Yang, J. J. Hnilo, M. Fiorino and G. L. Potter, NCEP-DOE AMIP-II Reanalysis (R-2), *Bull. Amer. Meteor. Soc.,* 83, 1631-1643, 2002.

Karoly, D. J., Southern Hemisphere circulation features associated with El Niño-Southern Oscillation events, *J. Climate,* 2, 1239-1252, 1989.

Karoly, D. J., The role of transient eddies in low-frequency zonal variations of the Southern Hemisphere circulation, *Tellus,* 42A, 41-50, 1990.

Karoly, D. J., P. Hope and P. D. Jones, Decadal variations of the Southern Hemisphere circulation, Int. *J. Climatol.,* 16, 723-738, 1996.

Kidson, J. W., Eigenvector analysis of monthly mean surface data, *Mon. Wea. Rev.,* 103, 177-186, 1975.

King, J. C., Recent climate variability in the vicinity of the Antarctic Peninsula, *Int. J. Climatol.,* 14, 357-369, 1994.

King, J. C., and S. A. Harangozo, Climate change in the western Antarctic Peninsula since 1945: Observations and possible causes, *Ann. Glaciol.,* 27, 571-575, 1998.

Kistler, R. et al., The NCEP-NCAR 50-year reanalysis: Monthly means CD-ROM and documentation, *Bull. Amer. Meteor. Soc.,* 82, 247-267, 2001.

Krivolutsky, A. A., Air depression over South Pole during Antarctic spring and its possible influence on ozone content in this region, *Advances in Space Research,* 24, 1627-1630, 1999.

Kushner, P. J., I. M. Held and T. L. Delworth, Southern Hemisphere atmospheric circulation response to global warming, *J. Climate,* 14, 2238-2249, 2001.

Kwok, R., and J. C. Comiso, Southern Ocean climate and sea ice anomalies associated with the Southern Oscillation, *J. Climate,* 15, 487-501, 2002.

MacAyeal, D. R., E. Rignot and C. L. Hulbe, Ice-shelf dynamics near the front of the Filchner-Ronne Ice Shelf, Antarctica, revealed by SAR interferometry: Model/interferogram comparison, *J. Glaciol.,* 44, 419-428, 1998.

Mechoso, C. R., J. D. Farrara and M. Ghil, Intraseasonal variability of the winter circulation in the Southern Hemisphere atmosphere, *J. Atmos. Sci.,* 48, 1387-1404, 1991.

Mo, K. C., and G. H. White, Teleconnections in the Southern Hemisphere, Mon. Wea. Rev., 113, 22-37, 1985.

Perrin, G., and I. Simmonds, The origin and characteristics of cold air outbreaks over Melbourne, *Aust. Meteor. Mag.,* 44, 41-59, 1995.

Peterson, R. G., and W. B. White, Slow oceanic teleconnections linking the Antarctic Circumpolar Wave with the tropical El Niño-Southern Oscillation, *J. Geophys. Res.,* 103, 24573-24583, 1998.

Pole, M. S., Can long-distance dispersal be inferred from the New Zealand plant fossil record?, *Australian Journal of Botany,* 49, 357-366, 2001.

Renwick, J. A., ENSO-related variability in the frequency of South Pacific blocking, *Mon. Wea. Rev.,* 126, 3117-3123, 1998.

Renwick, J. A., Southern hemisphere circulation and relations with sea ice and sea surface temperature, *J. Climate,* 15, 3058-3068, 2002.

Rignot, E., and S. S. Jacobs, Rapid bottom melting widespread near Antarctic ice sheet grounding lines, *Science,* 296, 2020-2023, 2002.

Rott, H., P. Skvarca and T. Nagler, Rapid collapse of northern Larsen Ice Shelf, Antarctica, *Science,* 271, 788-792, 1996.

Royer, J. F., S. Planton and M. Deque, A sensitivity experiment for the removal of Arctic sea ice with the French spectral general circulation model, *Climate Dyn.,* 5, 1-17, 1990.

Scambos, T. A., C. Hulbe, M. Fahnestock and J. Bohlander, The link between climate warming and break-up of ice shelves in the Antarctic Peninsula, *J. Glaciol.,* 46, 516-530, 2000.

Simmonds, I., Modes of atmospheric variability over the Southern Ocean, *J. Geophys. Res.,* 108, art. no. 8078, 2003.

Simmonds, I., and M. Dix, The use of mean atmospheric parameters in the calculation of modeled mean surface heat fluxes over the world's oceans, *J. Phys. Oceanogr.,* 19, 205-215, 1989.

Simmonds, I., and T. H. Jacka, Relationships between the inter-annual variability of Antarctic sea ice and the Southern Oscillation, *J. Climate,* 8, 637-647, 1995.

Simmonds, I., and D. A. Jones, The mean structure and temporal variability of the semiannual oscillation in the southern extratropics, Int. *J. Climatol.,* 18, 473-504, 1998.

Simmonds, I., and K. Keay, Mean Southern Hemisphere extra-tropical cyclone behavior in the 40-year NCEP-NCAR reanalysis, *J. Climate,* 13, 873-885, 2000a.

Simmonds, I., and K. Keay, Variability of Southern Hemisphere extratropical cyclone behavior 1958-97, *J. Climate,* 13, 550-561, 2000b.

Simmonds, I., K. Keay and E.-P. Lim, Synoptic activity in the seas around Antarctica, *Mon. Wea. Rev.,* 131, 272-288, 2003.

Simmonds, I., and R. J. Murray, Southern extratropical cyclone behavior in ECMWF analyses during the FROST Special Observing Periods, *Wea. Forecasting,* 14, 878-891, 1999.

Simmonds, I., R. J. Murray and R. M. Leighton, A refinement of cyclone tracking methods with data from FROST, *Aust. Meteor. Mag.,* Special Edition, 35-49, 1999.

Skvarca, P., W. Rack, H. Rott and T. I. Y. Donangelo, Climatic trend and the retreat and disintegration of ice shelves on the Antarctic Peninsula: An overview, *Polar Research,* 18, 151-157, 1999.

Smith, R. C., and S. E. Stammerjohn, Variations of surface air temperature and sea-ice extent in the western Antarctic Peninsula region, *Ann. Glaciol.,* 33, 493-500, 2001.

Thompson, D. W. J., and S. Solomon, Interpretation of recent Southern Hemisphere climate change, *Science,* 296, 895-899, 2002.

Thompson, D. W. J., J. M. Wallace and G. C. Hegerl, Annular modes in the extratropical circulation. Part II: Trends, *J. Climate,* 13, 1018-1036, 2000.

Torinesi, O., M. Fily and C. Genthon, 2003: Variability and trends of summer melt period of Antarctic ice margins since 1980 from microwave sensors. *Journal of Climate,* 16, 1047-1060.

Turner, J., S. A. Harangozo, G. J. Marshall, J. C. King and S. R. Colwell, Anomalous atmospheric circulation over the Weddell Sea, Antarctica during the austral summer of 2001/02 resulting in extreme sea ice conditions, *Geophys. Res. Lett.,* 29, art. no. 2160, 2002.

Turner, J., T. A. Lachlan-Cope, J. P. Thomas and S. R. Colwell, The synoptic origins of precipitation over the Antarctic Peninsula, *Antarctic Science,* 7, 327-337, 1995.

van den Broeke, M., The semi-annual oscillation and Antarctic climate. Part 4: A note on sea ice cover in the Amundsen and Bellingshausen Seas, *Int. J. Climatol.,* 20, 455-462, 2000.

van den Broeke, M. R., On the interpretation of Antarctic temperature trends, *J. Climate,* 13, 3885-3889, 2000.

Venegas, S. A., The Antarctic Circumpolar Wave: A combination of two signals?, *J. Climate,* 2003. (submitted).

Walland, D. J., and I. H. Simmonds, Sensitivity of the Southern Hemisphere semiannual oscillation in surface pressure to changes in surface boundary conditions, *Tellus,* 50A, 424-441, 1998.

Watkins, A. B., and I. Simmonds, Sensitivity of numerical prognoses to Antarctic sea ice distribution, *J. Geophys. Res.,* 100, 22,681-22,696, 1995.

Watkins, A. B., and I. Simmonds, Relationships between Antarctic sea-ice concentration, wind stress and temperature temporal variability, and their changes with distance from the coast, *Ann. Glaciol.,* 27, 409-412, 1998.

Watkins, A. B., and I. Simmonds, A late spring surge in the open water of the Antarctic sea ice pack, *Geophys. Res. Lett.,* 26, 1481-1484, 1999.

White, W. B., and R. G. Peterson, An Antarctic circumpolar wave in surface pressure, wind, temperature and sea-ice extent, *Nature,* 380, 699-702, 1996.

White, W. B., and I. Simmonds, Modulation of extratropical cyclone activity by the Antarctic Circumpolar wave in the Southern Ocean, *J. Climate,* 2003. (submitted).

Xie, S., C. Bao, Z. Xue, L. Zhang and C. Hao, Interaction between Antarctic sea ice and ENSO events, Proceedings of the NIPR Symposium on Polar Meteorology and Glaciology, 8, 95-110, 1994.

Yuan, X. J., and D. G. Martinson, Antarctic sea ice extent variability and its global connectivity, *J. Climate,* 13, 1697-1717, 2000.

Yuan, X. J., D. G. Martinson and W. T. Liu, Effect of air-sea-ice interaction on winter 1996 Southern Ocean subpolar storm distribution, *J. Geophys. Res.,* 104, 1991-2007, 1999.

Ian Simmonds, School of Earth Sciences, University of Melbourne, Victoria, Australia 3010 Email: simmonds@unimelb.edu.au

RESPONSE OF WINTERTIME ANTARCTIC TEMPERATURES TO THE ANTARCTIC OSCILLATION: RESULTS OF A REGIONAL CLIMATE MODEL

Michiel R. van den Broeke

Utrecht University, Institute for Marine and Atmospheric Research, Utrecht, Netherlands

Nicole P. M. van Lipzig*

Royal Netherlands Meteorological Institute, De Bilt, Netherlands

We describe Antarctic wintertime (July 1980–1993) surface temperature changes in response to the Antarctic Oscillation (AAO), using output of a regional atmospheric climate model. During conditions of AAO positive polarity (strong large scale westerly circulation), the south Atlantic storm centre is displaced eastward which cuts off the atmospheric branch of the Weddell Gyre. This causes surface temperatures in the Weddell Sea and over the Antarctic Peninsula to rise by up to 8 K, which has important implications for the formation of sea ice and deep water in the Weddell Sea and the viability of the Antarctic Peninsula ice shelves. On the other hand, a cooling of up to 5 K occurs locally in East Antarctica in places where near-surface easterly winds have decreased over the perennial ice. This is due to reduced downward mixing of warm air in the stable boundary layer. An amplified temperature response is found over the flat ice shelves, where the damping effect of the katabatic wind-forcing on surface temperature changes is absent. This contributes up to 3 K to the surface warming of the Antarctic Peninsula ice shelves and up to 8 K cooling over the ice shelves of in western Dronning Maud Land.

1. INTRODUCTION

Temperature trends in Antarctica have not been uniform in the last decades. A strong warming in the Antarctic Peninsula (2.5°C in the last 45 years) [*King and Harangozo*, 1998] has led to the rapid disintegration of the northern ice shelves [*Vaughan and Doake*, 1996, likely caused by increased meltwater ponding and associated ice shelf weakening [*Scambos et al.*, 2000]. On the other hand, a cooling between 1986 and 2000 has been signaled in the Dry Valleys in Victoria Land [*Doran et al.*, 2002] and in other regions in East Antarctica.

These regional differences have been explained in terms of Southern Hemisphere large scale circulation variability. The two leading modes of large scale variability in high southern latitudes [*Connolley*, 1997] are the Antarctic Oscillation (AAO) [*Thompson and Wallace*, 2000] and changes in the amplitude of the semi-annual oscillation (SAO) [*Van Loon*, 1967; *Simmonds and Walland*, 1998; and *Walland and Simmonds*, 1999]. There is an appreciable influence of the SAO on the distribution of sea ice around Antarctica [*Enemoto and Ohmura*, 1990]. Comparing pre and post 1975 amplitude of the SAO we see a decrease of SAO strength in the latter decades. This has led to changes in the annual cycle of pressure, temperature, wind speed and sea ice cover in Antarctica [*Harangozo*, 1997; and *Van den Broeke*,

---

*Presently at: British Antarctic Survey, Cambridge, UK

10.1029/079ARS04

2000a] as well as an outspoken regional differentiation in Antarctic temperature trends [*Van den Broeke*, 2000b]. On the other hand, regional differences in Antarctic near-surface temperature trends have also been ascribed to a persistent positive polarity of the AAO in recent decades, possibly related to springtime ozone depletion in the lower stratosphere [*Thompson and Solomon*, 2002]. There are indications that a weak SAO is another manifestation of the positive phase of the AAO [*Burnett and McNicoll*, 2000].

The interaction of the large scale circulation with the near-surface climate is not simple in Antarctica, because the Antarctic surface layer is characterised by strong horizontal and vertical temperature gradients [*Connolley*, 1996] and persistent katabatic winds [*Parish and Bromwich*, 1987]. All are ultimately forced by the surface radiation deficit over the ice sheet and interact in a complex fashion: a temperature deficit develops in the near-surface air when heat is exchanged with the cold surface. The negatively buoyant air flows down the slope of the ice sheet, resulting in the well known Antarctic katabatic winds. These winds, in their turn, generate turbulence that mix relatively warm air toward the surface. That is why in infrared satellite imagery regions of active katabatic winds are visible as warm surface signatures [*Bromwich*, 1989; and *Heinemann*, 2000]. They can also be detected through their elevated 10-m snow potential temperature [*Van den Broeke et al*, 1999]. Therefore, if one aims to understand Antarctic surface layer temperature change, one must also understand changes in the wind field and vice versa.

Given the above, it is not surprising that the effect of large scale circulation variability on the near surface climate in Antarctica depends strongly on the local geographical and climatological setting. In this paper we use output of a regional atmospheric climate model (RACMO/ANT1), which enables us to present in detail the surface layer wintertime temperature changes that occur in the Antarctic Peninsula region and other parts in Antarctica in response to the AAO.

2. MODEL AND METHODS

The regional climate model RACMO/ANT1 is based on the ECHAM4 model, a mix of the European centre for Medium-Range Weather Forecasts (ECMWF) and Max Planck Institut für Meteorologie global circulation models. The RACMO/ANT1 model domain of 122 x 130 grid points covers entire Antarctica and part of the surrounding oceans and corresponds to the area shown in

Figure 1. At the lateral boundaries, RACMO is continuously forced by ERA15 data (ECMWF reanalysis, 1980–1993). Sea ice and sea surface temperatures are prescribed from ERA15. Horizontal resolution is app. 55 km, which enables a reasonably accurate representation of the steep coastal ice slopes and the ice shelves fringing the coast. In the vertical, 20 hybrid levels are used. An additional layer at 6–7 m above the surface was included to better capture the strong temperature and wind speed gradients near the surface. Several improvements were made to improve the model performance over Antarctica, especially with regard to the physical representation of the snow surface. For example, the deep snow temperature initialisation was performed using the first two harmonics of the observed temperature cycle instead of a simple cosine function with a no flux boundary condition in the deep snow. For a more detailed description of the model the reader is referred to Van Lipzig, 1999.

The performance of RACMO/ANT1 is a great improvement over earlier models [*Van Lipzig et al.*, 1998; 1999]. In a detailed comparison with station data, Van Lipzig (1999) showed that annual mean temperature, wind speed and directional constancy are simulated with RMSE's of 1.5 K, 2 m s^{-1} and 0.12. Clearly, the more sophisticated TKE-based turbulence parameterisation in RACMO/ANT1 solved the problem of the much too low Antarctic surface temperatures in ERA15, which resulted from the decoupling of the surface layer from the overlying boundary layer due to a too strong suppression of mixing under stable conditions.

In this paper, the lowest terrain following model level (6–7 m above the surface) is referred to as the surface layer, while model layer 11 (with a height of about 6 km above the sea surface, less over the ice sheet surface) is chosen as being representative for the free atmosphere. To distinguish between the positive/negative polarities of the AAO we use the detrended 1980-93 July AAO index [*Thompson and Wallace*, 2000]. The timeseries of the AAO index is based on the first principal component of the NCEP/NCAR Reanalysis (NRA) 850-hPa extratropical height field (20–90°S). Because the index shows a spurious trend that is associated with unphysical atmospheric mass loss over Antarctica in the NRA, we used the index time series detrended for the period of interest (1980–1993).

Figure 2 shows the detrended July averaged AAO index together with the RACMO free-atmosphere zonal wind velocity U_{LSC} over the Southern Ocean (14 July months in the period 1980–1993). The ocean values of U_{LSC} represent the average over model level 11 grid

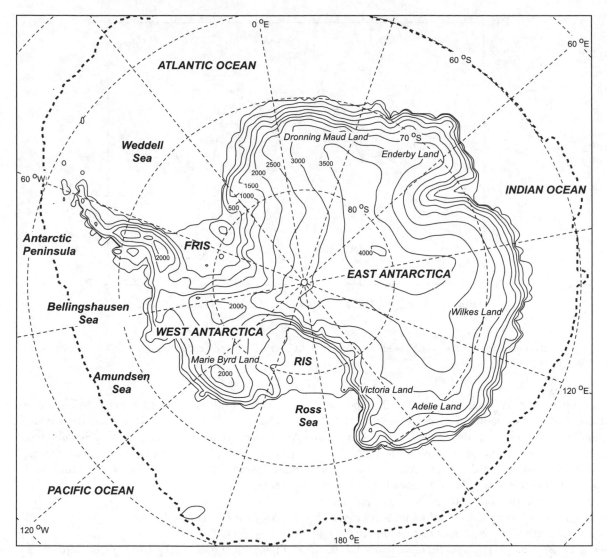

Fig. 1. Model domain, model topography and some topographical features of Antarctica. Stippled areas: model ice shelves; shaded: average July model sea ice extent (1980–93). Surface elevation (m asl) is contoured every 500 m.

points (about 6 km above the surface) that are situated between 600 and 800 km from the Antarctic coastline. The Antarctic coastline we defined as the first model grid point with a glacier mask, which includes the ice shelves. Figure 2 shows that 80% of the variability in U_{LSC} is explained by the AAO index (r = 0.89). In other words, the AAO index is a good measure for the strength of the free atmosphere zonal circulation (or 'polar vortex') over Antarctica. Note also the very high variability in polar vortex strength, with monthly mean values in U_{LSC} ranging from 9 to 20 m s^{-1}.

To distinguish between free atmosphere and surface layer changes, we partition potential temperature Θ (z)

into a background temperature profile Θ_0 (z), representative of free atmosphere conditions, and a potential temperature perturbation near the surface Δ_Θ (z):

$$\Theta\ (z) = \Theta_0\ (z) + \Delta_\Theta\ (z)$$

Note that Δ_Θ (z) is negative in the stable boundary layer over the ice sheet but may become positive in areas where cold continental air is being advected over relatively warm seawater. We assume that Θ_0 has a linear lapse rate $\gamma_\Theta = \partial\Theta_0/\partial z$ that is constant with height:

$$\Theta_0\ (z) = \Theta_0\ (0) + \gamma_\Theta\ z$$

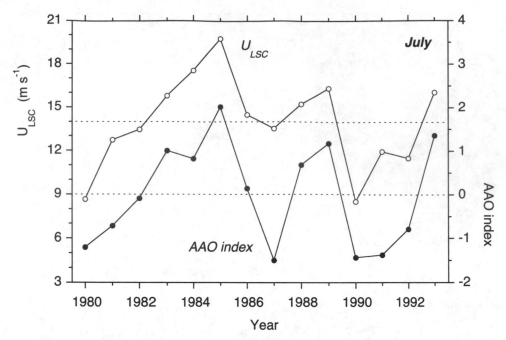

Fig. 2. July mean large scale zonal wind U_{LSC} from RACMO/ANT1 and detrended AAO index for the period 1980–1993 (see text).

γ_Θ is obtained by making a linear fit with respect to z to the potential temperature at model levels 9 to 11.

Finally, we constructed July ensembles for years with *AAO positive polarity* and *AAO negative polarity* (N=7), based on the sign of the deviation from the average in Figure 2. In the following, the difference between these ensembles is presented as *AAO positive polarity* minus *AAO negative polarity*, with a confidence level based on a two-sided t-test with 12 degrees of freedom. In the following figures, regions where changes reach the 95% confidence level are hatched.

3. RESULTS

3.1 Large Scale Circulation Changes

Figure 3 shows 500 hPa height fields (Z_{500}) for *AAO positive polarity* (dashed lines) and *AAO negative polarity* (solid lines). We chose this level because 500 hPa is the first standard pressure level that does not intersect the surface of Antarctica. It also is the main depression steering level and therefore crucial for the near-surface wind field. In the *AAO negative polarity* ensemble, two regional minima are found, one over the Ross Ice Shelf and one east of the Filchner Ronne Ice Shelf. Apart from a general lowering of the 500 hPa level, the second minimum has disappeared in the *AAO positive polarity*

ensemble, resulting in a more zonal flow pattern especially over the Antarctic Peninsula and the Weddell Sea. Meridional gradients in Z_{500} have generally increased over the ocean, indicative of a stronger mid-tropospheric polar vortex with relatively weak north-south air exchange.

3.2 Surface Pressure Changes

Figure 4a shows *AAO positive polarity* (dashed lines) and *AAO negative polarity* (solid lines) ensembles of surface pressure p_s. Because over the ice sheet this variable becomes small, we only present p_s over sea. In both ensembles the circumpolar pressure trough is visible with the South Atlantic, Indian and Pacific climatological storm centres off the Antarctic coast. In the *AAO positive polarity* ensemble the storm centres have deepened by 4 to 8 hPa, assumed a flatter shape and moved closer to the continent. As a result, the meridional component of the geostrophic flow over the ocean has generally decreased. The South Atlantic and Pacific storm centres have shifted about 20 longitudinal degrees to the east. As we will see later, this has important consequences for the regional climate.

Figure 4b shows the *AAO positive polarity - AAO negative polarity* difference in surface pressure (dashed regions indicate changes that reached the 95% confi-

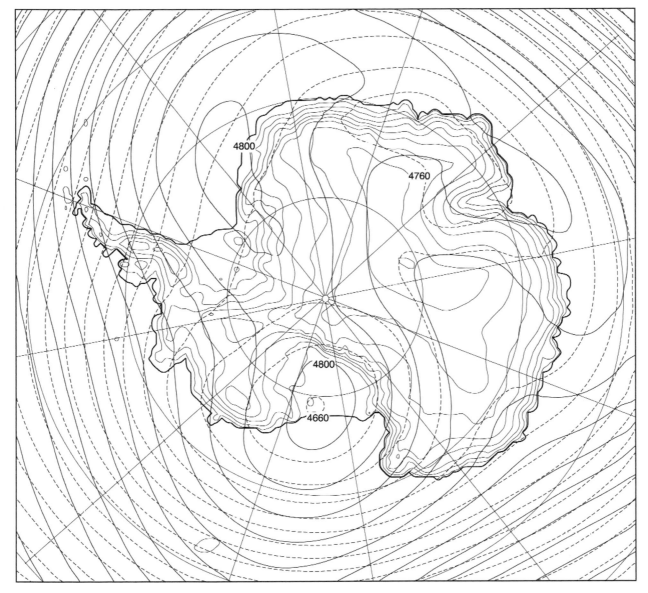

Fig. 3. July 500 hPa height fields (Z_{500}) for *AAO positive polarity* (dashed lines) and *AAO negative polarity* (solid lines).

dence limit). p_s has fallen over almost the entire model domain, as much as 11 hPa over central East Antarctica. This pattern of change represents a large scale intensification of westerly geostrophic winds around Antarctica and a weakening of the easterlies south of the circumpolar pressure trough including the continent (see next section).

Superimposed on this, there are important longitudinal asymmetries. The 20° eastward shift of the South Atlantic storm centre in the *AAO positive polarity* ensemble (Figure 4a) leads to relatively small values of surface pressure change over the Weddell Sea. However, the concurrent deepening and eastward shift of the Pacific

storm centre off the coast of Marie Byrd Land produces a dipole pattern which results in a strong north-westerly flow anomaly over the Weddell Sea and the Antarctic Peninsula. This has a large impact on the regional circulation and temperature, as will be discussed next.

3.3 Surface Layer Wind Changes

Plate 1a shows the July *AAO negative polarity* ensemble mean surface layer wind vector and wind directional constancy dc, the latter being defined as the ratio of the mean to vector mean wind speed.

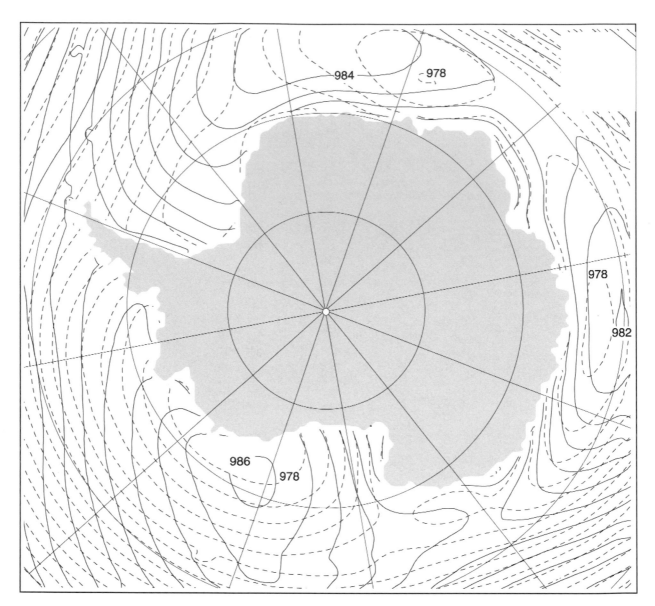

Fig. 4*a*. July surface pressure fields for *AAO positive polarity* (dashed lines) and *AAO negative polarity* (solid lines). Contour interval is 2 hPa.

In good agreement with observations, strong and persistent katabatic winds are modelled over the sloping ice sheet margins. The Coriolis force deflects the katabatic winds to the left of the topographic fall-line, but surface drag maintains a downslope component in the surface layer [*Van den Broeke et al.*, 2002]. With a directional constancy that exceeds 0.9, East Antarctic katabatic winds rival the trade winds as being the most directionally constant winds on Earth. Over West Antarctica and the Antarctic Peninsula the katabatic flow is less well developed because of the weaker surface layer temperature deficit (owing to more

clouds and a warmer atmosphere). Over the domes of the interior East Antarctic ice sheet, far away from the coastal depressions and in absence of the katabatic pressure gradient force, winds are weak and *dc* is low.

In the wintertime Antarctic surface layer the katabatic pressure gradient force dominates the downslope momentum budget, but recent studies have shown that up to 50% of the total downslope pressure gradient force over the ice sheet comes from the large scale pressure gradient force [*Parish and Cassano*, 2001; and Van den Broeke and Van Lipzig, 2002]. This explains for example

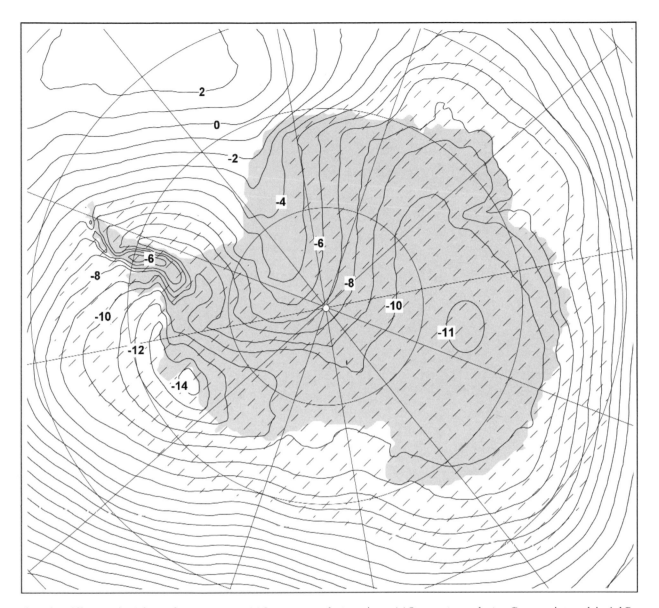

Fig. 4*b*. Difference in July surface pressure: *AAO positive polarity* minus *AAO negative polarity.* Contour interval is 1 hPa. Hatches indicate area where confidence is greater than 95%.

the persistent and strong surface layer easterlies that occur on the plateau in Wilkes Land [*Allison et al.*, 1993]. Over the ice shelves, where the surface slope is very small (in the order of 1/1000) but the surface layer temperature deficit can become very large, katabatic forcing can still be a significant term in the surface layer momentum budget, but only when the large scale flow is weak [*Kottmeier*, 1986; and *King*, 1993].

Over flat surfaces like sea and sea ice, the surface layer flow is driven by the large scale pressure gradient force and/or thermal wind effects. Clear examples in Plate 1*a*

are the strong and directionally constant westerlies north of the circumpolar pressure trough and the easterlies south of it. In between is a belt with low *dc*, where surface layer winds turn from westerly to easterly and the vector mean wind speed becomes small (but not the absolute mean wind speed!). Other manifestations of persistent surface layer circulations that are forced by the large scale pressure gradient force are the southerly winds that blow along the western boundaries of the Ross and Filchner-Ronne ice shelves. These circulations represent the atmospheric branches of the Ross and

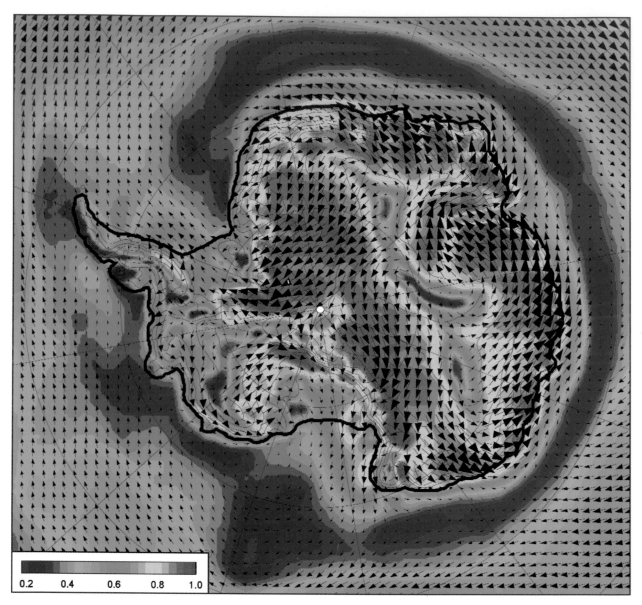

Plate 1*a*. July surface layer wind vector (arrows, maximum 14.7 m s^{-1}) and directional constancy (background colours) for *AAO negative polarity*.

Weddell Gyres that advect cold ice shelf air northward, promoting sea ice and deep water formation in the Ross and Weddell Seas [*Schwerdtfeger*, 1975]. They are primarily forced by the large scale pressure distribution (see solid lines in Figure 4*a*).

Plate 1*b* shows the *AAO positive polarity-low* anomaly field of surface layer vector wind (arrows) and the magnitude of the change (background colours). Over the ocean, the sign of the wind speed change reflects the stronger circumpolar vortex under AAO positive polarity conditions: north of the circumpolar pressure trough surface layer westerlies are enhanced by up to 4 m s^{-1} (and directional constancy has increased by up to 0.3, not shown). South of the circumpolar pressure trough, the circumpolar easterlies have weakened, especially over the Fimbul Ice Shelf around 0° E/W, which is associated with the eastward shift of the South Atlantic storm centre. Easterlies have also decreased significantly over the ocean north of Wilkes Land, in coastal West Antarctica and in the Weddell Sea.

Over the ice sheet the katabatic pressure gradient force dominates the momentum budget in the surface layer. This means that the sign of vector wind speed change in Plate 1*b* is determined by the orientation of the geostrophic flow anomaly with respect to that of the ice sheet slope (which determines the direction of the katabatic pressure gradient force). This explains the dipole patterns centred at the main ice divides of East and West Antarctica in Plate 1*b*. For example, in Dronning Maud Land west of the main ice divide, the westerly geostrophic wind anomalies oppose the katabatic pressure gradient force and surface layer easterlies have weakened by up to 1.5 m s^{-1}. Just east of the divide the situation is reversed. Here, the geostrophic wind anomaly supports the katabatic pressure gradient force, resulting in increased surface layer easterlies by up to 1 m s^{-1}. Similar considerations apply to the dipole pattern in Marie Byrd Land in West Antarctica and for smaller topographic features such as Berkner Island on the Filchner Ronne Ice Shelf.

An anomalous wind response is found over the Antarctic Peninsula. On the west coast of the Antarctic Peninsula, the weak south-westerly flow has traded places with stronger north-westerlies under strong AAO conditions, and wind speed has increased as a result. On the east coast, the migration of the South Atlantic storm centre has led to the cessation of southerly outflow from the Filchner-Ronne Ice Shelf, and wind speeds have decreased in a significant fashion.

3.4 Surface Potential Temperature Changes

As discussed in section 2, we interpret changes in surface potential temperature Θ (z_s) as a superposition of changes in the background surface potential temperature Θ_0 (z_s) and the surface potential temperature perturbation Δ_Θ (z_s) (Plates 2*a-c*). The latter may be interpreted loosely as a changed surface 'inversion' strength, although they are not equal. Also included in Plates 2*a-c* are the contours of 25 and 50% sea ice cover anomalies (dashed lines). Because sea ice is prescribed in RACMO/ANT1 as a layer of constant thickness of 1 m, this change can also be interpreted as a sea ice thickness change.

Plate 2*a* shows a general background cooling over East Antarctica under *AAO positive polarity* conditions. A notable feature is the zonally variable pattern in coastal East Antarctica which is clearly linked to changes in the shape and position of the storm centres: limited/enhanced cooling is found in areas where geostrophic wind anomalies have a northerly/southerly component. As a result, the strongest background cooling (up to 5 K) is found in Enderby Land, Wilkes Land/Victoria Land and Marie Byrd Land (although the change is not significant in the latter region). In between we find areas with little or only weak background cooling.

In strong contrast to the background cooling in East Antarctica is the strong and significant warming in the Weddell Sea, over the Antarctic Peninsula and parts of West Antarctica. The pressure changes in Figure 4*b* show a pronounced northerly circulation anomaly over these regions, representing the weakening of cold air advection from the Filchner-Ronne ice shelf, the atmospheric branch of the Weddell Gyre. The strong sensitivity of the Weddell Gyre to the AAO is interesting given its hemispheric (and possibly global) climatic importance through the formation of sea ice and deep water in the Weddell Sea.

Plate 2*b* presents the change of surface potential temperature perturbation Δ_Θ (z_s). The interpretation of Plate 2*b* is straightforward for regions where sea ice cover has changed in a significant fashion: decreased/increased sea ice cover effectively enhances/reduces upward heat transport with associated increase/decrease of the surface temperature deficit. Over open sea, where surface temperature remains unchanged, the change in Δ_Θ (z_s) must balance the change in Θ_0 (z_s). This effect is especially strong near the sea ice edge, in regions with significant changes in cold air outflow.

Over the land ice, turbulence is mainly generated by wind shear. A decrease in near-surface wind speed as depicted in Plate 1*b* therefore leads to a drop in surface temperature via a diminished turbulent heat transport toward the surface. As a result, areas in East Antarctica where surface layer wind speed has decreased in a significant fashion (Dronning Maud Land west of the main ice divide, Enderby Land and Wilkes Land) show an

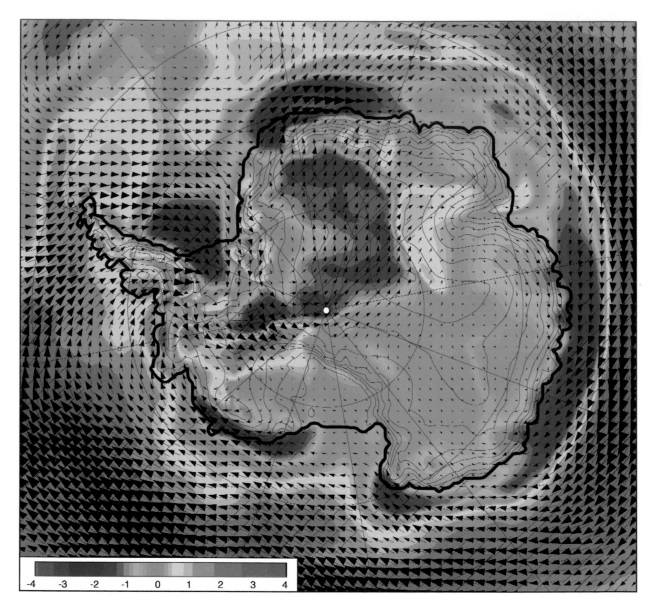

Plate 1*b*. Difference in July surface layer wind vector (arrows) and wind speed (colours): *AAO positive polarity* minus *AAO negative polarity*. Hatches indicate area where confidence is greater than 95%.

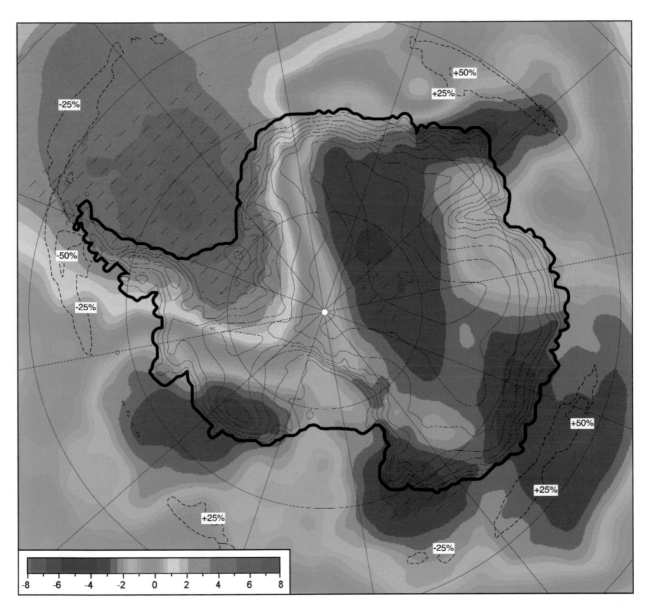

Plate 2a. Difference in July surface background potential temperature $\Theta_0(z_s)$: *AAO positive polarity* minus *AAO negative polarity*. 25 and 50% sea ice changes are indicated by dashed contours. Hatches indicate area where confidence is greater than 95%.

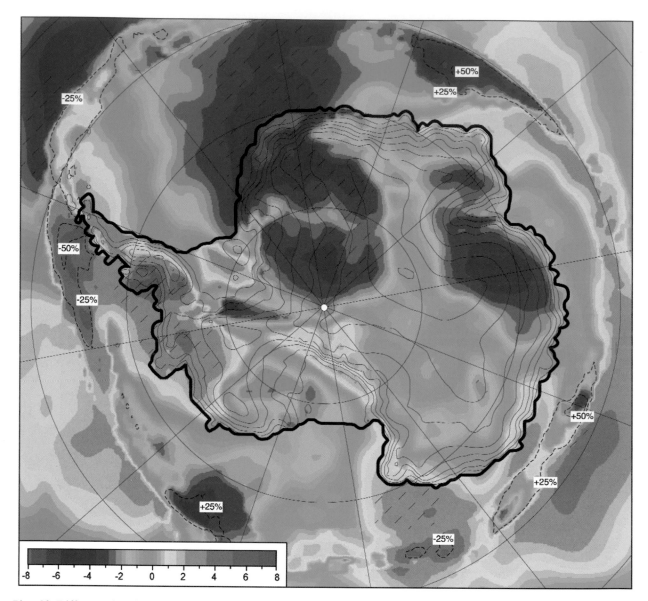

Plate 2*b*. Difference in July surface potential temperature perturbation $\Delta_\Theta(z_s)$: *AAO positive polarity* minus *AAO negative polarity*. 25 and 50% sea ice changes are indicated by dashed contours. Hatches indicate area where confidence is greater than 95%.

intensification of the surface temperature deficit (negative change in Δ_Θ (z_s)). Note that katabatic wind dynamics represent a negative feedback on changes in Δ_Θ (z_s): a stronger surface temperature deficit means a stronger katabatic wind which enhances mixing that in turn reduces the surface temperature deficit.

Over the flat sea ice and ice shelves this negative feedback is largely absent so that Δ_Θ (z_s) changes are larger in both directions. For instance, over the Antarctic Peninsula the temperature rise over the northern ice shelves is enhanced by 2–3 K through an increase in Δ_Θ (z_s), while the 8 K cooling over the narrow ice shelves around 0° E/W results entirely from a drop in Δ_Θ . The west coast of the Antarctic Peninsula experiences a sharp increase of Δ_Θ (z_s) as a direct result of the advection of warm and cloudy air under conditions of positive polarity AAO, resulting in a sharp increase of cloudiness and precipitation [*Van Lipzig and Van den Broeke*, 2002].

Plate 2c shows the total surface potential temperature change, i.e. the summation of Plates 2a and 2b. East Antarctic cooling is largely restricted to the grounded ice. An exception is the sea ice around 0° E/W where a southerly circulation anomaly leads to a decreased cloud cover (not shown) and an enhanced surface temperature deficit.

The very pronounced warming over the Antarctic Peninsula extends over the Filchner Ronne Ice Shelf and parts of West Antarctica and gradually decreases eastward and westward over the sea ice. The strongest warming of up to 8 K is found over areas that have experienced both significant background warming and a decreased surface inversion strength, i.e. on the Ronne Ice Shelf west of Berkner Island and large parts of the Antarctic Peninsula including the northern ice shelves. This is an interesting result in the light of the recent rapid disintegration of the northern Antarctic Peninsula ice shelves. The present picture is consistent with earlier findings that the temperature in the Antarctic Peninsula region is very sensitive to circulation changes [Marshall and King, 1998; and Van den Broeke, 2000]. RACMO/ANT1 output has now clarified this pattern for July, and future studies will address changes in the full annual cycle.

4. CONCLUSIONS

Output of the regional atmospheric climate model RACMO/ANT1 is used to study the influence of the Antarctic Oscillation (AAO) on the wintertime (July) near surface climate of Antarctica. In spite of the relatively short time series (14 years, 1980–1993), strong and significant signals are found. When the AAO polarity is positive (high AAO index), we find that:

- surface pressure is significantly below average over Antarctica. This enhances near-surface westerlies north of the circumpolar pressure trough and opposes easterlies south of it, but with important regional deviations;

- the south Atlantic storm centre is shifted 20° eastward while the Pacific storm centre has deepened; this causes a northerly geostrophic wind anomaly over the Antarctic Peninsula and the Weddell Sea basin, leading to a complete shut-down of the atmospheric branch of the Weddell Gyre. This causes a very pronounced surface warming of up to 8 K over the Antarctic Peninsula, the Filchner-Ronne Ice Shelf and the Weddell Sea;

- in response to intensified westerlies, the katabatic easterlies over West and East Antarctica become weaker, but not uniformly so: dipole patterns of surface layer wind speed change are found at the main ridges of the ice sheets and smaller scale features like Berkner Island in response to the change of large scale pressure gradient relative to the direction of the surface slope;

- significant surface cooling (up to 8 K) occurs in East Antarctica, which can be explained in terms of decreased meridional advection, lowering the background temperature over Antarctica, and decreased sensible heat exchange with the surface, increasing the magnitude of the surface temperature deficit;

- over the ice shelves, the surface temperature response is amplified in absence of the negative feedback that katabatic winds have on surface temperature changes.

The most striking result of this study is the complete shut-down of the cold-pump in the Weddell Sea under conditions of *AAO positive polarity*, leading to anomalous warming in the Antarctic Peninsula and Weddell Sea. This must strongly reduce the formation of sea ice and deep bottom water in the Weddell Sea, which are known to influence climate on the hemispheric and possibly global scale. The amplified warming at the surface of the Antarctic Peninsula ice shelves is interesting in the light of the recent rapid disintegration of some these ice shelves [*Vaughan and Doake*, 1996].

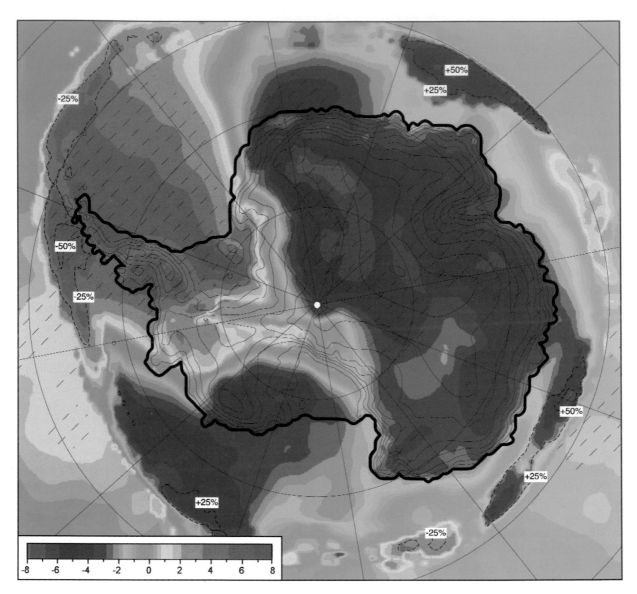

Plate 2c. Difference in July surface potential temperature $\Theta(z_s)$: *AAO positive polarity* minus *AAO negative polarity*. 25 and 50% sea ice changes are indicated by dashed contours. Hatches indicate area where confidence is greater than 95%.

Acknowledgments. Erik van Meijgaard is thanked for supervising the runs with RACMO/ANT. This is EPICA publication no. 64. This work is a contribution to the "European Project for Ice Coring in Antarctica" (EPICA), a joint ESF (European Science Foundation)/EC scientific programme, funded by the European Commission and by national contributions from Belgium, Denmark, France, Germany, Italy, the Netherlands, Norway, Sweden, Switzerland and the United Kingdom.

REFERENCES

Allison, I., G. Wendler and U. Radok, Climatology of the East Antarctic ice sheet (100°E to 140°E) derived from automatic weather stations, *J. Geophys. Res.* 98(D5), 8815-8823, 1993.

Bromwich, D. H., Satellite analyses of Antarctic katabatic wind behavior, *Bull. Am. Meteorol. Soc.* 70, 738-749, 1989.

Burnett, A. W. and A. R. McNicoll, Interannual variations in the southern hemisphere winter circumpolar vortex: relationships with the semiannual oscillation. *J. Climate* 13, 991-999, 2000.

Connolley, W. M., The Antarctic temperature inversion, *Int. J. Climatol.* 16, 1333-1342, 1996.

Connolley, W. M., Variability in annual mean circulation in southern high latitudes, *Climate Dynamics* 13, 745-756, 1997.

Doran, P. T., J. C. Priscu, W. B. Lyons, J. E. Walsh, A. G. Fountain, D. M. McKnight, D. L. Moorhead, R. A. Virginia, D. H. Wall, G. D. Clow, C. H. Fritsen, C. P. McKay and A. N. Parsons, Antarctic climate cooling and terrestrial ecosystem response. *Nature* 415, 517 – 520, 2002.

Enomoto, H. and A. Ohmura, The influence of atmospheric half-yearly cycle on the sea ice extent in the Antarctic, *J. Geophys. Res.* 95 (C6), 9497-9511, 1990.

Harangozo, S. A., Atmospheric meridional circulation impacts on contrasting winter sea ice extent in two years in the Pacific sector of the Southern Ocean, *Tellus* 49A, 388-400, 1997.

Heinemann, G., On the streakiness of katabatic wind signatures on high-resolution AVHRR satellite images: results from the aircraft-based experiment KABEG, *Polarforschung* 66, 19-30, 2000.

King, J. C. and S. A. Harangozo, Climate change in the western Antarctic Pensinsula since 1945: observations and possible causes, *Ann. Glac.* 27, 571-575, 1998.

King, J. C., Control of near-surface winds over an Antarctic ice shelf, *J. Geophys Res.* 98, 12949-12953, 1993.

Kottmeier, C., Shallow gravity flows over the Ekström Ice Shelf, *Boundary-Layer Meteorol.* 35, 1-20, 1986.

Marshall, G. J. and J. C. King, Southern hemisphere circulation anomalies associated with extreme Antarctic Peninsula winter temperatures, *Geophys. Res. Lett.* 25, 2437-2440, 1998.

Parish, T. R. and D. H Bromwich, The surface windfield over the Antarctic ice sheets, *Nature* 328, 51-54, 1987.

Parish, T. R. and J. J. Cassano, Forcing of the wintertime Antarctic boundary layer winds from the NCEP–NCAR global reanalysis, *J. Appl. Meteorol.* 40(4), 810-821, 2001.

Scambos, T. A., C. Hulbe, M. Fahnestock and J. Bohlander, The link between climate warming and break-up of ice shelves in the Antarctic Peninsula, *J. Glaciol.* 154, 516-530, 2000.

Schwerdtfeger, W., The effect of the Antarctic peninsula on the temperature regime of the Weddell sea, *Mon. Wea. Rev.* 103, 45-51, 1975.

Simmonds, I. and D. J. Walland, Decadal and centennial variability of the southern semiannual oscillation simulated in the GFDL coupled GCM, *Climate Dynamics* 14, 45-53, 1998.

Thompson, D. W. J., and J. M. Wallace, Annular modes in the extratropical circulation. Part I: Month-to-month variability. *J. Climate* 13, 1000-1016, 2000.

Thompson, D. W. J. and S. Solomon, Interpretation of recent Southern Hemisphere climate change. *Science* 296, 895-899, 2002.

Van den Broeke, M. R., J.-G. Winther, E. Isaksson, J. F. Pinglot, L. Karlöf, T. Eiken and L. Conrads, Climate variables along a traverse line in Dronning Maud Land, East Antarctica, *J. Glaciol.* 45, 295-302, 1999.

Van den Broeke, M. R., The semi-annual oscillation and Antarctic climate, part 5: Impact on the annual temperature cycle as derived from NCEP re-analysis, *Climate Dynamics* 16, 369-377, 2000a.

Van den Broeke, M. R., On the interpretation of Antarctic temperature time series, *J. Climate* 13, 3885-3889, 2000b.

Van den Broeke, M. R., N. P. M. van Lipzig and E. van Meijgaard, Momentum budget of the East Antarctic atmospheric boundary layer: results of a regional climate model, *J. Atmos. Sci.* 59(21), 3117-3129, 2002.

Van den Broeke, M. R. and N. P. M. van Lipzig, Factors controlling the near surface wind field in Antarctica, *Mon. Wea. Rev.* in press, 2002.

Van Lipzig, N. P. M., E. van Meijgaard and J. Oerlemans, Evaluation of a regional atmospheric climate model for January 1993, using in situ measurements from the Antarctic, *Ann. Glaciol* 27, 507-514, 1998.

Van Lipzig, N. P. M., E. van Meijgaard and J. Oerlemans, Evaluation of a regional atmospheric model using measurements of surface heat exchange processes from a site in Antarctica, *Mon. Wea. Rev.* 127, 1994-2011, 1999.

Van Lipzig, N. P. M., *The surface mass balance of the Antarctic ice sheet: a study with a regional atmospheric model,* PhD thesis, Utrecht University, 1999 [can be obtained from IMAU, PO Box 80005, 3508 TA Utrecht, The Netherlands].

Van Lipzig, N. P. M. and M. R. van den Broeke, A model study on the relation between atmospheric boundary-layer dynamics and poleward atmospheric moisture transport in Antarctica, *Tellus* 54A, 497-511, 2002.

Van Loon, H., The half-yearly oscillation in the middle and high southern latitudes and the coreless winter, *J. Atmos. Sci.* 24, 472-486, 1967.

Vaughan, D. G. and C. S. M. Doake, Recent atmospheric warming and retreat of ice shelves on the Antarctic Peninsula, *Nature* 379, 328-331, 1996.

Walland, D. J. and I. Simmonds, Baroclinicity, meridional temperature gradient and the southern semiannual oscillation, *J. Climate* 12, 3376-3382, 1999.

Michiel R. van den Broeke, Institute for Marine and Atmospheric Research Utrecht, P.O. Box 80005, 3508 TA Utrecht, The Netherlands.

Nicole P. M. van Lipzig, British Antarctic Survey, High Cross, Madingley Rd., CB3 0ET Cambridge, UK.

Glaciological Climate Relationships

ANTARCTIC PENINSULA CLIMATE VARIABILITY
ANTARCTIC RESEARCH SERIES VOLUME 79, PAGES 61-68

SPATIAL AND TEMPORAL VARIATION OF SURFACE TEMPERATURE ON THE ANTARCTIC PENINSULA AND THE LIMIT OF VIABILITY OF ICE SHELVES

Elizabeth M. Morris[1] and David G. Vaughan

British Antarctic Survey, Natural Environment Research Council, Cambridge, United Kingdom

Mapping surface air temperature in the Antarctic Peninsula region is made unusually difficult by: the scarcity of meteorological stations, strong climatic gradients and recent rapid regional warming. We have compiled a database of 534 mean annual temperatures derived from measurements of snow temperature at around 10-m depth and air temperature measured at meteorological stations and automatic weather stations. These annual temperatures were corrected for interannual variability using a composite record from six stations across the region. The corrected temperatures were then analysed using multiple linear regression to yield altitudinal and temporal lapse rates. A subset of 508 values were then used to produce a map of temperature reduced to sea level and for a specific epoch (2000 A.D.). The map shows the dramatic climate contrast (3–5°C) between the east and west coast of the Antarctic Peninsula in greater detail than earlier studies and also indicates that the present limit of ice shelves closely follows the –9°C (2000 A.D.) isotherm. Furthermore, the limit of ice shelves known to have retreated during the last 100 years is bounded by the –9°C and –5°C (2000 A.D.) isotherms, suggesting that the retreat of ice shelves in the Antarctic Peninsula region is consistent with a warming of around ~ 4°C.

1. INTRODUCTION

It has been long-understood, and widely discussed [e.g. *Schwerdtfeger*, 1974], that the high topography over the Antarctic Peninsula (Figure 1) forms a significant climatic barrier between the warm Bellingshausen Sea and the colder Weddell Sea. However, the sparse network of climatological observing stations in the Antarctic Peninsula region is sufficient to reveal only the broadest details of this climatic divide. An earlier attempt to map mean annual air temperature supplemented the station data with mean annual air temperatures inferred from ice cores and boreholes [*Reynolds*, 1981], and has helped us to understand the nature of the climatic boundary [e.g. *King and Turner*, 1997] and to interpret the geographical limit of

ice shelves around the Antarctic Peninsula as a thermal condition [*Vaughan and Doake*, 1996]. However, there are significant limitations to *Reynolds'* analysis; for example, he noted that "if more temperature data were available it would be better if a temporal regression were to be incorporated into the statistical analysis".

We now have the opportunity to improve *Reynolds'* map by including new station data, substantially more borehole temperatures, data from automatic weather stations, and by removing interannual and trend anomalies from the data. We thus present an updated map of mean annual temperature over the Antarctic Peninsula and discuss its significance.

2. DATA

A database of mean annual surface temperature for the Antarctic Peninsula region (between 40°W and 105°W,

---

[1]Now at: Scott Polar Research Institute, University of Cambridge, United Kingdom

10.1029/079ARS05

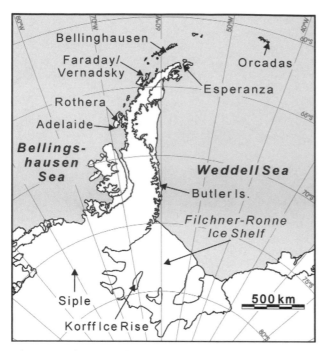

Fig. 1. Location map of the Antarctic Peninsula

2.1 Expedition Data

Measurements of air temperature in the Antarctic Peninsula were sparse before the mid-20th century, when the first scientific bases were established. However, *Jones* [1990] listed mean annual temperatures from 10 expeditions since 1898 that wintered on or near the Antarctic Peninsula. We have discarded values based on less than 12 months of observation but have used seven mean values based on yearlong measurements at 6, 8 or more observations at fixed times during the day (Table 1).

2.2 Station Data

Since the mid-20th century, surface air temperature has been measured at scientific stations in the Antarctic Peninsula region, although not all stations achieve an unbroken record. The SCAR READER project (Reference Antarctic Data for Environmental Research) has created a high quality, long-term data set of mean surface and upper air meteorological measurements from *in situ* observations. The data set, which is updated regularly as the project progresses, is available from the READER web site hosted by the British Antarctic Survey (www.antarctica.ac.uk). Table 2 lists the meteorological stations and the period during which some or all years have a percentage of observations high enough to calculate an accurate mean annual temperature. We have used the original READER criteria that, for each of 12 months, the data must be >80% complete and there must be no breaks of more than 5 days, to select data for inclusion in our database. (The first criterion has now been made more stringent for READER data by raising the required percentage to >90%). We have selected 210 separate years of mean annual surface air temperatures from 22 stations. Note that the focus of this paper is on using all sources of mean annual temperature data; we are not here concerned with other station data, such

and between 60°S and 83°S) has been constructed using 239 measurements of mean annual air temperature from expeditions, meteorological stations and automatic weather stations, and 295 measurements of snow temperature at around 10 m depth [*Morris et al.*, 2002]. We have not used proxy temperature data from ice cores, for example, oxygen isotope measurements, because the mean value of the proxy variable over an annual layer may not be an accurate indication of mean annual temperature in this area [*Vaughan et al*, in press]. Ice core data are indeed useful for studying temperature variations over different timescales, but in this paper we are primarily concerned with producing a map of mean annual temperature based on values derived from direct measurements of temperature.

TABLE 1. Mean annual surface air temperatures measured by expeditions [*Jones*, 1990]. All sites were on land, except for *Belgica*, which was trapped in sea ice to the west of the Antarctic Peninsula in 1898 and for which we give a mean position.

Site	Latitude/°S	Longitude/°W	Elevation/ m.a.s.l.	Date or Period	Number of months
Belgica	70.62	88.58	5	1898	12
Snow Hill Island	64.50	56.93	13	1902-03	18
Port Charcot	65.07	64.03	9	1904	12
Water Boat Point	64.8	62.72	3	1921	12
Winter Island	65.25	64.27	34	1935	12
Barry Island	68.13	67.10	15	1936	12
Stonnington Island	68.18	67.03	15	1940	12

TABLE 2. Details of records of mean annual air temperature from meteorological observing stations included in the database.

Station	Latitude/°S	Longitude/°W	Elevation/m.a.s.l.	Period	Number of complete years
Siple	75.93	84.25	1054	1979-86	5
Orcadas	60.75	44.72	6	1996-2000	3
Deception	63.98	60.57	8	1959-66	8
Esperanza	63.4	56.98	13	1997-99	3
San Martin	68.13	67.10	4	1996-99	4
Faraday/Vernadsky	65.25	64.26	11	1951-95	45
Arturo Prat	62.5	59.68	5	1996-2000	5
O'Higgins	63.32	57.90	10	1996-2000	5
Marsh	62.2	58.96	10	1996-2000	4
Adelaide	67.77	68.93	26	1963-74	12
Bellingshausen	62.20	58.96	16	1970-2000	13
Marambio	64.24	56.66	198	1996-2000	5
Rothera	67.57	68.12	16	1978-2000	22
Great Wall	62.22	58.96	10	1996-2000	4
Ferraz	62.08	58.39	20	1995-99	4
Jubany	62.24	58.66	4	1996-99	4
Arctowski	62.16	58.47	2	1978-96	15
Almirante Brown	64.88	62.88	7	1952-83	16
Hope Bay	63.4	56.98	10	1953-59	7
Matienzo	64.97	60.05	32	1962-75	10
Admiralty Bay	62.08	58.42	11	1951-60	10
Petrel	63.47	56.22	18	1968-75	6

as summer temperatures which are, of course, valuable for studies of other aspects of the Antarctic Peninsula climate

2.3 Automatic Weather Stations

The mean annual air temperatures recorded by an automatic weather station (AWS) are unlikely to be as accurate as those derived from manned stations but AWSs do yield data at remote locations that would otherwise not be sampled. We have included 22 measurements of mean annual surface air temperature from the 6 Automatic Weather Stations included in the READER database at the time of writing (Table 3). The same criteria for inclusion are used as for the station records.

2.4 Ten-Metre Snow Temperatures

Mean annual air temperature at the snow surface is commonly estimated by assuming that it is the same as the snow temperature measured at a depth of 10 m [*Paterson*, 1994]. The accuracy of this assumption depends on the stratigraphy of the snow, the shape of the annual cycle of air temperature and any long-term temperature trend. In a previous paper [*Morris and Vaughan*,

1992], we described a technique to remove the effects of measurement depth and date from borehole temperatures. We have applied the same technique to 10-m temperatures in this study. We assume that the accumulation, densification and metamorphosis of polar snow are controlled by the local climate, so that the snow cover at sites with the same temperature history, $T_s(t)$, develops broadly the same physical properties. In particular, we suppose that the thermal diffusivity in the upper 10 m of snow is similar for sites of similar $T_s(t)$. In this case, a standard set of snow temperature curves, $T(d,t)$, will apply for all sites with the same surface temperature history. The mean annual surface temperature, T_m, is then obtained by subtracting a correction, $aF(T_s,d_1,t)$, from the measured borehole temperature, $T(d_1,t_1)$. The amplitude of the first harmonic of the annual temperature wave, a, is known, and ranges from about 4°C at the northern end of the Antarctic Peninsula to 10°C at the South Pole. On the southern part of the Peninsula and on Filchner-Ronne Ice Shelf a is about 11°C. The function F is known at sites where snow temperatures have been measured at different depths throughout the year. For the northern part of the Antarctic Peninsula we estimated F from data recorded at Maudheim Station [*Dalrymple et al.*, 1966] and for southern areas, data recorded at Plateau

TABLE 3. Automatic Weather Stations

Station	Latitude/°S	Longitude/°W	Elevation/m.a.s.l.	Period	No. of complete years
Bonaparte Point	64.8	64.1	8	1992	1
Butler Island	72.21	60.17	91	1990-98	7
Limbert	75.42	59.95	40	1996-97	2
Racer Rock	64.07	61.61	17	1990-91	2
Siple	75.92	84.25	1054	1986-91	6
Uranus Glacier	71.43	68.93	780	1987-95	4

Station [*Weller and Schwerdtfeger*, 1977]. Borehole temperatures are normally made in the austral summer when the snow at 10-m depth is beginning to warm up after the arrival of the winter cold wave through the snow. Thus the corrected mean annual surface temperature is of the order of 0 - 0.5°C warmer than the borehole temperature [*Morris and Vaughan*, 1992].

We have extended the database of borehole measurements used in *Morris and Vaughan* [1992] by adding published and unpublished data corrected for depth and time of year. This yields a total of 285 estimates of mean annual surface air temperature. These are supplemented by a further 11 sites for which the date of the borehole measurement is unknown but which can be used for spatial analysis.

3. ANALYSIS

Broadly speaking, the 10-m temperature data give most information on spatial variability and the station data give most information on temporal variability. The key to our approach is, however, that both types of data are included in a unified analysis, aimed at producing a map of mean annual temperature for a specific epoch.

3.1 Assumptions

We assume that the mean annual surface air temperatures, T_s, can be represented by

$$T_s = \Phi \varphi + \Lambda \lambda + Z z + a_0 t + a_1(t) \qquad (1)$$

where for each measurement, φ is latitude, λ, longitude, z, elevation above mean sea level and t is date of measurement. The constants, Φ, Λ, Z and a_0 denote spatial and temporal lapse rates. The assumption of constant rate of change in mean annual temperature over the period covered by the database (102 years) is a simplification, but we do not consider that the amount of data available, especially from the earlier years, justifies using a more

complex expression in equation (1). Use of a single function, $a_1(t)$, to describe the interannual variability implies that this is constant over the domain. This assumption is supported by recent findings [*King and Comiso*, in press] that there is a good correlation (mostly > 0.5) between winter temperatures as expressed by passive microwave emissivity across most of the Antarctic Peninsula and the meteorological temperatures at Faraday (now re-named Vernadsky).

To estimate $a_1(t)$, the inter-annual variability, we first estimated mean annual temperature for all years (not only those which have data that satisfy the criteria we set for inclusion in the mean annual temperature database) for each of the stations with long records (Orcadas, Faraday/Vernadsky, Bellingshausen, Rothera, San Martin and Esperanza). We then calculated the deviation from the trend, for each of these stations. The mean of the surface temperature anomalies is used as an estimate, $\langle a_1(t) \rangle$, for $a_1(t)$. This was then used to calculate a temperature, $T = T_s - \langle a_1(t) \rangle$, for each value of T_s in the database. All station data for a given year are corrected by the same mean anomaly.

The above process distinguishes between mean annual temperature data which are adequate for calculating mean anomalies and those which are good enough to be included in the database. For example, 98 estimated mean annual temperatures for Orcadas (where records began in 1904) were used in the calculation for $\langle a_1(t) \rangle$. However, only 3 of these years have 12 months of more than 80% complete data and so only 3 mean annual temperatures for Orcadas are included in the database.

Borehole temperatures will also show inter-annual variations but these will be damped and lagged with respect to the variations in snow surface temperature. We have estimated the mean annual 10-m temperature anomaly series $a_2(t)$ by using the Faraday/Vernadsky temperature record as the upper boundary condition for a physics-based distributed snow model which can be used to predict the temperature variation at depth given the physical properties of the snow cover. Each mean annu-

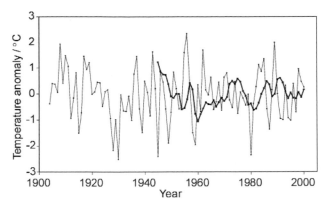

Fig. 2. Composite record of (i) inter-annual variability derived from long-term station data and (ii) smoothed and lagged record used to correct 10-m temperature data.

al surface air temperature, T_e, estimated from a borehole measurement was corrected to give a temperature, $T = T_e - a_2(t)$. Figure 2 shows $\langle a_1(t) \rangle$ and $a_2(t)$.

3.2 Linear Regression

We performed a multivariate linear regression analysis on the uncorrected and corrected mean annual temperatures (Table 4). Discarding expedition data before 1904 decreases the apparent temporal trend slightly but does not significantly alter the spatial lapse rates.

Correcting the data for inter-annual variability reveals a long-term time trend for the whole Antarctic Peninsula region of $(2.0 \pm 0.9)°C$ over the last 100 years and improves the value of r^2 to 0.93. However, noting that *Morris and Vaughan* [1994] found different lapse rates for sites west and east of the topographic divide, we also separated our data into western and eastern groups. The southern limit was set at 80°S. This improved the correlation for the western sites and increased the temporal trend to $(3.8 \pm 0.7)°C$ over the last 100 years. This is in agreement with *Morris and Vaughan's* previous value for western sites of $(4.0 \pm 7.6)°C$ century[-1] (note the greatly

improved uncertainty). The trend for the eastern sites $(-1.3 \pm 1.5)°C$ century[-1] gives no evidence for warming.

Because of the strong winter inversion over Filchner-Ronne Ice Shelf, mean annual temperature increases with altitude up to about 700 m.a.s.l. Separating eastern sites above 700 m.a.s.l. and eastern sites below 700 m.a.s.l. improves the correlation but in neither case is there any significant warming or cooling. *Morris and Vaughan* [1992] did find a warming of $(2.7 \pm 0.8)°C$ century[-1] for eastern sites below 700 m.a.s.l. but they used data from the eastern side of the Filchner-Ronne Ice Shelf so the domain is not the same

3.3 Spatial Variation

We produced a grid representing the spatial variation of mean annual air temperature corrected to sea level and a specific epoch (2000 A.D.) using an Arc/Info subroutine "Topogrid" to interpolate between the 534 corrected values in the data base.

Using this grid we could identify four 10-m temperatures (one temperature measured in 1962 by *Robin* and three temperatures measured by *Walton* in 1975 [*Reynolds*, 1981]) which did not fit the broad-scale pattern. This could be because of unusual conditions at the site, but there is also the possibility of instrumental error.

We also identified an apparent error in the position of the 10-m temperature measured in 1974 on Butler Island and corrected it to agree with the position of the Butler Island AWS.

One 10-m temperature comes from a site (the summit of Korff Ice Rise) where the presence of a boundary layer inversion means that using the global elevation lapse rate to correct it to sea level leads to an anomalous value. Furthermore, the 21 ten-metre temperatures warmer than $-7°C$ may not be good estimates of mean annual temperature because of melt effects [*Reynolds*, 1981].

We therefore repeated the interpolation using 508 values i.e. excluding the 26 values discussed above. Plate 1(a)

TABLE 4. Lapse rates, trends and correlation coefficients from multivariate regression of Antarctic Peninsula temperatures. Uncertainties are give at 1-sigma level

Set		Φ /°C deg[-1]	Λ /°C deg[-1]	Z /°C m[-1]	a /°C a[-1]	r^2
{A}	All temperatures	1.47 ± 0.03	-0.12 ± 0.02	-0.0044 ± 0.0002	0.037 ± 0.010	0.91
{B}	All {A} after 1904	1.47 ± 0.03	-0.11 ± 0.02	-0.0044 ± 0.0002	0.035 ± 0.010	0.92
{C}	All {B} corrected for variability	1.48 ± 0.03	-0.10 ± 0.02	-0.0045 ± 0.0002	0.020 ± 0.009	0.93
{D}	All {C} W of divide and N of 80°S	0.54 ± 0.05	0.15 ± 0.02	-0.0072 ± 0.0002	0.038 ± 0.007	0.95
{E}	All {C} E of divide and N of 80°S	1.43 ± 0.06	-0.007 ± 0.032	-0.0033 ± 0.0004	-0.013 ± 0.015	0.93
{F}	All {E} above 700 m.a.s.l.	0.63 ± 0.08	0.23 ± 0.04	-0.0031 ± 0.0006	0.0036 ± 0.015	0.95
{H}	All {E}.below 700 m.a.s.l.	1.47 ± 0.06	0.07 ± 0.05	-0.0058 ± 0.0017	0.013 ± 0.018	0.96

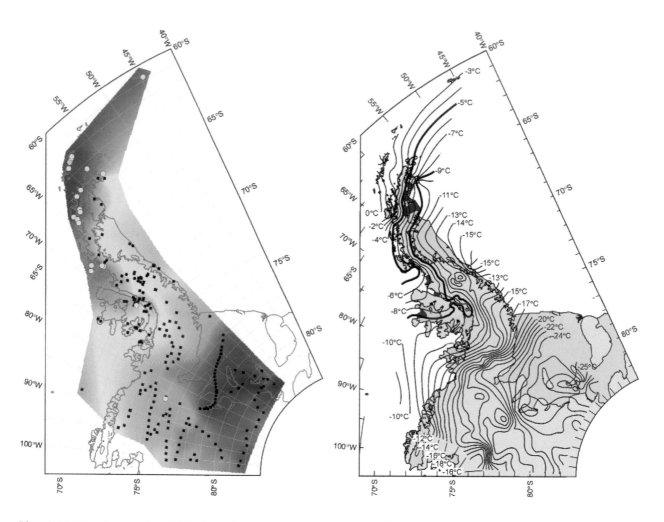

Plate 1 (a) Map showing the distribution of measurements of mean annual temperature used in the compilation overlaid on a colour image of the interpolated mean annual temperature. Grey circles show the station positions and black squares sites for all other data (b) Contours of interpolated mean annual temperature. Grounded ice and ice-free ground is shown in grey, and extant ice shelves are shown in blue. Portions of ice shelves that have been lost through climate-driven retreat are shown in red. Note: the Mhller (67°13'S, 66°50'W) and Jones (67°30s, 66°55'W) Ice Shelves were too small to be clear at this scale—their loss is indicated by red dots of nominal size.

shows the complete field of temperature corrected to sea level and Plate 1(b) shows contours of the same grid. These maps of spatial variability in mean annual temperature reveals some interesting points: First, while the juxtaposition of isotherms and the spine of the Antarctic Peninsula confirms the effect of topography in creating a climatic barrier, we find that, compared to *Reynolds'* map, the climatic divide between the east and west coast of the Antarctic Peninsula is less steep. We find no evidence that the climatic divide is displaced to the east of the topographic divide as suggested by *Reynolds*. Second, as a result of the correction for inter-annual variability, two of the three features referred to as "anomalies" by *Reynolds* have disappeared. We do not find higher than expected temperatures over James Ross Island or a 2°C difference between the west coast of Adelaide Island and north-east Marguerite Bay. On the other hand, we, also, find a warm basin east of Alexander Island. In our case the warm "tongue" lies along the low-lying George VI Ice Shelf, supporting *Reynolds'* suggestion that channelling of the weather along George VI Sound contributes to the relatively warm area east of Alexander Island.

3.4 Temporal Variation

With the exception of Siple Station, which has only a 5-year record, the meteorological stations in the Antarctic Peninsula region are in the north of our region and close to sea level. The combination of the Adelaide and Rothera records gives the most southerly record of mean annual temperature. Thus the station data only give a very limited view of the geographical extent of the recent warming.

We have used linear regression analysis of data covering the entire region as far as 83°S to estimate a constant warming rate for the entire area over the last 96 years. Importantly, this value, $(3.5 \pm 1.0)°C$ century^{-1}, is not significantly different to the trends reported for each of the long-term stations: Faraday/Vernadsky, $(5.7 \pm 2.0)°C$ century^{-1}; Bellingshausen, $(3.7 \pm 2.1)°C$ century^{-1}; Esperanza, $(3.4 \pm 1.3)°C$ century^{-1}; Orcadas, $(2.0 \pm 1.0)°C$ century^{-1} [*Vaughan et al.*, in press] over varying periods of the last century. The residuals from the linear regression analysis do not vary with time, so there is no evidence from our data of an increasing warming rate over the period.

Vaughan et al. pointed out that there are no coastal station data of sufficient duration to show whether the coast between the Antarctic Peninsula and the Ross Sea is

warming at a similar rate to the Antarctic Peninsula. We note however, that a warming is apparent at the seven 10-m temperature sites in Ellsworth Land that were visited in 1961/62 during the Antarctic Peninsula Traverse [*Shimizu*, 1964] and were revisited in 1997/98 – direct evidence that the warming is not confined to the northern part of the Antarctic Peninsula, but does extend south into West Antarctica. This point will be discussed in greater detail elsewhere.

4. ICE-SHELF DISTRIBUTION

The possibility that climate controls the viability of ice shelves has been widely discussed [*Mercer*, 1978; *Reynolds*, 1981; *Vaughan and Doake*, 1996]. The new map of mean annual temperature presented here allows us to examine in more detail the present position and rate migration of a limit of viability.

Plate 1(b) shows the distribution of ice shelves, which remain around the Antarctic Peninsula in 2002 and those which have shown climatically driven retreat since the beginning of the 20th Century [*Ward*, 1995; *Vaughan and Doake*, 1996; *Cooper*, 1997; *Scambos et al.*, 2000; *Fox and Vaughan*, in press; *Scambos et al.*, in press]. We note the following:

1. All the ice shelves that have been mapped on the Antarctic Peninsula, since the beginning of detailed exploration (c. 1898), have occurred to the south of the –5°C (2000 A.D.) isotherm.
2. The present (2002) limit of ice shelf distribution is well-approximated by the –9°C (2000 A.D.) isotherm.
3. All the ice shelves that lie between these two isotherms have shown significant progressive retreat or total loss, while none of the ice shelves south of the –9°C (2000 A.D.) isotherm have been reported as showing any progressive retreat.

In Section 3.4 we estimated that regional warming on the Antarctic Peninsula has occurred at a rate of (3.5 ± 1.0) °C century^{-1} over the last 96 years. This is entirely consistent with an apparent migration of the limit of viability between the present-day –5°C and –9°C isotherms for the period (little more than a century) for which we have maps of ice shelf distribution.

The connection between the boundary of retreating ice shelves and the mean annual air temperature contour suggests that there is an atmospheric (rather than perhaps a purely oceanographic) control on ice shelf retreat. We note that the amplitude of the first harmonic of the annual temperature wave (section 2.4) ranges from 4°C in the north to 11°C in the south of the region and suggest therefore that the –9°C contour indicates the onset of summer melt.

5. SUMMARY

We have compiled a new database of mean annual surface temperatures over the Antarctic Peninsula, and from this we have evaluated the spatial and temporal variability. The resulting map of mean annual air temperature, reduced to sea level, shows very strong climatic gradients both from north to south and from east to west, although the east/west boundary is not so localized as previous estimates suggested [e.g. *Reynolds*, 1981]. Our regression analysis shows a single value for warming for the whole region of $(3.5 \pm 1.0)°C$ century^{-1}, which is similar to those reported for the long-term meteorological stations.

The present limit of ice shelves is well approximated by the –9°C (2000 A.D.) isotherm. The ice shelves which have retreated since the early 1900s all lie between the –9°C and –5°C (2000 A.D.) isotherms. This distribution supports the hypothesis of a climate-controlled limit of viability for ice shelves that is marching south as temperature rises.

Acknowledgments. We wish to thank Veronika Hawenek for assistance with construction of the 10-m temperature database. Gareth Marshall and Stephen Colwell provided quality-controlled air temperature data from the READER Antarctic meteorological database.

REFERENCES

Cooper, A.P.R., Historical observations of Prince Gustav Ice Shelf, *Pol. Rec. 33*, 285-294, 1997.

Dalrymple, P.C., H. Lettau, and H. Wollaston, South Pole micrometeorology program (data analysis), in *Studies in Antarctic Meteorology*. M.J. Rubin, (Ed.) *Antarc. Res. Ser., 9*, 13-57, 1966.

Fox, A.J., and D.G. Vaughan, The retreat of Jones Ice Shelf, Antarctic Peninsula, *Geophys. Res. Letter,* in press 2003.

Jones, P.D., Antarctic temperatures over the present Century—a study of the early expedition record. *J. Clim. 3*, 1193-1203, 1990.

King, J.C., and J.C. Comiso, The spatial coherence of interannual temperature variations in the Antarctic Peninsula, *Geophys. Res. Letter,* in press 2003.

King, J.C., and J. Turner, *Antarctic Meteorology and Climatology*, pp. 409. Cambridge University Press, New York, 1997.

Mercer, J.H., West Antarctic ice sheet and CO2 greenhouse effect: a threat of disaster, *Nature, 271*, 321-325, 1978.

Morris, E.M., V. Hawanek, and D.G. Vaughan, Compilation of mean annual surface temperatures in the Antarctic Peninsula region, British Antarctic Survey. ES4/8/2/2002, Cambridge. 2002.

Morris, E.M., and D.G. Vaughan, Snow surface temperatures in West Antarctica, in *The Contribution of the Antarctic Peninsula to Sea Level Rise. Ice and Climate Special Report 1*, E.M. Morris (Ed.), British Antarctic Survey, Cambridge. 17-24, 1992.

Morris, E.M., and D.G. Vaughan, Snow surface temperatures in West Antarctica. *Ant. Sci. 6*(4), 529-535, 1994.

Paterson, W.S.B., *The Physics of Glaciers*, pp. 480, Elsevier, Oxford, 1994.

Reynolds, J., The distribution of mean annual temperatures in the Antarctic Peninsula, *Br. Antarct. Surv. Bull. 54*, 123-133, 1981.

Scambos, T., C. Hulbe, M. Fahnestock, and J. Bohlander, The link between climate warming and break-up of ice shelves in the Antarctic Peninsula, *J. Glaciol. 46*(154), 516-530, 2000.

Scambos, T., C. Hulbe, and M. Fahnestock, Climate-induced ice shelf distintegration in Antarctic, in press, *J. Geophys. Res.* 2003.

Schwerdtfeger, W., The Antarctic Peninsula and the temperature regime of the Weddell Sea, *Ant. J. U.S. 9*(5), 213-214, 1974.

Shimizu, H., Glaciological studies in West Antarctica, 1960-1962, in *Antarctic Snow and Ice Studies*, M. Mellor, (Ed.) *Antarc. Res. Ser., 2*, 37-64, 1964.

Vaughan, D.G., and C.S.M. Doake, Recent atmospheric warming and retreat of ice shelves on the Antarctic Peninsula, *Nature, 379*, 328-331, 1996.

Vaughan, D.G., G.J. Marshall, W.M. Connolley, C.L. Parkinson, R. Mulvaney, D.A. Hodgson, J.C. King, C.J. Pudsey, and J. Turner, Recent rapid regional climate warming on the Antarctic Peninsula, *Climatic Change,* in press, 2003.

Ward, C.G., The mapping of ice front changes on Mhller Ice Shelf, Antarctic Peninsula, *Ant. Sci., 7*(2), 197-198, 1995.

Weller, G., and P. Schwerdtfeger, Thermal properties and heat transfer properties of low temperature snow, in *Meteorological Studies at Plateau Station, Antarctica*, J.A. Businger, (Ed.), *Antarc. Res. Ser., 25*(2), 27-34, 1977.

E.M. Morris, Scott Polar Research Institute, University of Cambridge, Lensfield Road, Cambridge, CB2 1ER, United Kingdom.

D.G. Vaughan, British Antarctic Survey, High Cross, Madingley Road, Cambridge, CB3 0ET, United Kingdom.

ANTARCTIC PENINSULA CLIMATE VARIABILITY
ANTARCTIC RESEARCH SERIES VOLUME 79, PAGES 69-78

IMPACT ASSESSMENT OF REGIONAL CLIMATIC WARMING ON GLACIERS AND ICE SHELVES OF THE NORTHEASTERN ANTARCTIC PENINSULA

Pedro Skvarca and Hernán De Angelis

División Glaciología, Instituto Antártico Argentino, Buenos Aires, Argentina

The impact of regional climatic warming on grounded and floating ice masses in the northeastern Antarctic Peninsula is being assessed. Associated with the increasing atmospheric warming trend of the last two decades large thinning rates and negative mass balance were measured on a glacier with termini on land on Vega Island. Clear signals of glacier surface lowering were also detected on nearby James Ross Island, where monitoring of tidewater calving glaciers with sequential satellite imagery indicates an increasing rate of retreat during the last decade. However, the most striking evidence of climatic impact on ice masses is the abrupt disintegration of sections of the northern Larsen Ice Shelf that occurred in early 1995 and 2002, respectively, in coincidence with the two warmest summers recorded in the region. The final stage of these sudden and catastrophic events, a consequence of ice-shelf imbalance, has been triggered by a surplus of surficial meltwater, product of the warmest summers. From late 1975 to early 2002 the area of Larsen Ice Shelf north of Jason Peninsula has been reduced by 12,260 km^2, of which 76 percent was lost since the end of January 1995. The disintegration of these sections of the ice shelf has not only altered dramatically the geography of the northeastern Antarctic Peninsula but also triggered the retreat of former tributary glaciers beyond their grounding lines, an early warning of possible subsequent global sea level rise.

1. INTRODUCTION

Of particular interest for ice-climate interaction studies at a global scale are those glaciers that terminate on land, because their activity can be directly linked to atmospheric changes. In Antarctica such glaciers are scarce. On the northeastern Antarctic Peninsula (AP) a few glaciers with termini on land are located on Vega Island (VI) and northern James Ross Island (JRI) (Figure 1). To investigate the response of glaciers in this region to climate a test site was selected on VI, a glacier informally called "Glaciar Bahía del Diablo" (GBD) (Figure 2). Furthermore, a base line for glacier fluctuation monitoring has been established for the nearby JRI, based on satellite images from different sources that extend back to 1975 [*Skvarca et al.*, 1995].

The prediction of *Mercer* [1978] that ice shelves around the AP would start disintegrating in response to the atmospheric warming became a reality. The most dramatic events of area loss have occurred during the last decade on the northern part of the Larsen Ice Shelf (LIS). We refer to the disintegration of ice shelves in Prince Gustav Channel (PGC) and Larsen A, occurred in 1994–95 [*Rott et al.*, 1996; *Vaughan and Doake*, 1996; *Rott et al.*, 1998], and Larsen (B) Ice Shelf, which collapsed in 2001–02. Recent evidence from marine sediment cores indicates that ice shelves in the PGC and Larsen A also retreated in the middle Holocene, according to *Pudsey and Evans* [2001] and *Domack et al.* [2001]. The latter attribute the readvance of Larsen (A) Ice Shelf to the extended cold period and persistent sea-ice conditions of the last 2500 years. However, a preliminary

10.1029/079ARS03

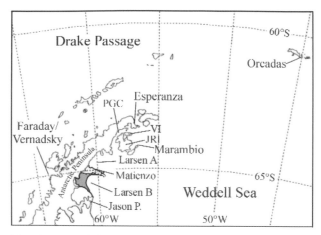

Fig. 1. Map of the northern Antarctic Peninsula indicating meteorological stations and sites of ice-climate interactions discussed in the text. The disintegrated part of Larsen (B) Ice Shelf in 2002 is shown in grey. JRI = James Ross Island, PGC = Prince Gustav Channel, VI = Vega Island.

analysis of marine sediment cores recovered in January 2002 reveals that the recent disintegration of Larsen (B) Ice Shelf is an unprecedented event in the past 9,000 years [*Domack et al.*, 2002]. The objective of this paper is to quantify and discuss the impact of recent atmospheric warming on glaciers and ice shelves of the northeastern AP, using updated climatic data and field measurements.

2. CLIMATIC CONDITIONS IN THE NORTH-EASTERN AP DURING THE 20TH CENTURY

For a better understanding of the drastic changes occurring in the northeastern AP, it is essential to analyse in detail the temporal pattern of climate variations during the last few decades as well as present-day climatic conditions. For investigation of recent climatic variability in this region, the temperature records of meteorological stations extending along a northeast-southwest transect have been analysed, including Esperanza Station (lat 63° 24' S; long 57° 00' W), Vega Island automatic weather station (VIAWS: lat 63° 49' S; long 57° 21' W), Marambio Station (lat 64° 14' S; long 56° 37' W; 198 m a.s.l.) and Base Matienzo automatic weather station (BMAWS: lat 64° 59' S; long 60° 04' W). These four stations are well distributed and are considered to be representative for the region where major changes in ice masses have been observed in the recent past. For comparison purposes and to extend the observational period for almost an entire century, the temperature record of Orcadas Station (lat 60° 44' S; long 44° 44' W), South Orkney Islands (Figure 1) is also being evaluated.

Orcadas Station provides the longest continuous instrumental climate-data record available in Antarctica (Figure 3). The long-term meteorological/climatological time serie from Orcadas Station is especially important for analysis of ice-climate interactions in the northeastern AP because the relatively high correlation with Esperanza Station and Marambio Station temperature records [*Skvarca et al.*, 1998] allows the information on climatic conditions in the region of interest to be extended back to 1903. The representativeness of selected meteorological stations for the northeastern AP region is evident, because the major climatic signals are present in the mean annual temperature (MAT) series (not shown), such as the low in 1980, the highs in 1989 and 1999, the latter being the extreme maximum MAT recorded in the northeastern AP since 1952.

The analysis of Orcadas Station mean decadal temperatures (MDT) shows a general warming trend from 1921–30 to the present despite a cold decade during 1971–80 (Figure 3). From Orcadas Station data *Hoffmann et al.* [1997] have computed a temperature increase of +2.1 °C between the decades 1921–30 and

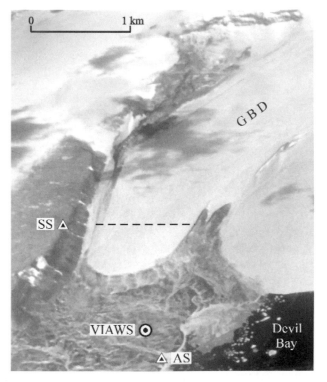

Fig. 2. Oblique aerial photograph of 27 March 1999, showing the lower part of "Glaciar Bahía del Diablo" on Vega Island. The dashed line indicates the profile shown on Figure 5 and VIAWS indicates the location of climate station. SS = survey station, AS = azimuth station.

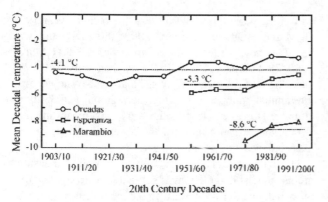

Fig. 3. Mean decadal temperatures at Orcadas, Esperanza and Marambio Stations during the 20th century.

1981–90; additional analysis extended to the decade 1991–2000 yields +2.2 °C. The MDT records indicate that the last two decades are the warmest at these stations during the period of available data (Figure 3). According to the meteorological/climatological record from Orcadas Station they are also the warmest of the 20th century. Moreover, the ice cores recovered from Dyer Plateau on AP reveal that recent decades are among the warmest since 1510 AD [*Thompson et al.*, 1994].

Mean summer temperature (MST) is a very important climatic parameter for the mass balance of glaciers and viability of ice shelves. Climatological records show that MST's in the Antarctic summer 2001–02 were the highest since the instrumental measurements began in the region. For example, +1.9 °C was the highest at Orcadas Station since 1903–04, the summer temperatures at Esperanza Station reached their historical maximum of +2.4 °C in 50 years, and the Marambio Station climatic record showed a maximum MST of +0.7 °C since the initiation of meteorological observations three decades ago (Figure 4). Furthermore, at Matienzo Station, which is located at the edge of LIS remnant section, the 2001–02 MST was positive for the first time, at +1.3 °C above zero. The analysis of available records also reveals that MAT trends are more significant toward the higher latitudes and increasing. For instance, the MAT trend at Marambio has increased from 0.056 °C a⁻¹ to 0.074 °C a⁻¹ in the recent five years. This is mainly due to the increase in mean autumn temperature trend (+0.12 °C a⁻¹). For the period considered (1971–2001), the summer warming trends also show a latitudinal increase, starting with +0.041 °C a⁻¹ at Orcadas Station, +0.060 °C a⁻¹ at Esperanza Station, and +0.074 °C a⁻¹ at Marambio Station. Despite the very large variability in both seasonal and annual temperatures, all records show a significant and consistent warming trend.

3. THINNING AND MASS BALANCE OF GLACIERS IN THE NORTHEASTERN AP

3.1 Glacier Thinning and Mass Balance on Vega Island

James Ross and Vega Islands are among the very few sites in Antarctica where glacier termini end on land. Glaciar Bahia del Diablo (GBD; lat 63° 49' S; long 57° 26' W) on VI (Figures 1; 2), was selected as a test site for glaciological studies in this region. GBD is a relatively small glacier, 14.3 km² in area and an accumulation area ratio (AAR) of 0.34.

As part of the local triangulation network, an azimuth station (AS) and survey station (SS) were established in early 1980s nearby the coast and on the southern margin of GBD, at an altitude of 4 m and 200 m above the sea level, respectively (Figure 2). In repeated surveys different distance meters-theodolites were used at SS to measure the glacier surface elevations by positioning the reflectors at selected points on the ablation area of GBD. The same survey points were located each time by using the corresponding azimuth angles and distances. As a control, during each survey a point on the glacier was also measured from AS.

Surveys made in Summers 1982–85 showed almost no change in elevation of the glacier surface. However, measurements carried out 13 years later (in 1998) revealed an average surface lowering of 13.1 m, at a rate of –1.0 m a⁻¹ (Figure 5). From 1998 to 2000, the thinning rate has increased to –1.5 m a⁻¹ due to a few consecutive warm summers, but the rate decreased to –0.40 m a⁻¹ in 2000–01 due to –2.6 °C lower summer temperatures and higher precipitation in Summer 2000–01. MST's are

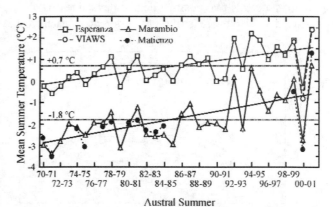

Fig. 4. Mean summer temperatures at Esperanza, VIAWS, and Marambio Stations from 1970–71 to 2001–02. Note that MST's at Matienzo Station are incomplete. Updated after *Skvarca et al., 1999b, Polar Research 18 (2)*. Reproduced by permission of the Norsk Polarinstitutt.

responsible for glacier-surface melt and run-off. The climatic record at Esperanza Station located 50 km north of test site, shows MST's at or above 0 °C since 1980, with the exception of Summer 2000–01 which was the coldest of the last two decades (Figure 4). Marambio Station and Esperanza Station MST's show a large interannual variability which is reflected in the rate of glacier thinning. The temperature record available from VIAWS, which is located in front of the terminus of GBD (Figure 2), indicates that duration of melt season is also very variable, in agreement with the regional interannual climatic variability. According to Esperanza Station and Marambio Station temperature records, 1999 was the warmest during the last 30 years on the northeastern AP. The large thinning rates measured on GBD during the last two decades are a direct response to the increasing regional warming trend during this period. The temperature record of recently installed VIAWS shows a strong correlation with Esperanza Station and Marambio Station records (Figure 4), confirming that large interseasonal and interannual climate variability prevails elsewhere on the northeastern AP.

Mass-balance measurements were initiated on GBD in Summer 1999-00. Field measurements by the stratigraphic method yield a negative mass balance of –534 mm of water equivalent (w. eq.) for the balance year 1999-2000. In contrast, a budget of –52 mm w.eq. was measured for the balance year 2000–01. In comparison to the previous summer, the MST in 2000–01 was –2.6 °C lower (Figure 4), resulting in almost no melting during the summer season, a much lower thinning rate (–0.40 m a⁻¹) and a slightly negative mass balance, in sharp contrast to the previous balance year. The analysis of VIAWS temperature data reveals that positive degree-day sums (PDD) at Bahía del Diablo are strikingly different in each year

within the period 1999–2001. For the balance year 1999 (1 March 1999–29 February 2000) the PDD sum was 380.7 °C day, while for the balance year 2000 (1 March 2000–28 February 2001) PDD reached only 140.0 °C day. The first two years of mass balance also show a high interannual contrast, reflecting the regional variability in climate. However, despite the very low mean summer temperatures responsible for the absence of melting, the mass balance still remains negative.

The data obtained so far from direct field measurements suggest that GBD was subject to a strong negative mass balance during the last two decades, in response to the regional atmospheric warming.

3.2 Retreat and Surface Lowering of Glaciers on James Ross Island

On JRI, a preliminary glacier inventory consisting of 138 glaciers of different types has been compiled [Rabassa et al., 1982]. Most glaciers on JRI are tidewater calving glaciers which react to climatic changes in decadal to centennial timescales. However, several glaciers whose termini end on land are located in a comparatively ice free northwestern part of the island. Optical and radar images used for studying the variations of 39 glaciers on JRI from late 1975 to early 1993 indicate a general reduction of 33.1 km² within the 17.4 year period [Skvarca et al., 1995]. Recent analyses based on a Landsat Enhanced Thematic Mapper (ETM+) image of 21 February 2000, indicate a further reduction of 26.8 km² from 1993 to 2000. The JRI glaciers decreased about 60 km² in area during the past 25 years. The reduction rate has almost doubled from the period 1988–1993 (2.0 km² a⁻¹) to 1993–2000 (3.8 km² a⁻¹), in response to climatic warming. In addition, strong evidence that intense surface lowering has also been affecting JRI glaciers comes from comparing oblique aerial photographs of Glaciar IJR68 [Rabassa et al., 1982], acquired on February 1980 and March 2002 (not shown). Unfortunately, the magnitude of the thinning rate is unknown, but similar values to those measured on GBD might be anticipated because of its proximity to Vega Island.

4. RETREAT AND DISINTEGRATION OF NORTHERN LARSEN ICE SHELF

4.1 Areal Decrease

From 1975 to 1986 the LIS sections extending between Cape Longing and Robertson Island (Figure 6) lost about 540 km², but no significant changes were

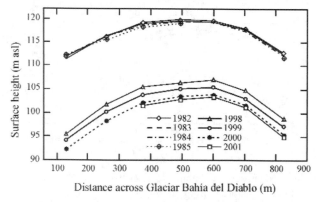

Fig. 5. Surface lowering of Glaciar Bahía del Diablo from 1982 to 2001 across a transverse profile (for location of the profile see Figure 2).

detected in PGC [*Skvarca*, 1993]. To measure the area loss from 1986 to 2002 we used the Landsat Thematic Mapper (TM) image mosaic of 1 March 1986, and the GPS ice-front survey carried out with Twin Otter aircraft on 13 March 2002. Along a few sections, the mapping was completed by coregistration of 5 March 2002 Moderate Resolution Imaging Spectroradiometer (MODIS) satellite image, provided by National Snow and Ice Data Center (NSIDC). The 1986 Landsat TM image mosaic, in Universal Transverse Mercator (UTM) projection (Figure 6), was georeferenced to coastlines, nunataks, and rock outcrops surveyed in the field with Differential Global Positioning System (DGPS) during several field campaigns.

The results of the March 2002 GPS mapping allowed us to measure an area loss of 1008 km^2 in PGC from 1986 to 2002. Visual observations during aerial surveys also revealed that the ice fronts of Sjögren and Boydell Glaciers had already retreated behind their previous confluence. Further south, that is within the Larsen Inlet (LI), 409 km^2 were lost during the same period. For Larsen (A) Ice Shelf the GPS survey, combined with MODIS image of 5 March 2002, showed a decrease of 2475 km^2 in area. At Larsen (B) Ice Shelf, the GPS survey indicated a loss of 7828 km^2 of shelf ice since 1986. In total, from 1 March 1986 to 5–13 March 2002 (16 years), the Larsen Ice Shelf north of Jason Peninsula (JP) decreased in area by 11,720 km^2. Adding the areal loss of 540 km^2 of the Larsen (A) Ice Shelf from 1975 to 1986, the total decrease for the period 1975–2002 amounts to 12,260 km^2. However, it is worth noting that 92% of this areal loss occurred from middle 1992 to early 2002, coincident with the increased warming trend in the region during the last decade, and 76% was lost mostly by disintegration since January 1995. At the end of the summer 2001–02 Larsen (B) Ice Shelf lost about 67% of its area compared to the maximum extent before the early 1995 calving event. About 33% of the ice shelf, which is nourished by Flask and Leppard Glaciers, and bounded at its southern margin by JP, still remains (Figure 6). A time-lapse history of Larsen (B) Ice Shelf areal changes, covering almost four decades (1963 to 2002), derived from different satellite images and airborne GPS surveys, is given in Table 1, updated and modified after Skvarca et al. [1999a].

4.2 February-March 2002 Larsen (B) Ice Shelf Event

A comparison of MODIS images acquired on 31 January 2002 and 17 February 2002 yields an areal loss of 611 km^2 for the Larsen (B) Ice Shelf [*Scambos et al.*,

this volume]. On 17 February 2002 the position of the ice front was also surveyed with airborne kinematic GPS (Figure 6). A comparison of this ice-front position with that surveyed on 18 October 2001 indicates a loss of 744 km^2 during 122 days (Table 1). From the October 2001 GPS survey and the MODIS image of 31 January 2002 is concluded that the disintegration process started in early February 2002, coincident with the unusually high mean monthly temperatures of +1.9 °C at Marambio Station and +1.7 °C at Matienzo Station. The MODIS image of 23 February 2002 indicates further disintegration of 164 km^2 and that of 5 March 2002 reveals the collapse of 1937 km^2 of the Larsen (B) Ice Shelf [*Scambos et al.*, this volume]. As expected, the disintegration has affected only those parts of Larsen B where meltwater was present in warm summers during the past decade. From the inland margin the breaking line extended along a distinctive boundary from Cape Disappointment (CD) toward Cape Framnes, the easternmost tip of the JP (Figure 6). On the remant 33% of Larsen B, corresponding to the inflow of Flask and Leppard Glaciers, the morphology was totally different to the rest of the ice shelf, with almost no meltwater features visible [*Skvarca et al.*, 1999a].

The second GPS mapping survey was carried out on 13 March 2002 over the disintegrated ice shelf (Figure 7), flying at low altitude along the new coastline northwest of Cape Framnes (Figure 6). This GPS survey allowed us to calculate that 2600 km^2 of ice shelf disintegrated between 17 February and 13 March 2002. Adding this areal loss to the 611 km^2 which had disintegrated during the first 17 days of February 2002, a total of 3211 km^2 is calculated for a period of 41 days. It is probable that the peak disintegration took place before the end of February 2002, when the strong melt season was interrupted abruptly by the initiation of a cold period in the region (Figure 8). The rate of area loss of the Larsen (B) Ice Shelf is twice as that of Larsen (A) Ice Shelf, occurred during the summer 1994–95. At that time 1600 km^2 disintegrated in almost the same span of time (39 days).

4.3 Climatic Setting Nearby Larsen B

Base Matienzo automatic weather station (BMAWS) is located on Larsen Nunatak at the Seal Nunataks (S-N) section of LIS. After the Larsen (B) Ice Shelf collapse, the station now lies at the edge of a narrow stripe of the shelf ice which still remains constrained by the nunataks between Larsen (A) and (B) ice shelf embayments (Figures 1; 6). The short-term record of BMAWS installed

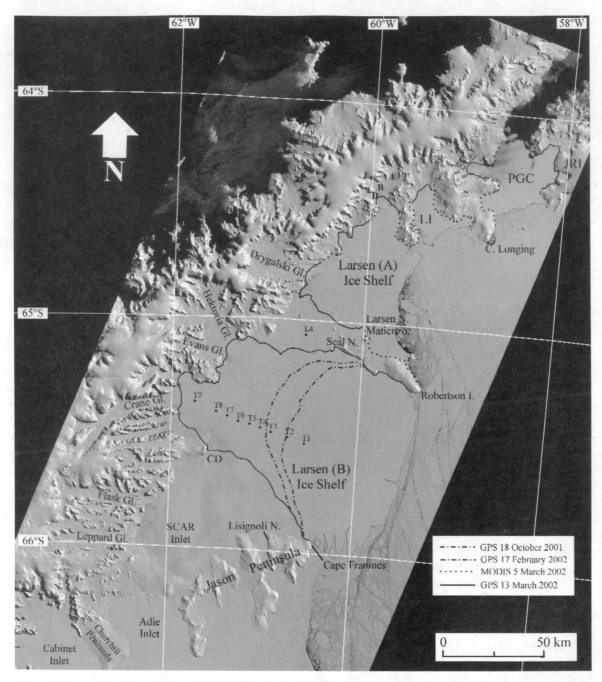

Fig. 6. Landsat TM mosaic of 1 March 1986 (band 4) in UTM projection of the northeastern AP. Airborne GPS surveys of the ice front in October 2001, and in February and March 2002 are indicated. The coastline derived from MODIS image of 5 March 2002 is indicated with a dotted line. Stakes along T-profile and L discussed in the text are also shown. JRI = James Ross Island, PGC = Prince Gustav Channel, Sj = Sjögren Glacier, Bo = Boydell Glacier, E = Edgeworth Glacier, B = Bombardier Glacier, D = Dinsmoor Glacier, LI = Larsen Inlet, Gl = Glacier, CD = Cape Disappointment.

TABLE 1. Areal extent and differences of Larsen (B) Ice Shelf from 1963 to 2002 derived from different source materials. Updated from *Skvarca et al.* [1999a]. Reproduced by permission of International Glaciological Society.

Date (d m yr)	Area (km²)	Δ Area (km²)	Data source
29 08 1963	10936		Argon
03 10 1975	11329	393	Kosmos KATE-200
01 03 1986	11560	231	Landsat 5
19 01 1988	11628	68	Landsat 4
08 01 1990	11695	67	Landsat 4
02 07 1992	11775	80	ERS/SAR[1]
26 08 1993	11770	–5	ERS/SAR[1]
28 01 1995	11816	46	ERS/SAR[1]
30 01 1995	9496	–2320	ERS/SAR[1]
08 03 1995	9496	0	ERS/SAR[1]
28 10 1995	9501	5	ERS/SAR[1]
29 02 1996	9483	–18	ERS/SAR[1]
01 11 1996	9391	–92	ERS/SAR[1]
02 03 1997	9397	6	ERS/SAR[1]
04 07 1997	9406	9	ERS/SAR[1]
25 04 1998	9326	-80	RADARSAT
Nov. 1998	8284	–1042	AVHRR[2]
06 02 1999	8084	–200	IAA-GPS[3] survey
11 10 1999	7616	–468	IAA-GPS[3] survey
24 01 2000	7334	–282	IAA-GPS[3] survey
18 10 2001	7252	–82	IAA-GPS[3] survey
31 01 2002	7119	–133	MODIS[4]
17 02 2002	6508	–611	MODIS[4] and IAA-GPS[3]
23 02 2002	6344	–164	MODIS[4]
05 03 2002	4407	–1937	MODIS[4]
13 03 2002	3908	–499	IAA GPS[3] survey

[1]ERS/SAR = European Remote Sensing Satellite/Synthetic Aperture Radar
[2]AVHRR = Advanced Very High Resolution Radiometer, [*Scambos et al., 2000*]
[3]IAA = Instituto Antártico Argentino; GPS = Global Positioning System
[4]MODIS = Moderate Resolution Imaging Spectroradiometer, Terra Satellite, [*Scambos et al.,* this volume]
Δ Area = area difference

in early 1999 in the same meteorological shelter where discontinued climate data were collected previously confirms an earlier conclusion that MSTs of Marambio are representative of the summer conditions on the northern LIS [*Skvarca et al.,* 1998]. However short, the recent continuous temperature record of BMAWS also indicates a very large seasonal variability in this region, for exam-

ple, –3.2 °C MST in 2000–01 and +1.3 °C MST in 2001–02 (Figure 4).

Analysis of BMAWS daily temperatures reveals that throughout most of February 2002 (when the ice shelf collapsed), air temperatures in the area were unusually high and persistently above 0 °C with only few short intervals below freezing (Figure 9). Similar conditions prevailed during December-January, with extended warm periods and low temperatures reaching only –2 °C. The absolute maximum temperature of +13.1 °C was recorded at Matienzo on 13 December 2001 at 11.30 h GMT. The mean temperature on that day was +8.1 °C (Figure 9); the average for December 2001 was +1.5 °C. The unusually high temperatures prevailing throughout the warmest summer recorded in the region produced more meltwater than usual over the ice-shelf surface. Note that melting was already initiated in October 2001 due to a warm period documented by BMAWS (Figure 8). There was almost no sea ice in the region during that period, and a wide polynia extended along the northern LIS front. In addition to the climate data, the available MODIS images reveal meltwater features over Larsen (B) Ice Shelf further south than previously detected. Photographs taken during the February GPS mapping surveys document the intense surface melting (Figure 10) and rifting. Blocks of ice thrusted above the surface of the ice shelf have been observed during the survey, indicating that a strong compression took place after the formation of rifts. Similar features were also seen over Larsen (A) Ice Shelf in October 1994, prior to its collapse [*Rott et al.,* 1996].

Fig. 7. View towards the disintegrated Larsen (B) Ice Shelf south of Seal Nunataks as observed during the GPS survey on 13 March 2002. From east to west are Gray, Bruce, and Bull Nunataks; further west lies Cape Fairweather and the new embayment east of Hektoria and Evans Glaciers.

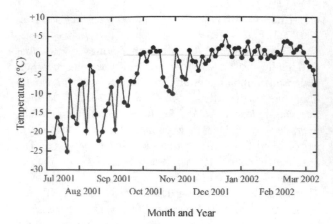

Fig. 8. Three-day average temperature at Base Matienzo AWS from 1 July 2001 to 10 March 2002.

5. DISCUSSION

The importance of meltwater in fracture toughness and disintegration of ice shelves has been discussed by *Doake and Vaughan* [1991]. According to *Weertman* [1973], crevasses may extend completely through the ice shelf and produce rifting, provided sufficient water is present within them. The occurrence, extent, and abundance of meltwater over northern LIS in response to the increasing summer warming trend in the region has been discussed elsewhere [*Skvarca*, 1993; *Skvarca et al.*, 1999a; *Rack*, 2000; *Scambos et al.*, 2000], and is illustrated in Figure 10. *Scambos et al.* [2000] tested a numerical model which supports the hypothesis that propagation of crevasses and rifts caused by surface meltwater is the main mechanism which causes the ice-shelf weakening and retreat. Following this model, we suggest that the sudden and abrupt collapse of Larsen (B) Ice Shelf has

been triggered by the surplus of meltwater produced by the warmest summer ever recorded in the region. This conclusion is strongly supported by the facts that almost no retreat occurred at Larsen (B) Ice Shelf in 2000–01 (see Table 1) during the coldest summer of the last decade (Figure 4), and that the southern section of Larsen (B) Ice Shelf, where little or no surface meltwater has been observed in the past, did not disintegrate.

In austral Spring 2001 the last glaciological field campaign was carried out over Larsen (B) Ice Shelf. Comparison of average velocities measured along the T-profile (Figure 6) in 1997–99 and 1999–01 shows an increase of 26%. This demonstrates that Larsen (B) Ice Shelf had accelerated considerably in its central part before disintegration, in comparison to only 10% increase detected from 1996–97 to 1997–99 [*Rack*, 2000]. Although lower, an increase in ice-flow velocity was also measured on Larsen (A) Ice Shelf prior to its collapse [*Bindschadler et al.*, 1994; *Rack et al.*, 1999; *Rack*, 2000]. As the velocity of section S-N is only about 24 m a^{-1} the acceleration of Larsen (B) Ice Shelf augmented the shearing along the band of rifts extending from the ice rise east of Cape Fairweather toward Bull Nunatak (Figure 7). As expected, the ice shelf separated along this band of rifts [*Rack et al.*, 1999]. In addition to the observed acceleration, a significant increase in longitudinal strain rates was measured along the T-profile. In comparison to the 1997–99 period, the strain rates averaged over the distance T4–T9 almost doubled during the period preceeding disintegration. Assuming pure strain, the ice shelf was thinning at an average rate of –0.27 m

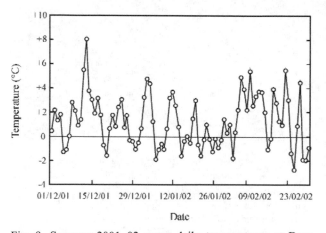

Fig. 9. Summer 2001–02 mean daily temperatures at Base Matienzo Automatic Weather Station (BMAWS).

Fig. 10. Complex pattern of meltwater features on ice shelf close to Base Matienzo AWS on 17 February 2002. To the north is the Weddell Sea where existed Larsen (A) Ice Shelf until 1995.

a^{-1} from 1996 to 2001. During the same period the surface mass balance along the T-profile was clearly positive, despite its decline toward 1999–01. During the 1990s, an average decrease of –2.0 m a^{-1} in ice thickness has been estimated for Larsen (B) Ice Shelf [*Rack*, 2000]. The average thinning rate due to strain along the T-profile is considerably lower, suggesting that the estimated change in ice thickness might be attributed to bottom melting.

Despite the positive surface mass balance in recent years Larsen (B) Ice Shelf was out of balance, partly due to thinning caused by increased strain rates but mainly due to loss of mass caused by melting beneath the ice shelf. However, meltwater produced during the warmest summers played a definitive role in the fracturing and rifting processes, and triggered without a doubt the final collapses.

Because of the recent disintegration of Larsen (B) Ice Shelf there is no more shelf ice to exert backpressure on major glaciers which drain the AP plateau, namely Hektoria, Evans, and Crane Glaciers (Figure 6). *Rott et al.* [2002] detected a dramatic acceleration of Drygalski Glacier only four years after the removal of Larsen (A) Ice Shelf, showing a high sensitivity of glacier flux to ice-shelf backpressure. A similar increase in ice velocity should be expected at former tributary glaciers after the removal of Larsen (B) Ice Shelf. Furthermore, since the disintegration of the ice shelf in PGC and the Larsen (A) Ice Shelf, a significant retreat of grounded ice has been already measured at Sjögren and Boydell Glaciers calving now into PGC (Figure 6), as well as at Dinsmoor, Bombardier, Edgeworth, and Drygalski Glaciers [*Rott et al.*, 2002]. The recent demise of Larsen (B) Ice Shelf will certainly add to the loss of grounded ice along the new coastline. Further retreat of grounded glaciers along the new coastline after the removal of LIS sections is expected to contribute to the global sea level rise, as formerly land-based glacier ice enters the ocean.

6. CONCLUSIONS

Evidence of significant changes on glaciers, either with termini ending on land or in tidewater, and ice shelves fringing the northeastern side of the Antarctic Peninsula have been discussed. The changes are associated with the increasing regional atmospheric warming. The last two decades have been the warmest recorded in this region. The most dramatic changes are caused by the retreat of ice fronts followed by disintegration of the northern sections of the Larsen Ice Shelf (PGC, LI, A and B). The total areal decay north of 66 °S from late 1975 to

early 2002 amounts to 12,260 km^2, with 11,300 km^2 lost in only one decade, the warmest one. What is striking is the rapidity of such catastrophic events: Larsen (A) Ice Shelf disintegrated in 39 days, whereas an area of Larsen (B) Ice Shelf twice as large collapsed in only 41 days. Both events were coincident with the two warmest summers in the region. Most of Larsen (B) Ice Shelf disintegrated within a month of unprecedented high mean temperatures of +1.7 °C. Disintegration is a consequence of unstable conditions within the ice shelf. However, the sudden collapse of Larsen (A) and (B) ice shelves was triggered by a surplus of surface meltwater, produced during the warmest summers. In regions of ice-shelf disintegration, several glaciers have accelerated significantly and started to retreat beyond their grounding lines. Glacier retreat is also expected to occur in the region of Larsen (B) Ice Shelf after its disintegration. Monitoring of such glaciers is essential, because their continued retreat will likely contribute to the global sea level rise. The impact of recent dramatic events on the local environment and on wildlife is still poorly known and remains to be investigated.

Acknowledgments. The temperature data of Argentine Antarctic Stations were kindly provided by Servicio Meteorológico Nacional, Fuerza Aérea Argentina. The authors would like to express their thanks to Evgeniy Ermolin, Andrés F. Zakrajsek, Teodoro Toconás, Juan C. Quinteros and all those who have contributed in the collection of field data. This paper is a contribution to the project "Ice-Climate Interaction and Dynamics of Glaciers on Antarctic Peninsula and Southern Patagonia" of the Instituto Antártico Argentino-Dirección Nacional del Antártico.

REFERENCES

Bindschadler, R.A., M.A. Fahnestock, P. Skvarca and T.A. Scambos, Surface-velocity field of the northern Larsen Ice Shelf, Antarctica. *Ann. Glaciol.*, 20, 319-326, 1994.

Doake, C.S.M. and D.G. Vaughan, Rapid disintegration of the Wordie Ice Shelf in response to atmospheric warming. *Nature*, 350 (6316), 328-330, 1991.

Domack, E.W., A. Leventer, R. Gilbert, S. Brachfeld, S. Ishman, A. Camerlenghi, K. Gavahan, D. Carlson and A. Barkoukis, Cruise reveals history of Holocene Larsen Ice Shelf. *EOS*, Transactions, AGU, 82 (2), 13, 16-17, 2002.

Domack, E.W., Duran, D., McMullen, K., Gilbert, R. and A. Leventer. Sediment lithofacies from beneath the Larsen B Ice Shelf: can we detect ice shelf fluctuation? *EOS* trans. AGU, 83 (47), Fall Meet. Suppl., Abstract C52A-04, 2002.

Hoffmann, J.A.J., S.E. Nuñez and W.M. Vargas, Temperature, humidity and precipitation variations in Argentina and the adjacent sub-antarctic region during the present century. *Meteorol. Zeitschrift, N.F.* 6, 3-11, 1997.

Mercer, J.H., West Antarctic ice sheet and CO_2 greenhouse effect: a threat of disaster. *Nature* 271, 321-325, 1978.

Pudsey, C.J. and J. Evans, First survey of Antarctic sub-ice shelf sediments reveals mid-Holocene ice shelf retreat. *Geology*, 29 (9), 787-790, 2001.

Rabassa, J., P. Skvarca, L. Bertani and E. Mazzoni, Glacier inventory of James Ross and Vega islands, Antarctic Peninsula. *Ann. Glaciol.*, 3, 260-264, 1982.

Rack, W., H. Rott, A. Siegel and P. Skvarca, The motion field of northern Larsen Ice Shelf, Antarctic Peninsula, derived from Satellite imagery. *Ann. Glaciol.* 29, 261-266, 1999.

Rack, W., Dynamic behaviour and disintegration of the northern Larsen Ice Shelf. Antarctic Peninsula. Doctoral thesis, Science Faculty, Univ. of Innsbruck, Austria, 166 pp., 2000.

Rott, H., P. Skvarca and T. Nagler, Rapid collapse of northern Larsen Ice Shelf, Antarctica. *Science* 271, 788-792, 1996.

Rott, H., W. Rack, T. Nagler and P. Skvarca, Climatically induced retreat and collapse of Northern Larsen Ice Shelf, Antarctic Peninsula. *Ann. Glaciol.* 27, 86-92, 1998.

Rott, H. W. Rack, P. Skvarca and H. De Angelis, Northern Larsen Ice Shelf, Antarctica: further retreat after collapse. *Ann. Glaciol.* 34, 277-282, 2002.

Scambos, T.A., C. Hulbe, M. Fahnestock and J. Bohlander, The link between climate warming and break-up of ice shelves in the Antarctic Peninsula. *J. Glaciol.* 46 (154), 516-530, 2000.

Scambos, T.A., C. Hulbe and M. Fahnestock, Climate-induced ice shelf disintegration in the Antarctic Peninsula, this volume.

Skvarca, P., Fast recession of the northern Larsen Ice Shelf monitored by space images. *Ann. Glaciol.* 17, 317-321, 1993.

Skvarca, P., H. Rott and T. Nagler, Satellite imagery, a base line for glacier variation study on James Ross Island, Antarctica. *Ann. Glaciol.*, 21, 291-296, 1995.

Skvarca, P., W. Rack and H. Rott, 34 year satellite time series to monitor characteristics, extent and dynamics of Larsen B Ice Shelf, Antarctic Peninsula. *Ann. Glaciol.* 29, 255-260, 1999a.

Skvarca, P., W. Rack, H. Rott and T. Ibarzábal y Donángelo, Evidence of recent climatic warming on the eastern Antarctic Peninsula. *Ann. Glaciol.* 27, 628-632, 1998.

Skvarca, P., W. Rack, H. Rott y T. Ibarzábal y Donángelo, Climatic trend, retreat and disintegration of ice shelves on the Antarctic Peninsula: an overview. *Polar Research* 18 (2), 151-157, 1999b.

Thompson, L.G., D.A. Peel, E. Mosley-Thompson, R. Mulvaney, J. Dai, P.N. Lin, M.E. Davis and C.F. Raymond, Climate since AD 1510 on Dyer Plateau, Antarctic Peninsula: evidence for recent climate change. *Ann. Glaciol.* 20, 420-426, 1994.

Vaughan, D.G. and C.S.M. Doake, Recent atmospheric warming and retreat of ice shelves on the Antarctic Peninsula. *Nature* 379, 328-331, 1996.

Weertman, J., Can a water-filled crevasse reach the bottom surface of a glacier ? *International Association of Scientific Hydrology,* Publication 95, 139-145, 1973.

Hernán De Angelis. División Glaciología, Instituto Antártico Argentino, Cerrito 1248, C1010AAZ, Buenos Aires Argentina. Email: hda@dna.gov.ar

Pedro Skvarca. División Glaciología, Instituto Antártico Argentino, Cerrito 1248, C1010AAZ, Buenos Aires Argentina. Email: pskvarca@dna.gov.ar

ANTARCTIC PENINSULA CLIMATE VARIABILITY
ANTARCTIC RESEARCH SERIES VOLUME 79, PAGES 79-92

CLIMATE-INDUCED ICE SHELF DISINTEGRATION IN THE ANTARCTIC PENINSULA

Ted Scambos

National Snow and Ice Data Center, University of Colorado, Boulder, Colorado

Christina Hulbe

Department of Geology, Portland State University, Portland, Oregon

Mark Fahnestock

Institute for the Study of Earth, Oceans, and Space, University of New Hampshire, Durham, New Hampshire

Climate warming in the Antarctic Peninsula has caused the disintegration of several ice shelves there. The rapid loss of 3320 km^2 from the northern Larsen B ice shelf in early 2002 typifies the pattern and pace of these events. Extended melt seasons leading to melt ponding on the shelf surfaces are the apparent cause of the breakups. Enhanced fracturing of pre-existing crevasses and shelf rifts driven by this meltwater occurs during warmer summers, leading to disintegration. Shelf breakups in the Peninsula over the last 20 years, coupled with geological evidence of their prior stability over the previous several millennia, imply that Peninsula climate is warmer now than at any time during this period. Stability of the remaining Peninsula shelves and other ice shelves can be assessed using remotely sensed indicators based on the presented model of ice shelf breakup. We find that, among 11 Antarctic ice shelves fringing the continent, only the currently retreating Larsen B, Wilkins, and George VI ice shelves, and the northernmost portions of the Larsen C shelf (northeast of the Alexander Peninsula) have the firn characteristics and melt season length we associate with impending breakup.

INTRODUCTION

Previous studies have considered the climatic limit of ice shelves [*Mercer*, 1978; and *Doake and Vaughan*, 1991] and the possible role of meltwater in ice shelf breakup [*Weertman*, 1973; *Hughes*, 1981; and *van der Veen*, 1998], but the rapidity of ice shelf retreat in the Peninsula, and particularly the events of 1995 and 2002, has been a continuing surprise. Although some Peninsula ice shelves have been retreating since the 1970s (*e.g.*, the Wilkins and George VI shelves), beginning around 1986 an increased rate of retreat was observed for all the north- ernmost ice shelves on the eastern and western Peninsula coast [*Vaughan et al.*, 1996; *Rott et al.*, 1998; *Skvarca et al.*, 1999a; *Scambos et al.*, 2000]. Early retreat events were small, trimming a few kilometers off the fronts of the shelves in warmer summers. However, a series of very warm summers throughout the Peninsula in the 1990s was followed by much larger, more rapid breakup events—events that could truly be called 'disintegrations'. Figure 1 summarizes retreat totals for the larger Peninsula shelves since 1980, along with the temperature changes there.

The significance of climate-change-induced ice shelf breakup is in the effect it may have on grounded ice

10.1029/079ARS07

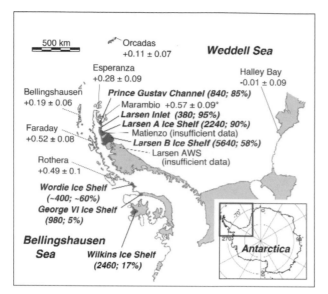

Fig. 1. Map of the Antarctic Peninsula showing climate trends for selected stations and total breakup extents in square kilometers for seven ice shelves fringing the Peninsula. Station trends are reported in °C per decade, spanning the last ~45 years [*Doake et al.,* 1998], except for Marambio, where records extend back only to 1971 [*Skvarca et al.,* 1999b]. Breakup areas (in km^2) and percentages [*Vaughan and Doake,* 1996; *Scambos et al.,* 2000; *Rott et al.,* 1996; *Rott et al.,* 1998] indicate loss of shelf by retreat-style calving since 1980, exclusive of areas that would be expected to calve under steady-state conditions.

upstream. *Mercer* [1978] stated that glaciers formerly fronted by an ice shelf would rapidly increase in speed following breakup, and change the mass balance of the ice sheet system. This in turn would result in sea level rise. In this scenario, removal of the shelf decreases longitudinal stresses in the glacier system, leading to the increased flow. This idea was supported in modeling studies by *Thomas et al.* [1979]. Some subsequent modeling studies [*Hindmarsh,* 1993] and observations [*Vaughan et al.,* 2001b] have cast doubt on the stress decrease, or the speed increase, effects. However, recent observational evidence has shown that, for the feeder glaciers in the former Larsen A area, a rapid increase in

flow speed *has* occurred, with ice flow up to triple the speed measured prior to breakup—well in excess of that required to balance mass input [*Rott et al.,* 2002]. This increase may possibly be a result of percolating meltwater acting on the glaciers as well as the ice shelf [e.g., *Zwally et al.,* 2002], or a loss of buttressing longitudinal stress as originally thought. In either case, the concern over changes in sea level due to ice shelf loss is underscored by recent events and their aftermath.

Here we review the shelf breakup and disintegration events of the 1990s, and revisit a previously described mechanism for their breakup [*Scambos et al.,* 2000] with new observational evidence. We also use the proposed mechanism to develop a means to estimate the likelihood that other shelf areas in Antarctica are poised for breakup.

RECENT SHELF BREAKUP EVENTS

In late January 1995, the Prince Gustav and Larsen A ice shelves disintegrated over the course of just a few days [1][*Rott et al.,* 1996]. These shelves had slowly lost more than half of their historic extent prior to this through smaller breakups. The 1995 events, however, introduced a new pattern of ice shelf breakup, in which thousands of square kilometers of shelf (~2030 km^2 total) rapidly disaggregated to sliver-shaped, few-kilometer-to-sub-kilometer icebergs. Although somewhat similar patterns of shelf fracturing during slow retreat had been observed before [e.g., *Skvarca,* 1993; *Lucchitta and Rosanova,* 1998], the speed and magnitude of the Larsen A event awakened glaciologists to the fact that climate-related processes could lead to near-instantaneous ice shelf losses. This rapidity and style of retreat was repeated a few years later in a breakup event of the Wilkins Ice Shelf in March, 1998 [*Scambos et al.,* 2000]. Figure 2 summarizes these events.

After the events of January, 1995 the National Snow and Ice Data Center (NSIDC) began monitoring ice shelf activity in Antarctica using visible, near-infrared, and thermal satellite images (NSIDC's satellite image archive is available at http://nsidc.org/iceshelves). This time series is useful for constraining the timing and causes of Peninsula ice shelf breakups, and for evaluating the stability of other shelves. Ice shelf retreat events, showing the sliver-iceberg calving style, were found to occur when ponded melt was present at or near the retreating ice shelf front. Disintegration occurred in ice that underlay the area of ponds. Timing of these events, in mid- to late austral summer, also implicates surface melt. This

---

[1]We use the informal geographic nomenclature adopted by *Vaughan and Doake* [1996] to divide the Larsen Ice Shelf. The Larsen 'A' extends from the Sobral Peninsula to Robertson Island; the Larsen 'B' refers to the shelf between Robertson Island and Jason Peninsula; the Larsen 'C' is the section between Jason Peninsula and Gipps Ice Rise. A far southern section, Larsen 'D', is not discussed in this paper.

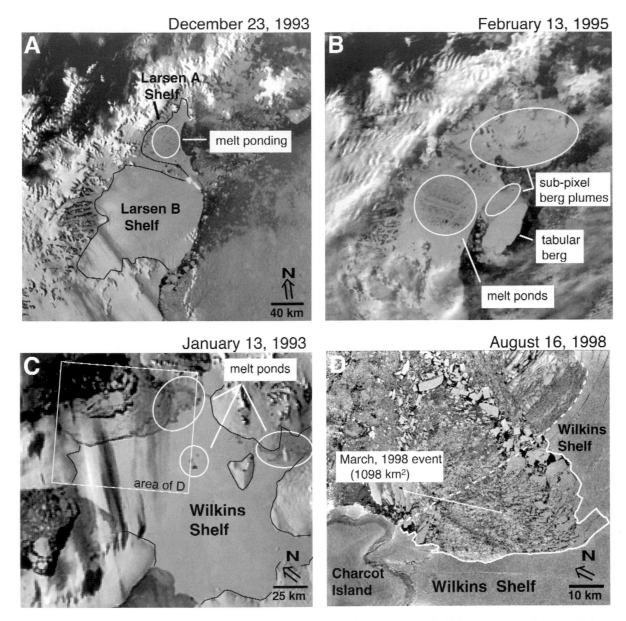

Fig. 2. Satellite images of the Larsen A and Wilkins ice shelf breakups. A: AVHRR (Advanced Very High Resolution Radiometer) image of the Larsen A area prior to breakup, showing ice extent on Larsen A and B ice shelves, and surface melting on Larsen A. B: AVHRR image acquired just after disintegration of the Larsen A, showing berg plumes from the collapsed shelf ice, and the calving of the last major iceberg of the Larsen B. Note the southern berg plume from the northernmost Larsen B. C: AVHRR image of the Wilkins ice shelf showing melt ponding areas. D: Synthetic Aperture Radar (SAR) image from Radarsat 1 showing detail of breakup area. Breakup was observed first in earlier AVHRR images.

association of melt ponds with late-summer breakups was observed in a series of minor events on the Wilkins and Larsen B in the late 1990s, and with retrospective images acquired by NSIDC, in the George VI, Larsen A, Larsen Inlet, and earlier Wilkins events.

At the time of the Larsen A disintegration, a large tabular iceberg, calved from the Larsen B (designated 'A32' by the U.S. National Ice Center, initially 26 km wide and 70 km long, total 1720 km²). This continued a pattern of quasiperiodic calvings from this shelf, having a period of

50 years. Up to this time, the Larsen B, further south than the other retreating ice shelves of the eastern Peninsula coast, had not experienced sustained retreat. However, in addition to the large berg, the northernmost portion of the shelf front (~550 km²) broke up as smaller, elongate icebergs, in the manner of the retreating shelves to the north [*Rott et al.*, 1996]. At the end of these events the Larsen B had an extent similar to that in 1902, when it was first mapped, and to that inferred for ~1950, when an unobserved calving probably occurred [*Scambos et al.*, 2000]. This cycle is consistent with the ice shelf growth rate (in the absence of calving), which for the central Larsen B was greater than 440 ma⁻¹ [*Rack et al.* 2000], and would have been still higher in the area of the calved berg. If the periodic, 'stable', calving cycle had continued, the shelf would have advanced eastward from this limit (roughly the line between the eastern tips of Robertson Island and the Jason Peninsula) and calved another large berg around 2045.

In a study of the stress and strain field of the Larsen B shelf, this 1902/1995 line was identified by *Doake et al.* [1998] as very near the minimum stable ice front position, represented by the easternmost zone of transversely compressed ice in the shelf (the 'compressive arch'). *Doake et al.* predicted the shelf would enter a breakup phase within a few years if this arch were disrupted. Just a few weeks after publication, in February, 1998, a relatively small calving event (~125 km²), composed of sliver-shaped icebergs and sub-kilometer pieces, removed part of the compressive arch area. Following this event, the Larsen B rapidly began to shed mass, losing 1839 km² by the end of the next austral summer, and 477 km² by the following summer. Ponding occurred near the retreating front in both these seasons. However, the unique, catastrophic-style breakup seen in the Larsen A and Wilkins was not repeated until the austral summer of 2002.

The 2002 Breakup of the Larsen B Ice Shelf

The spatial and temporal association of melt ponds and breakup discussed above was clearly present again in a February-March 2002 event on the Larsen B (Plate 1). Better satellite coverage by the MODIS sensor (Moderate Resolution Imaging Spectrophotometer), both spatially and temporally, and the coincidence of both Argentine and British field groups in the area, provided more detailed observational data for this breakup. The differences in behavior between the northern and southern portions of the shelf are particularly illuminating.

On January 31 2002, MODIS images revealed that the northern two-thirds of the Larsen B shelf was again extensively covered with melt ponds, from the Seal Nunataks to the southern edge of the Crane Glacier outflow. At the front, the calving rate of elongate, front-parallel icebergs increased significantly, and an initial breakup event began. Between January 31 and February 17, 611 km² calved away, which, like earlier events, was most active in the area near ponds. The southern portions of the shelf, where ponding was nearly absent, calved fewer, larger icebergs. By February 23 an additional 164 km² had calved. A March 5 image shows a loss of 1937 km² over the preceding 8 days, and a March 7 image reveals the loss of an additional 522 km². By March 17, a total of 3320 km² of shelf had calved, with the last ~2500 km² occurring as a catastrophic disintegration. Further retreat at a much slower pace occurred over the winter months, primarily in the Hektoria/Evans and Crane outlet glacier areas, totaling 251 km² by the end of December, 2002. The lost shelf area for the year lay almost completely within the region of scattered melt ponds indicated in the January 31 image.

Extreme disruption of the shelf ice during the disintegration phase of the breakup is indicated by the brilliant bluish color in the March 7 image, typical of the color of interior glacier ice. Floating ice in this area is fragmented to below the resolution of the satellite sensor, in this case 250 m. Aerial images during the breakup [Skvarca, personal communication] show that the shelf fragmented at all scales from ~10 km down to sub-meter. This image also shows englacial debris, previously encased within the shelf, as dark bands trending parallel to the shelf ice flow direction, providing further evidence of extreme disruption.

CLIMATE WARMING, MELT SEASON LENGTH, AND MELT PONDS

Retreat of the ice front past the 'compressive arch', while possibly a contributing factor to the speed of breakup of the ice shelves, does not provide a mechanism for the increased calving rate—i.e., what *causes* the ice to calve up to and past the 'compressive arch' line? A likely root cause would be some effect from the Peninsula's profound climate warming. But there are several effects of warming on ice that might play a role in ice shelf retreat: ice becomes softer, it may resist fracturing less, and, of course, it melts. As developed and supported below, observational evidence points to the effects of surface meltwater as the cause of breakup.

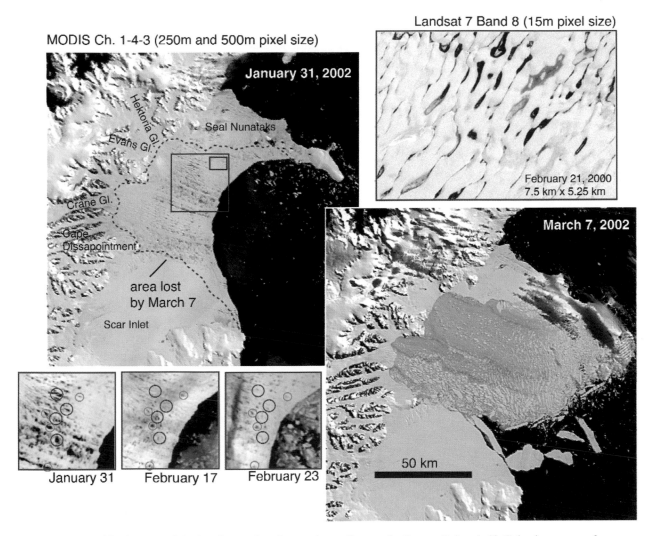

Plate 1. Satellite images of the breakup and surface melt ponding on the Larsen B ice shelf. Color images are from Channel 1 (red, 250 m pixel scale), Channel 3 (blue, 500 m pixel scale) and Channel 4 (green, 500 m pixel scale) of the MODIS sensor on the Terra satellite. Ponded meltwater shows as dark blue spots in the January 31 image, organized into linear patterns by subtle flow-related features on the shelf surface. The dashed blue line shows the extent of disintegration on March 7, 2002. Inset images in lower left track the disappearance of melt ponds in the weeks prior to breakup, using Channel 1 of MODIS. Circled in green are ponds that survived the 23-day period after January 31; in red and blue are ponds that disappeared by February 17 and February 23, respectively. The Landsat 7-derived inset in the upper right provides a high-resolution view of the surface (15 m pixel scale) in a previous summer.

Surface air temperatures measured at several long-occupied research stations in the area show a mean air temperature rise of ~2.5°C over the last 5 decades [*Morris and Vaughan*, this volume; *Skvarca et* al., 1999b; *Skvarca et al.*, this volume; *Vaughan et al.*, 2001a; *Comiso*, 2000] (Figure 1). A number of proxy observations also indicate warming. Sea ice extent in the Bellingshausen Sea west of the Peninsula has decreased by 20% over the last two decades [*Jacobs and Comiso* 1997; *Parkinson*, 2002] and both sea surface temperatures and temperatures in the mid-ocean have risen [*Reynolds and Smith*, 1994; *Gille*, 2002]. The air temperature rise and sea ice decline is observed for every season of the year, and is actually most profound for the winter months [*King*, this volume]. However, in summer, the warming has resulted in a gradual increase in the length of the melt season, rising approximately 1 day per year on average over the period 1978–2000 [*Fahnestock et al.*, 2002]. A summary of the warming trend in the region of the Larsen B, and related effects discussed below, are shown in Plate 2.

Causes of this warming are elusive. A complex relationship between temperatures, winds, and sea ice in this area and the Southern Oscillation has been identified [*Vaughan et al.*, 2001a; and *Kwok and Comiso*, 2002], but a causal sequence among these climate indicators is not yet established. Recently, circulation effects resulting from the seasonal loss of ozone have been proposed as a likely contributing factor [*Thompson and Solomon*, 2002].

As the melt season lengthened, ponded meltwater began to appear on the northernmost shelves during mid- to late summer. Melt ponds have been observed on the northern George VI shelf since the 1930s [*Reynolds*, 1981], but for the Larsen Ice Shelf they are apparently a more recent occurrence. A review of late summer Landsat and AVHRR images of the Larsen B from 1975 to the present tracks the frequency of melt ponding (Plate 2); however, the image record is sparse for the earlier years. Comparing this record with mean summer air temperatures for nearby stations indicates that melt ponding on the Larsen B is associated with mean summer temperatures exceeding –1.5°C at the Marambio or Matienzo weather stations. Melt ponds can form in sub-zero mean air temperatures because air temperature is generally lower than surface skin temperature, and because a strong melt–albedo feedback helps melt additional water once melting begins. During cooler periods within the summer, thin ice layers may form over growing ponds, insulating them but still permitting light (and

therefore energy) to enter the melt below. In a study of melt season length for the continent, melt onset (wet snow) was determined to coincide with a mean monthly air temperature of –2.5°C, and severity of melt increased rapidly with temperature [*Zwally and Fiegles*, 1994].

Cooler years in the northern Larsen area (1977, 1986, 1994, and 2001) have little or no ponding. Mean summer temperatures rarely exceeded –1.5°C prior to 1977 [*Skvarca et al.*, 1999b and this volume], and so we infer that melt ponds on the Larsen B must have been rare until the late 1970s. Further, the extent of ponding on the Larsen B has increased in recent years. Landsat images in 1979 and 1988 indicate melt ponds were limited to the northwestern corner of the shelf, near the Hektoria and Evan outflow. During the warmer summers of the 1990s the area marked by melt ponds gradually expanded south to the Crane Glacier outflow and east. By January 1998, summer melt ponds were present throughout the northern two-thirds of the shelf, from Cape Disappointment north to the Seal Nunataks, and from near the grounding line to the ice front. The record extent for ponds was observed in 2002, just prior to the largest breakup event to date.

THE CLIMATE-INDUCED BREAKUP PROCESS

Given the close spatial and temporal association of surface melt ponding and ice shelf breakup in the Larsen A event, we used a numerical model of ice shelf flow to evaluate processes by which meltwater could directly change ice-shelf calving behavior [*Scambos et al.*, 2000] (Figure 3). The model incorporated ice thickness, input flux, the stress-strain relation for ice, and geometry of the shelf at several stages of retreat to determine stress field and flow speed. Comparison of our modeled flow speed to ice flow speed observations for the Larsen A and B [*Bindschadler et al.*, 1994; *Rott et al.*, 1998; *Rack et al.*, 2000] validated the model.

As part of the model, it was necessary to infer a temperature profile for the interior of the ice shelf. We determined that matching the observed surface velocity field required a polythermal internal temperature profile, with mid-shelf temperatures near -13.5°C. For comparison, the mean annual temperature at the shelf surface is ~-9°C [*Morris and Vaughan*, this volume]. We explain the cold interior as a result of influx from ice accumulated at higher altitude on the peninsula. Given the cold interior required by our numerical model, and the rapidity of the warming trend and breakups, we doubt hypotheses for breakup based on warmth of the interior ice.

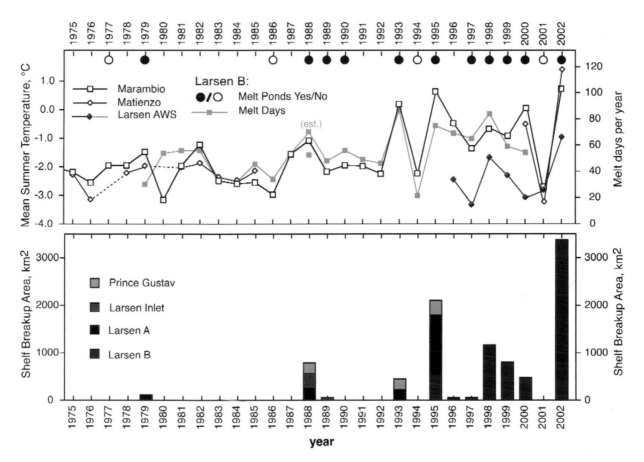

Plate 2. Trend of climate, melting, and retreat in the northeastern Antarctic Peninsula ice shelves over the last three decades. Mean summer temperatures (December, January, and February) are shown for three stations: Marambio (64° 14' S, 56° 37' W); Matienzo (64° 59' S, 60° 04' W) and Larsen AWS (66° 57' S, 60° 54' W) [*Skvarca et al.,* 1999b; *Skvarca and De Angelis,* this volume]. Melt days per year for a 25 km by 25 km area in the central Larsen B are derived from satellite passive microwave observation [*Fahnestock et al.,* 2002]. For the melt season of 1988, an estimate of 20 additional days was added to account for a lack of satellite coverage from December 3, 1987 to January 12, 1988. The summertime presence of melt ponds on the Larsen B was determined from Landsat, AVHRR, and MODIS satellite images. Breakup events are shown when the timing is known to within a single year. Several minor breakups of the Larsen A, Larsen Inlet, and Prince Gustav that occurred in the 1980s cannot be precisely timed. The Larsen B iceberg calving event of January 1995 is not included because it may not be related to climate-driven retreat.

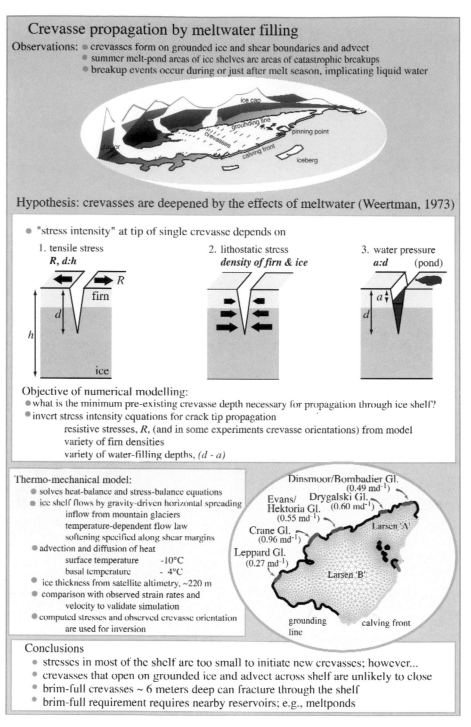

Fig. 3. Review and summary of model presented in *Scambos et al.* [2000]. Top cartoon: perspective view of Larsen-like shelf showing geographical features important to the model. Middle panel of cartoons: 1. With low, but non-compressive, resistive stresses, pre-existing crevasses neither close nor grow. This situation is typical for many ice shelves. 2. Lithostatic pressure limits the penetration of air-filled crevasses. 3. With sufficient water depth (i.e., if the value of a is small), the lithostatic pressure is overcome. Combined with the resistive stress, the extra outward pressure due to water-filling can induce the fracture tip to propagate downward. Bottom panel cartoon; model grid and input glaciers.

A process in which surface or near-surface fractures are extended downward by water pressure from infilling meltwater, previously discussed by Weertman [1973], Hughes [1983], and van der Veen [1998], was selected as the most likely connection between melt ponding due to climate warming and ice-shelf retreat (Figure 3).

In the simulations of Larsen A and B [*Scambos et al.,* 2000] pre-existing crevasses deeper than 6 to 22 meters (depending on firn density), if filled 90% or more with water, were able to penetrate the ice shelf completely. This occurs because the extra pressure due to meltwater filling increases the stress intensity at the crevasse tip. If the sum of stresses acting on the tip from ice dynamics and water pressure meets or exceeds the fracture toughness of the ice (typically 30 to 80 kPa) [*Van der Veen,* 1998] the crevasse propagates downward. Propagation halts when the lower surface of the shelf is breached, or when tensile stress drops below the threshold value. The latter case would result from lowering of the water level in the crevasse. Thus a reservoir of meltwater, e.g., surface ponds, is required to maintain a water level of at least 90% of the crevasse depth as the crevasse grows.

Penetration of meltwater into cold interior ice to the point of breaching the shelf requires that the process happen quickly—within hours to days for ~0.5m-wide cracks. However, once the shelf is breached, the fractures are likely to remain open, or at least structurally weak. Refrozen fresh-water ice in the crack will remain warmer, and therefore softer, than the surrounding shelf ice for many years. More importantly, seawater from beneath the shelf is likely to infiltrate the cracks, creating permanently weak vertical zones of partially frozen brine.

The melt-enhanced fracturing model requires pre-existing shallow surface fractures within the shelf. In mid-shelf ice where stresses are low, crevasses are only likely to form near shear margins or at the calving front. However, pre-existing crevasses formed at or above the grounding line can move by ice flow into low-stress regions within the shelf. In these regions there is near-zero compression, and so any advected crevasses will not close completely (until deeply buried by accumulation). In our model, meltwater acts on these existing, 'dormant', unhealed crevasses. Ice shelf rifts, formed by extending crevasses from shear-inducing obstructions [*MacAyeal et al.,* 1998], are another potential source of surface fractures in mid-shelf that could be augmented by meltwater infiltration. In many cases, several rifts are formed by the obstruction, but only a few are 'extended' across the shelf surface to become berg-forming rifts.

With meltwater enhancement, these shorter rifts might be extended both down and laterally into the shelf. A third source of fractures is present near the shelf front or rift edges. Forces deriving from the high freeboard typical of ice shelves induce rift- or front-parallel fractures within about one-half ice thickness of the ice edge [*Reeh,* 1968; *Fastook and Chapman,* 1989]. It is possible that the weight of ponded water itself may cause new stresses and cracking [*Bindschadler et al.,* 2002, Figure 7] but we believe this is a minor factor in the rapid breakup process.

Evidence Supporting the Model From the Larsen A and Larsen B Breakup Events

The change in calving style associated with rapid breakup, in which the majority of ice is calved in small (~10 km^2 and less), often sliver-shaped icebergs, points directly to a change in fracturing style as the cause of collapse. Major calving events from stable shelves tend to be dominated by one or a few large tabular bergs [*Lazzara et al.,* 1999]. The January 1995 event on the Larsen B, consisting of a large tabular berg and a region of sliver-shaped crevasses, represents an example of a transition from stable to unstable behavior.

Lost shelf areas from splinter-style calving in the Larsen A and B events lie almost exactly within the region of scattered melt ponds—in the case of the Larsen B, this is indicated by comparison of the January 31 MODIS image with that of March 7 (Plate 1). In addition, greater disruption of the shelf (indicated by fewer white bergs within the blue sub-pixel fragments in the March 7 image) occurred in the northwestern region of the shelf, where seasonal ponding had been observed for the greatest number of years. In the southernmost area of breakup, bergs were larger.

It may be noted from Figure 2 that surface melt ponds on the Wilkins Ice Shelf are not present directly over the breakup areas. However, even as early as January 1972, firn cores in the Wilkins encountered a 'water table' of fresh water just a few meters below the surface [*Vaughan et al.,* 1993]. Given the much longer melt seasons of recent years, we suspect that much of the Wilkins may be a snow swamp in warmer summer periods, and therefore capable of acting as a meltwater reservoir for the fracture growth mechanism.

Timing of the events, on both seasonal and inter-annual scales, also suggests that surface melt is the root cause. All four of the major shelf retreats in the last ten years (the Prince Gustav, the Larsen A, the Wilkins, and the

Larsen B) and most of the splinter-style retreat events that can be timed by satellite imagery, occurred in mid- to late austral summer. Further, the largest breakup events occur during the warmest summers (Plate 2). These associations imply ponded meltwater, and not associated winter re-freezing or some other cause with less seasonality (e.g., sub-shelf ocean currents), is responsible. Further, the sequence of ice shelves losing mass, beginning with the Wordie, Prince Gustav, and Larsen Inlet, and continuing with the Wilkins, Larsen A, and finally the Larsen B, follows the trend expected for an air-temperature related change [*Vaughan and Doake*, 1996; *Morris and Vaughan*, this volume]. Stable tabular-berg calving events do not show this summer-linked sea-sonality, nor the co-location with melt ponds [*Lazzara et al.*, 1999; *Jacobs et al.*, 1986].

A Landsat 7 image of the Larsen B acquired during a previous warm summer season appears to show that our modeled process is indeed operating on the shelf (upper right inset on Plate 1). Here we see lakes with elongate furrows extending out of them, essentially water-filled cracks. In some areas of this image, drained lakes and cracks appear, with sharp-edged fissures parallel or sub-parallel to the nearby, still-filled, furrows.

Further evidence comes from a close inspection of sub-scenes from the 2002 MODIS satellite image time series (lower left insets on Plate 1). Although lower in resolution than Landsat, the images show a gradual reduction in the number of dark ponds during the weeks just prior to the final disintegration of the ice shelf. Nearby areas on either side have a few ponds persisting through the period—so we do not attribute the disappearance to freezing or snowfall on the pond surfaces. We attribute the changes to shelf cracking and draining of ponded water just prior to breakup.

This model of cracking and draining also explains the sediment patterns on the seafloor beneath the former extent of the Larsen A [*Gilbert and Domack*, in press]. Although coarse-grained gravel and cobbles dominate much of the sub-shelf sediments, a substantial fraction of sediment in some locations is finer-grained sands and silts. These are interpreted as wind-blown surface sediments that accumulate in the upper ice shelf firn during ice flow. Concentration and deposition of these fines occurs when melt ponds accumulate meltwater from larger surface areas and then suddenly drain through the ice sheet. Gilbert and Domack [in press] used radioisotopes to determine that the fines under Larsen A had been deposited on the seabed primarily in the 1990s.

Other Factors Affecting Ice Shelf Susceptibility and the Breakup Mechanism

The persistence of the George VI shelf, which is only slowly retreating despite decades of abundant melt ponding [*Reynolds*, 1981] suggests that shelves can tolerate extensive melt accumulation if new crevasses cannot form and old crevasses close due to compression. Confined shelves, like George VI, or shelves where melting occurs in regions without near-surface crevasses, will not be as susceptible to the rapid breakup or catastrophic disintegration seen in the other Peninsula shelves.

The Amery and Fimbul ice shelves are examples where extensive ponding occurs in many or all summers without shelf breakup. In both cases, melt ponds currently form in areas behind major obstructions to ice flow, such as ice rises or islands, where along-flow and across-flow tension is low [e.g., *Young and Hyland*, 2002].

While capable of explaining many observations, the melt-driven fracturing model alone is not sufficient to explain the most dramatic aspects of the January 1995 and March 2002 events. Neither the pre-existing crevasse density nor the volume of meltwater stored in surface ponds is likely to have been great enough to crack these shelves as rapidly and as finely as the final catastrophic collapses require. Instead, cumulative effects of melt-driven fracturing may result in ice shelves so unstable that additional processes beyond typical calving can occur. In a separate study, an explanation is developed for the final disruption of the shelves based on a domino-like tipping of narrow intrashelf blocks [*MacAyeal et al.*, 2003]. These blocks are carved by closely spaced fractures (closer than the ice shelf thickness, creating tall narrow blocks) resulting from melt-driven fracturing. This tipping action contributes additional spreading force within the shelf. These forces can greatly exceed the normal shelf driving force once tipping proceeds to moderate angles.

REMOTE DETECTION OF PRE-BREAKUP CONDITIONS ON ICE SHELVES

The relationship between air temperature rise, increased surface melting, melt ponding, and ice shelf disintegration provides a set of remotely observable surface characteristics that can be used to evaluate any Antarctic ice shelf for susceptibility to the melt ponding-fracture enhancement breakup process.

Melt season length can be monitored by passive microwave emission changes [*Fahnestock et al.*, 2002].

To generate the melt pools needed in our model of shelf breakup, a long melt season is needed. However, to support ponded water on the surface, the shelf firn must also be impermeable. This occurs after repeated extensive melting events, in which surface melt percolates down into porous snowpack and refreezes. This eventually densifies the firn to the point of impermeability.

This process has a profound effect on the post-melt-season radar backscatter of the upper few meters of the firn [*Fahnestock et al.,* 1993; *Long and Drinkwater,* 1999]. In areas of near-zero melting, snow is an absorber of radar-wavelength (~decimeter) energy, allowing it to penetrate deep within the snowpack with little backscatter to the radar receiver. However, with even modest amounts of melt, radar backscatter increases significantly due to coarse-grained, radar-reflective ice layers formed by refreezing of surface melt. Such regions are termed the 'percolation facies' of the ice sheet [*Benson,* 1962]. With increasing melt, percolation layering increases, and so radar backscatter increases, making some portions of the Greenland and Antarctic ice sheets among the highest backscattering surfaces on Earth. Backscatter intensity peaks when the firn saturates with melt layers, creating a solid, impermeable layer of ice. At this point, increasing melt smoothes the upper surface, resulting in specular reflection of most of the radar energy. For non-nadir-looking radar instruments, this reduces backscatter by directing energy away from the receiver.

We assume this *geographic* pattern of backscatter versus melt intensity is an *evolutionary* one as well for ice shelves in a warming climate.

The general relationship of radar backscatter to surface melt season length clearly indicates the importance of firn impermeability for Antarctic ice shelves (Figure 4). Although significant melt duration occurs around much of the Antarctic perimeter, only the Peninsula shelves are at or approaching firn saturation. The three actively retreating shelves (Larsen B, Wilkins, and George VI) have the longest melt seasons on the continent, and significantly reduced backscatter. Regions adjacent to these areas, i.e., the southern Larsen B and northeastern Larsen C, have firn characteristics similar to the active-retreat shelves. With further warm summers, these areas will likely disintegrate via splinter-style calving, possibly catastrophically. The more southern parts of the Larsen C are nearly ice-layer saturated, but do not yet show reduced backscatter.

The relationship between melt season length and backscatter is likely dependent on accumulation rate as well, with areas of greater accumulation requiring longer melt seasons to achieve ponding or a given backscatter level. Much of the perimeter of Antarctica has a moderate accumulation rate, between 200 and 600 kg/m^2 [*Vaughan et al.,* 1999]. Further study of the melt—backscatter relationship may be able to identify a secondary dependence on accumulation and perhaps other factors.

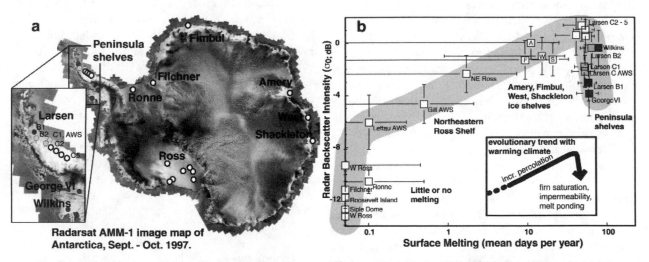

Fig. 4. a: RAMP image of Antarctica showing approximate backscatter intensity, with shelf locations used in b plotted as circles; b: Mean melt days per year versus backscatter intensity for shelf and ice sheet areas. Backscatter is the mean value from 100 km^2 regions at each circle in (a). Mean melt days over 21 years (1978–1999) are determined from passive microwave data [*Fahnestock et al.,* 2002]. RAMP image copyright Canadian Space Agency. Error bars indicate one-sigma errors.

DISCUSSION

It follows from melt onset data and our melt-fracture-enhancement breakup model that the climate stability limit of ice shelves is near the −1.5°C summertime isotherm. This value is distinct from previous estimates of the stability limit, and not just a refinement. Previous studies inferred either a 0°C summertime (or January) limit [*Mercer*, 1978; *Robin and Adie*, 1964] or a −5°C mean annual temperature limit [*Reynolds*, 1981] from geography: no ice shelves existed in areas exceeding these values. One explanation for the −5°C limit, as mentioned above, was a decrease in fracture toughness of ice as it nears the melting point [*Doake and Vaughan*, 1991]. For the 0°C limit, one hypothesis was that the ice shelf becomes nearly isothermal due to melt percolation into the annual winter cold wave. As a temperate ice mass, the shelf ice is much softer, and becomes unstably deformable, [*Mercer*, 1978].

For the model presented here, mean annual temperature may be unimportant – which in turn is very important for evaluating ice shelf stability. Given our model, mean summer temperature and melt season length, impermeability of the upper firn, and the proximity of melt ponds to shallow crevasses are paramount. This places several ice shelves closer to a threshold of retreat and breakup than previously thought. Much of the Antarctic perimeter is fringed with ice shelves having low mean annual temperatures (−15 to −25°C) but still experiencing a significant summer melt season. Summer climate warming and melt season increases at the rate occurring in the Antarctic Peninsula could render several ice shelves susceptible to breakup in just a few decades. As illustrated in Figure 4, the Amery, Fimbul, West, and Shackleton ice shelves are closest to this limit. To date, none of these shelves shows anything but ordinary cyclical calving behavior [e.g., *Fricker et al.*, 2002].

Glaciological and geological evidence suggests that the current warming causing the shelf retreats in the Peninsula is unusual to unprecedented in the Holocene epoch. Formation of the largest retreating shelves requires several centuries, and so summer temperatures must have been below the breakup threshold for at least that period [*Scambos et al.*, 2000]. Sediments in the seabed beneath the former shelves provide a longer-term perspective. Coarse till-like layers of uniform provenance are interpreted as sub-glacial tills from a thicker ice sheet in the past. Layers of fine-grained, biogenic-rich material with high depositional rates, or dropstones of mixed provenance, are interpreted to mean that bays were open-water or seasonally sea-ice-covered in the

past. In the case of the Prince Gustav and Larsen A, a few such layers within sediment cores indicate these areas were likely open water 2000 to 5000 YBP, and possibly 700 to 1500 YBP as well [*Pudsey and Evans*, 2001]. However, sediment cores acquired from the embayed area in front of the Larsen B in 2002 show no such layers in preliminary analysis [*Domack et al.*, 2002; *Gilbert and Domack*, in press]. This is interpreted to mean that, until the present warming, this shelf had not disintegrated since the Last Glacial Maximum, 12,000 YBP.

SUMMARY

The northernmost ice shelves in the Antarctic Peninsula are retreating via a distinctive calving style, culminating in rapid disintegration. These events are closely associated in time and space with areas of greater melt, and in particular melt ponding. A model for these events based on fracture deepening from meltwater infiltration is supported by several lines of evidence.

Glaciers formerly feeding the Larsen A increased flow speed by up to a factor of three since its breakup [*Rott et al.*, 2002]. This observation raises the significance of ice shelf breakup from that of a passive indicator of climate change to a possible step in raising sea level. Of particular concern is the Ross Ice Shelf, into which substantial portions of the West and East Antarctic ice sheets flow, each containing ice equal to several meters of sea level rise.

The presented model leads to remotely observable parameters that allow us to evaluate susceptibility of the Ross and any other ice shelf to climate-induced breakup. Further development of this monitoring tool, and further evaluation of ice shelves in more detail, is planned.

The present warming in the Peninsula does not pose much of a threat. Ice mass stored on the Peninsula is insignificant in terms of sea level rise. This is not the case with other ice shelf areas. As our Figure 4 demonstrates, summer melt season increases at the rate seen for the Peninsula area (one day per year) would bring several much larger shelves to the brink of the inferred threshold within just a few decades. Once at this threshold, the Peninsula shelves demonstrate that collapse occurs within about a two decades.

Acknowledgments. T. Scambos and M. Fahnestock were supported by NASA Grant NAG5-11308 and NAG5-3438; C Hulbe by NSF-OPP grant 01-25754. Extensive, helpful discussions with Douglas MacAyeal, David Vaughan, and Eugene Domack contributed to the ideas presented here, and we are grateful to S. Das, J. Ferrigno, and S. Jacobs for their comments in review of the manuscript. We thank J. Bohlander for data processing of the ice shelf images and Radar sat data.

REFERENCES

Benson, C., Stratigraphic studies in the snow and firn of the Greenland ice sheet, *U. S. Army Snow, Ice, and Permafrost Research Establishment Research Report 70*, 1962, 120 pp.

Bindschadler, R. A., and P. Vornberger, Interpretation of SAR imagery of the Greenland Ice Sheet using co-registered TM imagery, *Remt. Sens. Env.*, 42, 167–175, 1992.

Bindschadler, R. A., M. A. Fahnestock, P. Skvarca, and T. A. Scambos, Surface-velocity field of the northern Larsen Ice Shelf, Antarctica, *Ann. Glaciol.* 20, 319–326, 1994.

Bindschadler, R. A., T. A. Scambos, H. Rott, P. Skvarca, and P. Vornberger, Ice dolines on Larsen Ice Shelf, Antarctica, *Ann. Glaciol.*, 34, 283–290, 2002.

Comiso, J., Variability and trends in Antarctic surface temperatures from in situ and satellite infrared measurements, *J. Climate*, 13, 1674–1696, 2000.

Doake, C. S. M., and D. G. Vaughan, Rapid disintegration of the Wordie Ice Shelf in response to atmospheric warming, *Nature*, 350, 328–330, 1991.

Doake, C. S. M., H. F. J. Corr, J. H. Rott, P. Skvarca, and N. W. Young, Break-up and conditions for stability of the northern Larsen Ice Shelf, Antarctica, *Nature*, 391, 778–780, 1998.

Domack, E. W., D. Duran, K. McMullen, R. Gilbert, and A. Leventer, Sediment lithofacies from beneath the Larsen B Ice Shelf; can we detect ice shelf fluctuation?, *Eos*, 83, F301, 2002.

Fahnestock, M. A., R. Bindschadler, R. Kwok, and K. Jezek, Greenland Ice Sheet surface properties and ice dynamics from ERS-1 SAR imagery, *Science*, 262, 1530–1534, 1993.

Fahnestock, M. A., W. Abdalati, and C. Shuman, Long melt seasons on ice shelves of the Antarctic Peninsula: an analysis using satellite-based microwave emission measurements, *Ann. Glaciol.*, 34, 127–133, 2002.

Fastook, J. L., and J. E. Chapman, A map-plane finite-element model: three modeling experiments, *J. Glaciol.*, 35, 48–52, 1989.

Fricker, H. A., N. W. Young, I. Allison, and R. Coleman, Iceberg calving from the Amery Ice Shelf, East Antarctica, *Ann. Glaciol.*, 34, 241–246, 2002.

Gilbert, R. and E. Domack, The sedimentary record of disintegrating ice shelves in a warming climate, Antarctic Peninsula, *Geology, Geophysics, and Geosystems,* in press, 2003.

Gille, S. T., Warming of the southern ocean since the 1950s, *Science*, 295, 1275–1277, 2002.

Hindmarsh, R. C. A., Qualitative dynamics of marine ice sheets, in *Ice in the Climate System*, edited by W. R. Peltier, NATO ASI Ser., Ser. I, 12, pp. 67–99, 1993.

Hughes, T., On the disintegration of ice shelves: the role of fracture, *J. Glaciol.*, 29, 98–117, 1983.

Hulbe, C. L., and A. J. Payne, The contribution of numerical modelling to our understanding of the West Antarctic Ice Sheet, *Antarctic Research Series*, 77, 201–219, 2001.

Jacobs, S. S., D. R. MacAyeal, J. L. Ardai Jr., The recent advance of the Ross Ice Shelf, Antarctica, *J. Glaciol.*, 32, 464–474, 1986.

Jacobs, S. S., and J. C. Comiso, Climate variability in the Amundsen and Bellingshausen seas, *J. Climate*, 10, 697–709, 1997.

King, J. C., Recent climate variability in the vicinity of the Antarctic Peninsula, *Int. J. Climatology*, 14, 357–369, 1994.

King, J. C., J. Turner, G.J. Marshall, W. M. Connolley, and T. A. Lachlan-Cope, Antarctic Peninsula climate variability and its causes as revealed by analysis of instrument records, this volume.

Kwok, R., and J. C. Comiso, Southern ocean climate and sea ice anomalies associated with the southern oscillation, *J. Climate*, 15, 487–503, 2002.

Lazzara, M. A., K. C. Jezek, T. A. Scambos, D. R. MacAyeal, and C. J. Van der Veen, On the recent calving of icebergs from the Ross Ice Shelf, *Polar Geography*, 23, 201–212, 1999.

Lucchitta, B., and C. Rosanova, Retreat of northern margins of George VI and Wilkins ice shelves, Antarctic Peninsula, *Ann. Glaciol.*, 27, 41–46, 1998.

MacAyeal, D. R., E. Rignot, and C. L. Hulbe, Ice-shelf dynamics near the front of the Filchner-Ronne Ice Shelf, Antarctica, revealed by SAR interferometry: model/interferogram comparison, *J. Glaciol.*, 44, 419–428, 1998.

MacAyeal, D. R., T. A. Scambos, C. L. Hulbe, and M. A. Fahnestock, Catastrophic ice shelf breakup by an ice shelf fragment capsize mechanism, *J. Glaciol.,* in press, 2003.

Mercer, J. H., West Antarctic ice sheet and CO_2 greenhouse effect: a threat of disaster, *Nature*, 271, 321–325, 1978.

Morris, E. M., and D. G. Vaughan, Spatial and temporal variation of surface temperature on the Antarctic Peninsula and the limit of viability of ice shelves, this volume.

Parkinson, C., Trends in the length of the southern ocean sea ice season, 1979–99, *Ann. Glaciol.*, 34, 435–440, 2002.

Pudsey, C. J., and J. Evans, First survey of Antarctic sub-ice shelf sediments reveals mid-Holocene ice shelf retreat, *Geology*, 29, 787-790, 2001.

Rack, W., C. S. M. Doake, H. Rott, A. Siegel, and P. Skvarca, Interferometric analysis of the deformation pattern of the northern Larsen Ice Shelf, Antarctic Peninsula, compared to field measurements and numerical modeling, *Ann. Glaciol.*, 31, 205–210, 2000.

Reeh, N., On the calving of ice from floating glaciers and ice shelves, *J. Glaciol.*, 7, 215–232, 1968.

Reynolds, J. M., Lakes on George VI Ice Shelf, *Polar Record*, 20, 425– 432, 1981.

Reynolds, R. W., and T. M. Smith, Improved global sea surface temperature analyses using optimal interpolation, *J. Climate*, 7, 929–948, 1994.

Robin, G. de Q., and R. J. Adie, *Antarctic Research*, Butterworths, London, 1964.

Rott, H., P. Skvarca, and T. Nalger, Rapid collapse of northern Larsen Ice Shelf, Antarctica, *Science*, 271, 788–792, 1996.

Rott, H., W. Rack, T. Nalger, and P. Skvarca, Climatically induced retreat and collapse of northern Larsen Ice Shelf, Antarctic Peninsula, *Ann. Glaciol.*, 27, 86–92, 1998.

Rott, H., W. Rack, P. Skvarca, and H. De Angelis, Northern Larsen Ice Shelf, Antarctica: further retreat after collapse, *Ann. Glaciol.*, 34, 277–282, 2002.

Scambos, T. A., C. Hulbe, M. Fahnestock, and J. Bohlander, The link between climate warming and break-up of ice shelves in the Antarctic Peninsula, *J. Glaciol.*, 46, 516–530, 2000.

Skvarca, P., Fast recession of the northern Larsen Ice Shelf, Antarctic Peninsula monitored by space images, *Ann. Glaciol.*, 17, 317–321, 1993.

Skvarca, P., W. Rack, and H. Rott, 34 year satellite time series to monitor characteristics, extent, and dynamics of Larsen B Ice Shelf, Antarctic Peninsula, *Ann. Glaciol.*, 29, 255–260, 1999a.

Skvarca, P., W. Rack, H. Rott, and T. Ibarzábal y Donángelo, Climatic trend and the retreat and disintegration of ice shelves on the Antarctic Peninsula: an overview, *Polar Research*, 18, 151–157, 1999b.

Skvarca, P., and H. De Angelis, Impact assessment of climatic warming on glaciers and ice shelves on northeastern Antarctic Peninsula, this volume.

Thomas, R. H., T. O. Sanders, and K. E. Rose, 1979, Effect of warming on the West Antarctic Ice Sheet, *Nature,* 277, 355–358.

Thompson, D., and S. Solomon, Interpretation of recent southern hemisphere climate change, *Science*, 296, 895–899, 2002.

Van der Veen, C. J., Fracture mechanics approach to penetration of surface crevasses on glaciers, *Cold Regions Science and Technology*, 27, 31–47, 1998.

Vaughan, D. G., D. R. Mantripp, J. Sievers, and C. S. M. Doake, A synthesis of remote sensing data on Wilkins Ice Shelf, Antarctica. *Ann. Glaciol.*, 17, 211–218, 1993.

Vaughan, D. G., and C. S. M. Doake. 1996. Recent atmospheric warming and retreat of ice shelves on the Antarctic Peninsula. *Nature*,379, 328-331.

Vaughan, D. G., J. L. Bamber, M. Giovinetto, J. Russel, and A. P. R. Cooper, *J. Climate*, 12, 933–

Vaughan, D. G., G. J. Marshall, W. M. Connolley, J. C. King, and R. Mulvaney, Devil in the detail, *Science*, 293, 1777–1779, 2001a.

Vaughan, D. G., A. M. Smith, H. F. J. Corr, A. Jenkins, C. R. Bentley, M. D. Stenoien, S. S. Jacobs, T. B. Kellogg, E. Rignot, B. K. Lucchitta, A review of Pine Island Glacier, West Antarctica: hypothesis of instability vs. observations of change, *Ant. Res. Ser.,* 77, 237–256, 2001b.

Weertman, J., Can a water-filled crevasse reach the bottom surface of a glacier?, *Int. Assoc. of Scientific. Hydrology Publication,* 95, 139–145, 1973.

Young, N. W. and G. Hyland, Velocity and strain rates derived from InSAR analysis over the Amery Ice Shelf, Antarctica, *Ann. Glaciol.*, 34, 228–234, 2002.

Zwally, H. J., and Fiegles, S., Extent and duration of Antarctic surface melt, *J. Glaciol.*, 40, 463–476, 1994.

Zwally, H. J. , W. Abdalati, T. Herring, K. Larson, J. Saba, and K. Steffen, Surface melt-induced acceleration of Greenland ice-sheet flow, *Science, 297*, 218–222, 2002.

Mark A. Fahnestock, Institute for the Study of the Earth, University of New Hampshire, 39 College Road, Durham, NII, 03824-3525.

Christina L. Hulbe, Department of Geology, Portland State University, Portland, OR 97207.

Ted A. Scambos, National Snow and Ice Data Center, 1540 30th Street, CIRES, University of Colorado, Boulder, CO 80309-0449.

Terrestrial Archives of Paleoenvironmetal Change

THE LATE PLEISTOCENE AND HOLOCENE GLACIAL AND CLIMATE HISTORY OF THE ANTARCTIC PENINSULA REGION AS DOCUMENTED BY THE LAND AND LAKE SEDIMENT RECORDS – A REVIEW

Christian Hjort[1], Ólafur Ingólfsson[2], Michael J. Bentley[3], and Svante Björck[1]

Geomorphological and stratigraphical glacial- and climate history information from presently ice-free land areas in the Antarctic Peninsula region derive from glacially transported boulders, glacial striations, tills and moraine ridges, from raised beach-, littoral- and deeper marine sediments, and from lake sediments and moss-banks. Such sequences have been extensively ^{14}C dated, providing a rather coherent chronology for the last c. 10 ka (^{14}C kilo-years), with a less coherent pattern to 30 ka BP (before present) and beyond, complemented with occasional amino acid analyses and U/Th datings. These land data suggest that glaciation during the earlier part of the last (Wisconsinan/Weichselian) glacial cycle was more extensive than during the Last Global Glacial Maximum (LGM) around 18 ka BP. The earlier glaciation(s) was followed by an interstadial, probably during marine oxygen isotope stage (MIS) 3, and thereafter by the LGM glaciation. Deglaciation after LGM began with lowering of the ice surface on the southwestern Antarctic Peninsula around 14.5 ka BP, and with ice-front retreat prior to 14 ka BP on the continental shelf west of the peninsula. The latter was probably a result of increased calving and grounding-line retreat, due to the Northern Hemisphere-induced eustatic global sea level rise. Well inside the present coastlines deglaciation rates slowed down, with the earliest coastal areas becoming ice-free c. 9 ka BP and many more around 7.5 ka BP, but some areas considerably later. A mid-Holocene period of glacial readvances culminated shortly after 5 ka BP. It was closely followed by the Holocene climatic optimum 4.5 – 2.5 ka BP, roughly coinciding with the solar insolation maximum for these latitudes. Neoglaciation started c. 2.5 ka BP and Little Ice Age advances during the last 750 years have been identified, particularly in the South Shetland Islands.

1. INTRODUCTION

Much information on the geology and ecosystems of the Antarctic Peninsula region (Figure 1) was assembled by scientists onboard 19th Century sealing and research vessels, by men like Joseph D. Hooker and James Eights.

[1]Quaternary Geology, Lund University, Lund, Sweden
[2]University Courses on Svalbard (UNIS), Longyearbyen, Norway
[3]Dept. of Geography, University of Durham, Durham, UK.

However, it seems that the first one to take note of the fact that Antarctica had been even more glaciated before than it is today was *Henryk Arctowski* (1900), who in 1898 wintered with the *Belgica* on the western side of the Antarctic Peninsula. His brief notes were followed by more comprehensive observations on former glaciations, and on raised marine sediments, during the Swedish Antarctic expedition under Otto Nordenskjöld 1901–03, as summarized by *J.G. Andersson* (1906). Thereafter followed a long period, until after the 2nd World War, before Quaternary studies in this part of the world really got

10.1029/079ARS08

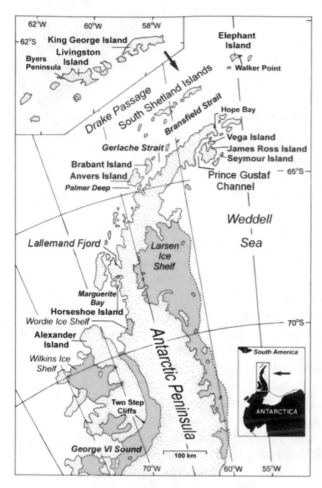

Fig. 1. Map of the Antarctic Peninsula region. The South Orkney Islands lie to the northeast, slightly outside the map.

underway. Although scattered notes of interest may be found in many earlier publications and reports, e.g. issued by the Falkland Islands Dependencies Survey (the predecessor to the British Antarctic Survey), most relevant work has been carried out during the last c. 35 years and with an almost exponential increase during the last 15 years. This applies to both land- and marine based studies.

Modern land based studies began with surveys of the glacial, marine and lacustrine history of the South Shetland Islands (e.g. *John and Sugden,* 1971; *Everett,* 1971; *John,* 1972; *Sugden and John,* 1973; *Curl,* 1980, *Birkenmajer,* 1981; *Barsch and Mäusbacher,* 1986). It continued with studies on ice-shelf related chronological problems south on Alexander Island (*Sugden and Clapperton,* 1981; *Clapperton and Sugden,* 1982) and with the glacial history of the islands in the northwestern Weddell Sea (*Elliot,* 1981; *Rabassa,* 1983). These papers

roughly set the baseline for the following brief review of what we today know about the glacial and climate history of the region, based on geological evidence in the presently ice-free land areas.

2. METHODS

Reconstructions of glacial, marine and other events of climatic significance in the Antarctic Peninsula region have made use of geomorphological and stratigraphical information from glacially transported boulders, glacial striations, tills and moraine ridges, raised beach-, littoral- and deeper marine sediments (often shell-bearing), plus lake sediments and moss-banks. These landforms and deposits permit the identification of different glacial events and their ice-flow directions, the study of the marine inundations and regressions which derive from the glacio-isostatic lowering of land and its subsequent rebound, and the evaluation of the "postglacial" Holocene climate variations. The latter studies are preferrably done by chemical and micropaleontological analyses of cores from lake sediments and moss-banks.

Absolute dating of the different events has usually been made by the ^{14}C-method (in later years with the AMS-version), sometimes complemented by tephro-, aminoacid- and lichenometric chronologies (e.g. *Birkenmajer,* 1981; *Clapperton and Sugden,* 1982; *Matthies et al.,* 1990; *Björck et al.,* 1991a; *Ingólfsson et al.,* 1992; *Martinez-Macciavello et al.,* 1996). U/Th dating has been attempted (*Barsch and Mäusbacher,* 1986), but the ESR and TL/OSL methods have to our knowledge not yet been tried in these parts of Antarctica. For TL/OSL, this is primarily due to the general lack of suitable eolian or waterlain sediments. More recently, cosmogenic exposure dating (using ^{10}Be and ^{26}Al) has been introduced (Bentley et al., submitted). All ages given below are in thousands of ^{14}C- years before present (ka BP). For the cosmogenic datings, the original ages in calendar years are also shown.

3. GLACIATION HISTORY AND THE RAISED MARINE SHORELINES

3.1 Pre-LGM Events

The altitude of some high and morphologically intact marine shorelines in the South Shetland Islands (> 200 m above present sea level; e.g. *John and Sugden,* 1971; *Birkenmajer,* 1997) may result from plate-tectonic related uplift. But like the shoreline at c. 80 m a.s.l. on James Ross Island in the northwestern Weddell Sea (*Hjort et al.*, 1997), some of them could instead indicate former heavy glacio-isostatic depression of the land. There is also a wide occurrence of glacially transported Antarctic Peninsula crystalline boulders, often at considerable altitudes, e.g. among the islands in the Weddell Sea (*Elliot*, 1981; *Rabassa*, 1983). In combination with glacially reworked shell-bearing marine sediments aged well beyond the limit of ^{14}C dating (*Clapperton and Sugden*, 1982; *Ingólfsson et al.*, 1992), these high shorelines and erratic boulders seem to indicate one or more late Pleistocene glacial event(s) predating and outsizing the Last Global Glacial Maximum (LGM, culminating around 18 ka BP and equivalent to marine oxygen isotope stage 2, MIS 2) in this region. Some erratic boulders well above the present trimlines on nunataks in the southern parts of the peninsula (*Carrara*, 1979, 1981; *Waitt*, 1983; Bentley et al.,submitted) may also belong to pre-LGM stages. However, no such older glacial event has yet been properly dated, although ^{14}C shell dates and amino acid values from just above a glaciomarine diamicton interpreted as a till, at St.Martha's Cove on James Ross Island, may put the latest such advance not too far beyond 30 ka BP (*Ingólfsson et al.*, 1992; J.M. Lirio, personal communication 2002). This indicates that it predated MIS 3, which B*erkman et al.* (1998) suggested was an important interstadial period also around Antarctica and which *Anderson and Andrews* (1999) regarded as a period with more extensive deglaciation around the Weddell Sea than during the Holocene. The latest more-extensive-than-the-LGM glaciation may thus have taken place during the earlier part of the Wisconsinan/Weichselian glacial cycle, during MIS 4 or 5.

3.2 LGM and the Subsequent Deglaciation

The LGM glaciation in the Antarctic Peninsula region may thus not have been the most extensive one during the last c. 100 ka glacial cycle, and perhaps the inland ice thickness and the offshore ice extension (particularly in the Weddell Sea sector) were less than postulated by e.g. *Bentley and Anderson* (1998). There are indications (like the extension of a thin till sheet visible on airphotos of the Prince Gustaf Channel area on James Ross Island) that the ice surface reached only a few hundred meters above present sea level in the lanes among the islands in the western Weddell Sea (*Hjort et al.*, 1997). It did, however, probably increase in thickness southwards, as indicated by the "postglacial" (Holocene) marine limit. This rises from c. 30 m in the north (James Ross Island, at c. 64°S; *Hjort et al.*, 1997) to 55 m in the south (Horseshoe Island in Marguerite Bay, at c. 68°S; *Hjort and*

Ingólfsson, 1990). The James Ross Island shoreline is ^{14}C dated to 7.5 ka BP, whereas the Horseshoe Island one has not yet been directly dated. However, as the latter stratigraphically post-dates the latest glacial overriding of the island and as deglaciation of Marguerite Bay was later than on James Ross Island (c. 6.5 ka BP; e.g. *Kennedy and Anderson*, 1989) the higher southern marine limit should in fact be younger than the northern one – what further supports the idea of a thicker ice in the south.

The deglaciation of the presently ice-free coastal areas has been ^{14}C dated with the help of subfossil mollusc shells from raised marine deposits, by remains from pioneer penguin rookeries, and by samples dating the onset of moss-bank- and lake sedimentation (see below). Recently some cosmogenic exposure ages ([10]Be, [26]Al) have contributed to the picture. The latter (Bentley et al., submitted) indicate that thinning of the ice stream in George VI Sound on the west side of the peninsula (c. 70°S, Figure 1) had begun by 14.5 ka BP (17.3 ka calendar years), but there are no indications of ice-surface lowering before c. 10 ka BP (11.5 ka calendar years) toward the higher and colder Weddell Sea side of the peninsula's ice divide.

The oldest ^{14}C ages, on subfossil mollusc shells from raised marine sediments, date the initial deglaciation of coastal areas on King George Island in the South Shetlands to 9-8 ka BP (*Sugden and John*, 1973; *Mäusbacher*, 1991). The deglaciation of northern James Ross Island has now been dated to around 7.5 ka BP, by *Hjort et al.*, (1997), who also discussed earlier published deglaciation dates from that area of 9-10 ka BP (*Zale and Karlén*, 1989; *Ingólfsson et al.*, 1992), now regarded as too old. The same deglaciation age (c. 7.5 ka BP; 8.2 ka calendar years) has recently been suggested for the east coast of Alexander Island, through exposure dating (Bentley et al., submitted). Shell dates (^{14}C) from there confirm deglaciation well before 6 ka BP (*Clapperton and Sugden*, 1982; *Hjort et al.*, 2001). This also agrees with ^{14}C dates of initial penguin colonization in Marguerite Bay, of 6.5-5.5 ka BP (*Emslie*, 2001).

Dating the onset of lake sedimentation and moss-bank growth indicates that once the glacier fronts had receded behind the present coastline and were not directly influenced by the eustatically induced early Holocene sea level rise, their continued retreat was often slow (*Barsch and Mäusbacher*, 1986; *Mäusbacher et al.*, 1989; *Mäusbacher*, 1991; *Ingólfsson et al.*, 1992; *Björck et al.*, 1991b, 1993, 1996a, 1996b; *del Valle and Tatur*, 1993; *Hjort et al.*,1997). On King George Island glaciers were at or within their present limits by 6 ka BP (*Martinez-Maciavello et al.*, 1996), and prior to 5.4 ka BP on northern James Ross Island (*Hjort et al.*, 1997). Some parts of Byers Peninsula on Livingston Island in the South Shetlands seem to have been deglaciated as late as 5-3 ka BP (*Björck et al.*, 1996b). Further north in that archipelago a minimum date for the deglaciation of some south-facing parts of Elephant Island is given as 5.5 ka BP by the onset of moss-bank growth there (*Björck et al.*, 1991b).

4. MID-HOLOCENE GLACIAL READVANCES

Rabassa (1983) described a mid-Holocene glacial readvance on James Ross Island, the shell-bearing sediments of which he named the Bahia Bonita Drift and ^{14}C-dated to around 5 ka BP. Later studies (*Hjort et al.*, 1997) have shown that this advance culminated around 4.6 ka BP and was associated with a marine shoreline at c. 15 m a.s.l.

Mäusbacher (1991) found evidence of increased glacial activity on King George Island at 5-4 ka BP and *Zale* (1994) suggested that a set of moraines in Hope Bay, at the northern tip of the Antarctic Peninsula, date from a glacial oscillation around 4.7 ka BP. The late deglaciation of parts of Byers Peninsula on Livingston Island (*Björck et al.*, 1996b) may also be an effect of such an oscillation, although no direct indication of any mid-Holocene glacial readvance has yet been found there. The present ice shelf in George VI Sound may have come into existence soon after 5.7 ka BP, as no molluscs younger than that age have been found in the ice-shelf moraines there (*Clapperton and Sugden*, 1982; *Hjort et al.*, 2001). It may thus originate from this mid-Holocene period of increased glacial mass balance.

5. THE HOLOCENE CLIMATIC OPTIMUM

The brief period of mid-Holocene glacial readvances was followed by the Holocene climatic optimum. It lasted about 2000 years, roughly between 4.5 and 2.5 ka BP, and culminated in the interval 4-3 ka BP. This comparatively warm and humid period is recorded by a number of biological and chemical proxies in lake cores (e.g. *Björck et al.*, 1991c, 1993; *Martinez-Machiavelli et al.*, 1996; *Mäusbacher et al.*, 1989; *Wasell*, 1993; *Zale*, 1993) and in moss-banks (*Fenton*, 1980; *Björck et al.*, 1991b). Paleoclimatic syntheses with discussions of this period in the Antarctic Peninsula region have been made by *Björck et al.* (1996b), and by *Jones et al.* (2000) who also suggested a longer optimum period (until c. 1.3 ka BP) in the South Orkney Islands. The latest review of the matter is one by Hodgson et al. (submitted).

6. NEOGLACIATION AND THE LITTLE ICE AGE

Several studies have demonstrated evidence of glacial expansion in the Antarctic Peninsula region during the last c. 2.5 ka (e.g. *John and Sugden*, 1971; *Sugden and John*, 1973; *Zale and Karlén*, 1989; *Clapperton*, 1990; *Lopéz-Martinez et al.*, 1996). For example, *Curl* (1980), *Birkenmajer* (1981), *Clapperton and Sugden* (1988) and *Björck et al.* (1996b) found evidence for neoglacial expansions in the South Shetlands in the form of moraines transgressing earlier Holocene raised beaches. Lichenometric dating (using the thalli diameter of *Rhizocarpon geographicum*; *Curl*, 1980; *Birkenmajer*, 1981) and a glacially transported whalebone on the Rotch Dome ice cap on Livingston Island (*Björck et al.*, 1996b) place several of these advances within the last c. 750 years - roughly in the Little Ice Age.

7. DISCUSSION AND CONCLUSIONS

Although many indications point that way, the question whether glaciation in the Antarctic Peninsula region was more extensive during an earlier phase of the last (Wisconsinan/Weichselian) glacial cycle than during the LGM is still open. Neither the glaciomarine sediments post-dating the latest pre-LGM glacial event, nor the high marine limits possibly associated with that event (e.g. the 80 m shoreline on James Ross Island), or the widespread glacially transported boulders at considerable altitudes, have yet been well dated. Nonetheless, geological data suggesting a larger than LGM earlier Wisconsinan/ Weichselian ice sheet exist on land, and also seem supported by marine data. For example, the youngest glaciomarine sediments in parts of both the eastern and western Weddell Sea seem to be older than 26 ka BP (*Elverhöi*, 1981; *Anderson and Andrews*, 1999). A distinct IRD (ice rafted detritals) peak there indicates a period of major deglaciation, more extensive than during the Holocene and lasting until c. 26 ka BP (*Anderson and Andrews*, 1999; *Anderson et al.* 2002), thus being placed in MIS 3. On land, the best way to aswer this open question would be through an extensive dating program, using methods reaching beyond the reach of ^{14}C, such as exposure dating glacial erratics and pre-LGM shorelines, and ESR-dating mollusc shells from the older raised marine sediments.

The deglaciation of the Antarctic Peninsula shelf areas (e.g. *Pope and Anderson*, 1992; *Pudsey et al.*, 1994; *Banfield and Anderson*, 1995; *Bentley and Anderson*, 1998; *Canals et al.*, 2000; *Domack et al.*, 2001; *Anderson*

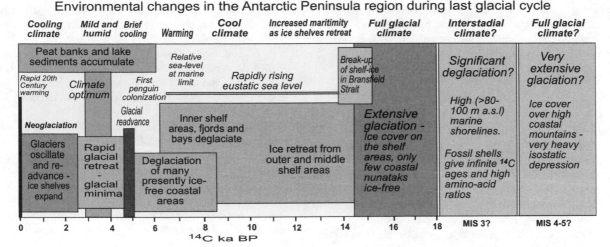

Fig. 2. Synthesis of the Late Pleistocene and Holocene glacial and climatic developments in the Antarctic Peninsula region.

et al., 2002) started before 14 ka BP and coincided with the rapid eustatic rise of global sea level following the LGM in the Northern Hemisphere. Therefore the glacial retreat from the shelves around the peninsula and the associated downdraw of ice on the peninsula itself do not nescessarily provide any regional or Southern Hemisphere climatic signal. But once the ice fronts had calved back to near the present coastlines the rate of deglaciation slowed down considerably, became more dependent on the regional climate, and distinct readvances took place well into the mid-Holocene. This, as compared to the Northern Hemisphere, considerably delayed deglaciation and meltwater production is also recorded in the Ross Sea area (e.g. *Licht et al.,* 1996). It may explain the continuation of global Holocene eustatic sea level rise well after 7 ka BP, when the northern ice sheets had largely vanished (*Ingólfsson and Hjort,* 1999; *Lambeck and Chappell,* 2001), and Antarctic deglaciation could thus be the reason for the northwest European Tapes- and Littorina transgressions between c. 7.5 and 5 ka BP.

The "post-glacial" Holocene climatic optimum in the Antarctic Peninsula region occurred as late as 4.5–2.5 ka BP. This is the same interval as for the "penguin optimum" in the Ross Sea area (*Baroni and Orombelli,* 1994), and this timing of the climatic optimum has also been documented in lake cores from the Bunger Oasis in East Antarctica (*Kulbe et al.,* 2001). It coincides roughly with the insolation maximum for these southern latitudes (e.g. *Berger,* 1978), but postdates most of the Northern Hemisphere Holocene optima by several millenia. Thus, according to the geological records on land and as sug-

gested by *Hjort et al.* (1998), neither the timing of the Holocene deglaciation, nor that of the climatic optimum in the Antarctic Peninsula region, show any convincing synchroneity with early-middle Holocene climatic events in the Northern Hemisphere. We are, however, aware that some marine data indicate a wider period of climatic optimal conditions around the peninsula, between c. 8 and 3 ka BP (*Domack et al.,* 2001), although in the near-coastal waters the highest productivity seems to have occurred c. 4 -3 ka BP (*Shevenell et al.,* 1996).

Also the change into Neoglacial conditions in the Antarctic Peninsula region lagged similar developments in the Northern Hemisphere by some thousand years. From marine data the cooling trend in the Northern Hemisphere started around 5 ka BP (e.g. *Koc et al.,* 1993), and from Greenland ice core data it began around 4 ka BP (*Johnsen et al.,* 2001), whereas the Antarctic cooling started around 2.5 ka BP. However, the distinct glacial readvances in the Antarctic Peninsula region during the last c. 750 years, particularly in the South Shetland Islands, seem roughly to parallel the Little Ice Age in the Northern Hemisphere. It may thus be a non-Milankovichean event.

We summarize our interpretations in Figure 2.

Acknowledgments. Financial support for our work has mainly been through the Swedish Natural Sciences Research Council, the National Environmental Research Council, UK, and the Universities of Lund and Göteborg in Sweden, and Edinburgh and Durham in the UK. Logistic support has been through the Swedish Polar Research Secretariat, the British

Antarctic Survey, the Alfred-Wegener Institut, the Instituto Antartico Argentino and the Programa Espanol de Investigation Antartida. We also acknowledge constructive reviews by John B. Anderson, Glenn W. Berger and Brenda Hall.

REFERENCES

Anderson, J.B. and Andrews, J.T., 1999: Radiocarbon constraints on ice sheet advance and retreat in the Weddell Sea, Antarctica. *Geology,* 27: 179-182.

Anderson, J.B., Shipp, S.S., Lowe, A.L., Smith Wellner, J. and Mosola, A.B., 2002: The Antarctic Ice Sheet during the Last Glacial Maximum and its subsequent retreat history: a review. *Quaternary Science Reviews,* 21: 49-70.

Andersson, J.G., 1906: On the geology of Graham Land. *Bull. Geological Institute, Uppsala University,* 7: 19-71.

Arctowski, H., 1900: Géographie physique de la région Antarctique visitée par l'expedition de la Belgica. *Bulletin de la Société Royale belge de Géographie,* 1900: 93-99.

Banfield, L.A. and Anderson, J.B., 1995: Seismic facies investigation of the Late Quaternary glacial history of Bransfield Basin, Antarctica. *Antarctic Research Series,* 68: 123-140.

Baroni, CA. and Orombelli, G., 1994: Abandoned penguin rookeries as Holocene paleoclimate indicators in Antarctica. *Geology,* 22: 23-26.

Barsch, D. and Mäusbacher, R., 1986: Beiträge zur Vergletscherungsgeschichte und zur Reliefentwicklung der Südshetland Insel. *Zeitschrift für Geomorphologie,* 61: 25-37.

Bentley, M.J. and Anderson, J.B., 1998: Glacial and marine geological evidence for the ice sheet configuration in the Weddell Sea-Antarctic Peninsula region during the Last Glacial Maximum. *Antarctic Science,* 10: 309-325.

Bentley, M.J., Fogwill, C.J., Kubik, P.W. and Sugden, D.E. submitted: Geomorphological evidence and cosmogenic ^{10}Be/^{26}Al exposure ages for the Last Glacial Maximum configuration and deglaciation history of the Antarctic Peninsula Ice Sheet. *Geology.*

Berger, A., 1978: Long-term variations of daily insolation and Quaternary climatic change. *Journal of Atmospheric Sciences,* 35: 2362-2367.

Berkman, P.A., Andrews, J.T., Björck, S., Colhoun, E.A., Emslie, S.D., Goodwin, I.D., Hall, B.L., Hart, C.A.P., Hirakawa, K., Igarashi, A., Ingólfsson, Ó., López-Martínez, J., Lyons, W.B., Mabin, M.C.A.G., Quilty, P.G., Taviani, M. and Yoshida, Y., 1998: Circum-Antarctic coastal environmental shifts during the Late Quaternary reflected by emerged marine deposits. *Antarctic Science,* 10: 345-362.

Birkenmajer, K., 1981: Lichenometric dating of raised marine beaches at Admirality Bay, King George Island (South Shetland Islands, West Antarctica). *Bulletin de l'Academie Polonaise des Sciences, Series des Sciences de la Terre,* 29(2): 119-127.

Birkenmajer, K., 1997: Quaternary geology at Arctowski Station, King George Island, South Shetland Islands (West Antarctica). *Studia Geologica Polonica,* 110: 91-104.

Björck, S., Sandgren, P. and Zale, R., 1991a: Late Holocene Tephrochronology of the Northern Antarctic Peninsula. *Quaternary Research,* 36: 322-328.

Björck, S., Malmer, N., Hjort, C., Sandgren, P., Ingólfsson, Ó., Wallén, B., Smith R.I.L. and Liedberg-Jönsson, B., 1991b: Stratigraphic and paleoclimatic studies of a 5,500 years old moss bank on Elephant Island, Antarctica. *Arctic and Alpine Research,* 23: 361-374.

Björck, S., Håkansson, H., Zale, R., Karlén, W. and Liedberg-Jönsson, B., 1991c: A late Holocene lake sediment sequence from Livingston Island, South Shetland Islands, with palaeoclimatic implications. *Antarctic Science,* 3, 61-72.

Björck, S., Håkansson, H., Olsson, S., Barnekow, L. and Jansens, J., 1993: Paleoclimatic studies in South Shetland Islands, Antarctica, based on numerous stratigraphic variables in lake sediments. *Journal of Paleolimnology,* 8: 233-272.

Björck, S., Håkansson, H., Olsson, S., Ellis-Evans, C.A., Humlum, O. and Lirio, J.M., 1996a: Late Holocene palaeoclimatic records from lake sediments on James Ross Island, Antarctica. *Palaeogeography, Palaeoclimatology, Palaeoecology,* 113: 195-220.

Björck, S., Hjort, C. Ingólfsson, Ó., Zale, R. and Ising, J., 1996b: Holocene deglaciation chronology from lake sediments. *In* López-Martínez, J., Thomson, M.R.A. and Thomson, J.W. (eds), *Geomorphological map of Byers Peninsula, Livingston Island.* BAS GEOMAP Series, Sheet 5-A, 1:25 000, with supplementary text, 65 pp. Cambridge, British Antarctic Survey.

Canals, M., Urgeles, R. and Calafat, A.M., 2000: Deep seafloor evidence of past ice streams off the Antarctic Peninsula. *Geology,* 28: 31-34.

Carrara, P., 1979: Former extent of glacial ice in Orville Coast region, Antarctic Peninsula. *Antarctic Journal of the United States,* 14: 45-46.

Carrara, P., 1981: Evidence for a former large ice sheet in the Orville Coast-Ronne Ice Shelf area, Antarctica. *Journal of Glaciology,* 27: 487-491.

Clapperton, C.A.M., 1990: Quaternary glaciations in the Southern Ocean and Antarctic Peninsula area. *Quaternary Science Reviews,* 9: 229-252.

Clapperton, C.A.M. and Sugden, D.E., 1982: Late Quaternary glacial history of George VI Sound area, West Antarctica. *Quaternary Research,* 18: 243-267.

Clapperton, C.A.M. and Sugden, D.E., 1988: Holocene glacier fluctuations in South America and Antarctica. *Quaternary Science Reviews,* 7: 185-198.

Curl, J.E., 1980: A glacial history of the South Shetland Islands, Antarctica. *Institute of Polar Studies Report,* 63. Columbus: Institute of Polar Studies, Ohio State University, 129 pp.

Domack, E., Levanter, A., Dunbar, R., Taylor, F., Brachfield, S., Sjunneskog, C.A. and ODP Leg 178 Scientific Party, 2001: Chronology of the Palmer Deep site, Antarctic Peninsula: a Holocene palaeoenvironmental reference for the circum-Antarctica. *The Holocene,* 11: 1-9.

Elverhöi, A., 1981: Evidence for a late Wisconsin glaciation of the Weddell Sea. *Nature,* 293: 641-642.

Elliot, D.H., 1981: Glacial geology of Seymour Island. *Antarctic Journal of the United States,* 16(5): 66-67.

Emslie, S.D., 2001: Radiocarbon Dated Abandoned Penguin Colonies in the Antarctic Peninsula Region. *Antarctic Science,* 13: 289-295.

Everett, K.R., 1971: Observations on the glacial history of Livingston Island. *Arctic,* 24:41-50.

Fenton, J.H.C.A., 1980: The rate of peat accumulation in Antarctic moss banks. *Journal of Ecology,* 68: 211-228.

Hjort, C. and Ingólfsson, Ó., 1990: Studies of the Glacial History in the Antarctic Peninsula Area. *In* Karlqvist, A. (ed.), *Swedish Antarctic Research Programme 1988/89 – A Cruise Report.* Stockholm, Swedish Polar Research Secretary, 76-80.

Hjort, C., Ingólfsson, Ó., Möller, P. and Lirio, J.M., 1997: Holocene glacial history and sea-level changes on James Ross Island, Antarctic Peninsula. *Journal of Quaternary Science,* 12: 259-273.

Hjort, C., Björck, S., Ingólfsson, Ó. and Möller, P., 1998: Holocene deglaciation and climate history of the northern Antarctic Peninsula region - a discussion of correlations between the Southern and Northern Hemispheres. *Annals of Glaciology,* 27: 110-112.

Hjort, C., Bentley, M.J. and Ingólfsson, Ó., 2001: Holocene and pre-Holocene temporary absense of the George VI Ice Shelf, Antarctic Peninsula. *Antarctic Science,* 13: 296-301.

Hodgson, D.A., Doran, P.T., Roberts, D., McMinn, A. & Vyverman, W., submitted: Antarctic Paleolimnology. *In* Smol, J.P. and Last, W.M. Eds. *Environmental Change in Arctic and Antarctic Lakes. Developments in Paleoenvironmental Research,* 8. Kluwer, Dordrecht.

Ingólfsson, Ó., Hjort, C., Björck, S. and Smith, R.I.L., 1992: Late Pleistocene and Holocene glacial history of James Ross Island, Antarctic Peninsula. *Boreas,* 21: 209-222.

Ingólfsson, Ó. & Hjort, C., 1999: The Antarctic contribution to Holocene global sea level rise. *Polar Research,* 18: 323-330.

John, B.S., 1972: Evidence from the South Shetland Islands towards a glacial history of West Antarctica. *Institute of British Geographers, Special Publication,* 4: 75-89.

John, B.S. and Sugden, D.E., 1971: Raised marine features and phases of glaciation in the South Shetland Islands. *British Antarctic Survey Bulletin,* 24: 45-111.

Johnsen, S.J., Dahl-Jensen, D., Gundestrup, D., Steffensen, J.P., Clausen, H.B., Miller, H., Masson-Delmotte, V., Sveinbjörnsdottir, A.E. & White, J., 2001: Oxygen isotope and palaeotemperature records from six Greenland ice-core stations: Camp Century, Dye-3, GRIP, GISP2, Renland and NorthGRIP. *Journal of Quaternary Science,* 16: 299-307.

Jones, V.J., Hodgson, D.A. & Chepstow-Lusty, A., 2000: Palaeolimnological evidence for marked Holocene environmental changes on Signy Island, Antarctica. *The Holocene,* 10: 43-60.

Kennedy, D.S. & Anderson, J.B., 1989: Glacial-marine sedimentation and Quaternary glacial history of Marguerite Bay, Antarctic Peninsula. *Quaternary Research,* 31: 255-276.

Koc, N., Jansen, E. & Haflidason, H., 1993: Paleoceanographic reconstructions of surface ocean conditions in the Greenland, Iceland and Norwegian Seas through the last 14 ka based on diatoms. *Quaternary Science Reviews,* 12: 115-140.

Kulbe, T., Melles, M., Verkulich, S.R. and Pushina, Z.V., 2001: East Antarctic climate and environmental variability over the last 9400 years inferred from marine sediments in the Bunger Oasis. *Arctic, Antarctic, and Alpine Research,* 33: 223-230.

Lambeck, K. & Chappell, J., 2001: Sea Level Change Trough the Last Glacial Cycle. *Science,* 292: 679-686.

Licht, K.M., Jennings, A.E., Andrews, J.T. & Williams, K.M., 1996: Chronology of late Wisconsin ice retreat from the western Ross Sea, Antarctica. *Geology,* 24: 223-226.

López-Martínez, J., Thomson, M.R.A., Arche, A., Björck, S., Ellis-Evans, J.CA., Hathway, B., Hernández-Cifuentes, F., Hjort, CA., Ingólfsson, Ó., Ising, J., Lomas, S., Martínez de Pisón, E., Serrano, E., Zale, R. and King, S., 1996: *Geomorphological map of Byers Peninsula, Livingston Island.* BAS GEOMAP Series, Sheet 5-A, 1:25 000, with supplementary text. Cambridge, British Antarctic Survey, 65 pp.

Martinez-Macchiavello, J.CA., Tatur, A., Servant-Vildary, S. and Del Valle, R., 1996. Holocene environmental change in a marine-estuarine-lacustrine sediment sequence, King George Island, South Shetland Islands. *Antarctic Science,* 8: 313-322.

Matthies, D., Mäusbacher, R. and Storzer, D., 1990: Deception Island tephra: a stratigraphical marker for limnic and marine sediments in Bransfield Strait area, Antarctica. *Zentralblatt für Geologie und Paläontologie,* 1: 153-163.

Mäusbacher, R., 1991: Die Jungkvartäre Relief- und Klimageschichte im Bereich der Fildeshalbinsel, Süd-Shetland-Inseln, Antarktis. *Heidelberger Geographische Arbeiten,* 89: 207 pp.

Mäusbacher, R., Müller, J. and Schmidt, R., 1989: Evolution of postglacial sedimentation in Antarctic lakes. *Zeitschrift für Geomorphologie,* 33: 219-234.

Pope, P.G. and Anderson, J.B., 1992: Late Quaternary glacial history of the northern Antarctic Peninsula's western continental shelf: evidence from the marine record. *American Geophysical Union, Antarctic Research Series,* 57: 63-91.

Pudsey, C.A.J., Barker, P.F. and Latrer, R.D., 1994: Ice sheet retreat from the Antarctic Peninsula Shelf. *Continental Shelf Research,* 14: 1647-1675.

Rabassa, J., 1983: Stratigraphy of the glacigenic deposits in northern James Ross Island, Antarctic Peninsula. *In* Evenson, E., Sclüchter, CA. and Rabassa, J. (eds.), *Tills and Related Deposits.* Rotterdam: A.A. Balkema Publishers, 329-340.

Shevenell, A.E., Domack, E.W. and Kernan, G.M., 1996: Record of Holocene palaeoclimate change along the

Antarctic Peninsula: evidence from glacial marine sediments, Lallemand Fjord. *Papers and Proceedings of the Royal Society of Tasmania*, 130: 55-64.

Sugden, D.E. and Clapperton, CA.M., 1981: An ice-shelf moraine, George IV Sound, Antarctica. *Annals of Glaciology*, 2: 135-141.

Sugden, D.E. and John, B.S., 1973: The ages of glacier fluctuations in the South Shetland Islands, Antarctica. *In* van Zinderen Bakker, E.M. (ed.), *Palaeoecology of Africa, the surrounding Islands and Antarctica*, 8: Cape Town, Balkema, 141-159.

Valle, del, R. and Tatur, A., 1993: Holocene evolution of landscape and biota on King George Island, Antarctica. *Verhandlungen Internationalen Verein fur Limnologie*, 25: 1128-1130.

Waitt, R.B., 1983: Thicker West Antarctic ice sheet and peninsula ice cap in late-Wisconsin time – sparse evidence from northern Lassiter Coast. *Antarctic Journal of the United States*, 18: 91-93.

Wasell, A., 1993: Diatom stratigraphy and evidence of Holocene environmental changes in selected lake basins in the Antarctic and South Georgia. Stockholm University: *Department of Quaternary Geology Research Reports*, 23: 1-15.

Zale, R., 1993: Lake sediments around the Antarctic Peninsula, archives of climatic and environmental change. *GERUM Naturgeogr.*, 17: 1-21.

Zale, R., 1994: [14]C age corrections in Antarctic lake sediments inferred from geochemistry. *Radiocarbon*, 36: 173-185.

Zale, R. and Karlén, W., 1989: Lake sediment cores from the Antarctic Peninsula and surrounding islands. *Geografiska Annaler*, 71(A): 211-220.

M.J. Bentley, Dept. of Geography, Durham University, South Road, Durham DH1 3LE, UK; email: m.j.bentley@durham.ac.uk

S. Björck and C. Hjort, Quaternary Geology, Dept of Geology, Lund University, Tornavägen 13, SE-223 63 Lund, Sweden; email: svante.bjorck@geol.lu.se, christian.hjort@geol.lu.se

Ó. Ingólfsson, University Courses on Svalbard (UNIS), Box 156, N-9170 Longyearbyen, Norway; email: olafur @unis.no

AN OVERVIEW OF LATE PLEISTOCENE GLACIATION IN THE SOUTH SHETLAND ISLANDS

Brenda L. Hall

*Institute for Quaternary and Climate Studies and Department of Geological Sciences,
University of Maine, Orono, Maine*

During the last glacial maximum (LGM), ice caps in the South Shetland Islands expanded onto the continental shelf, possibly no more than a few kilometers beyond the present coastline. Sea level may have been an important factor in controlling ice extent. Widespread preservation of pre-LGM raised marine platforms and beaches suggests that most LGM ice was cold-based. Deglaciation began prior to ~8500 ^{14}C yr B.P., although some areas may have remained ice-covered until mid-Holocene time.

INTRODUCTION

A key problem in global change research concerns the synchrony or asynchrony of climate variations north and south of the Antarctic Convergence [*Broecker*, 1998]. Evidence for the timing of climate oscillations in Antarctica is scarce and inconclusive. The Byrd ice core in West Antarctica shows behavior out-of-phase with that of the Northern Hemisphere during rapid climate events of the last glaciation [*Blunier et al.*, 1998; *Blunier and Brook*, 2001]. Moreover, oxygen-isotope values show a trend believed to represent regional warming several thousand years earlier at Byrd than in Greenland [*Sowers and Bender*, 1995]. In contrast, the Taylor Dome ice core, located in East Antarctica, indicates in-phase climate behavior [*Steig et al.*, 2000]. Resolving this question of synchrony or asynchrony is critical for understanding the basic mechanism behind climate fluctuations.

The most likely place for a change from in-phase to out-of-phase climate behavior would be at the Antarctic Convergence, particularly in areas of compressed oceanic and atmospheric gradients, such as the Drake Passage. Situated adjacent to the Drake Passage, the South Shetland Islands (SSI; 61°00'-63°30'S; 53°30'-62°45'W) are in an ideal location to test hypotheses concerning the global synchrony of climate change. But before one can address the global problems posed above, more basic questions must be answered. These include: How big was the ice in the SSI during the last glacial maximum (LGM)? Did it extend offshore? When did it begin to retreat? Was the ice temperate or polar? The purpose of this paper is to present a synthesis of available information, as well as some new data and interpretations from recent fieldwork that bear on these questions.

The SSI consist of eleven major and hundreds of minor islands and shoals, stretching over 230 km from the southwest to northeast (Figure 1). They are in a maritime polar zone in the path of the highest frequency of storm tracks in the Antarctic [*Lamb*, 1964]. The climate is wet and commonly windy. Annual precipitation averages 1170 mm/yr [*Loewe*, 1957] and the mean annual temperature is about -3 $^{\circ}$C [*Lamb*, 1964]. The island chain is heavily glaciated, with relatively few ice-free areas. Glaciers tend to terminate at the shore. Their extent likely is controlled largely by sea level and wave erosion at the base of the glacier cliff.

The SSI form part of the Scotia Arc. Bedrock consists mainly of Upper Paleozoic to late Cenozoic volcanic rocks and marine and terrestrial mudstones, sandstones, and conglomerates [*Smellie et al.*, 1984]. Early Cretaceous to early Tertiary plutons, composed of gabbro, quartz gabbro, tonalite, and granodiorite, occur in isolated locations.

Inherent logistical difficulties and the small number of ice-free areas have contributed to there being only a handful of glacial geologic studies in the SSI. *John and*

10.1029/079ARS09

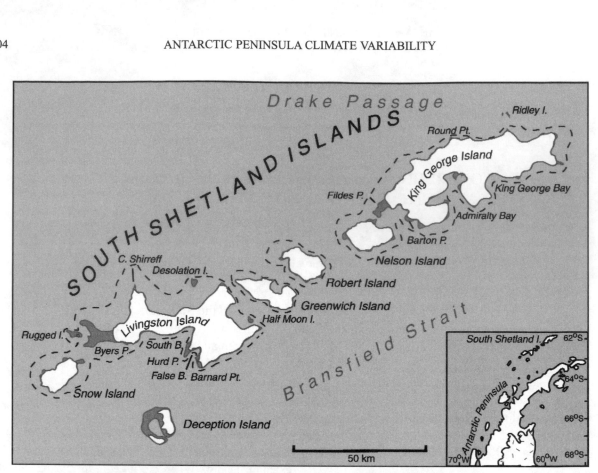

Fig. 1. Index map of the South Shetland Islands with ice caps shown in white. Dashed line shows schematic reconstruction of ice in the SSI at the LGM. Former ice extent on Deception Island is not shown, because there are not any data.

Sugden [1971] produced the most comprehensive work, visiting more than eighty sites in 1965/66. On the basis of geomorphologic features, particularly subglacial meltwater channels and deeply cut troughs, they concluded that the entire island chain had been covered by a large ice cap, centered off the northern coast on the broad continental shelf. This ice cap was thought to predate the LGM, largely because some features associated with it have been truncated by marine erosion. LGM ice was believed to have formed slightly smaller ice caps, centered over individual islands. Subsequent researchers generally have concurred with that opinion [*Curl*,1980; *Clapperton and Sugden*, 1988; *Mäusbacher*, 1991].

In 2001, we spent 2.5 months in the SSI, carrying out glacial geologic field work on King George and Livingston Islands. Characterization of the deposits was based on hand-dug excavations and surface expression. Elevations were obtained using a Garmin GPS unit with built-in barometer, as well as by surveying with an auto level and staff. Our work concentrated on three sites: Fildes, Barton, and Byers Peninsulas. Fildes Peninsula is a northeast-southwest trending ice-free area at the westernmost tip of King George Island. The peninsula has a mix of rugged topography and broad platforms extend-

ing to ~180 m elevation. A central valley separates two highlands (in the north and south). The coastline is heavily indented with coves and displays numerous sea cliffs and stacks. Nearby Barton Peninsula occupies ~15 km² between Marian and Potter Coves. Tidewater glaciers draining inland ice occupy the heads of both of these coves. Similar to Fildes Peninsula, the coastline is rugged, with needle-like sea stacks. Situated 100 km to the west on Livingston Island, Byers Peninsula rises from sea level to 265 m elevation. Unlike the other two areas, the coast here is marked by broad bays with large sandy beaches. Although most of the peninsula consists of gently undulating platforms, terrain is more mountainous in the northwest.

GLACIAL FEATURES

Erosional Landforms

Cross-cutting all but the lowest exposed platform on both Fildes Peninsula and adjacent Ardley Island is a series of anastomosing channels (~100 m wide—commonly 10–20 m, 1.5 km long) first noted by *John and Sugden* [1971; Figures 2a,3]. Although orientation and slope are generally to the southeast, many channels trend oblique to topography and rise up and down over divides. Their geometry and slope both suggest that they are subglacial meltwater features. Because they rise towards the northwest coast, the channels are inferred to have formed under an ice cap centered just northwest of Fildes Peninsula. Similar channels occur at nearby Rip Point on Nelson Island [*John and Sugden*, 1971], as well as on the upper platforms of Barton Peninsula subparallel to Marian and Potter Coves. I believe that these also formed from subglacial meltwater. However, v-notched channels that follow the topography on Barton Peninsula most likely were downcut by subaerial streams.

Large, deep transverse troughs have been cited as evidence for ice movement across the island chain [*John and Sugden*, 1971]. Inter-island troughs head on the northwest platform in keeping with the pattern of ice flow suggested by the meltwater channels. Other troughs, such as Maxwell and Admiralty Bays (Figures 1, 2a), are developed on the south side of the islands. Although possibly tectonic in origin, these latter two bays, in particular, show clear evidence of glacial erosion. Longitudinal profiles indicate that the bays are overdeepened with prominent sills near their mouths, similar to fjords [*Griffiths and Anderson*, 1989; *Domack and Ishman*, 1993; *Yoon et al.*, 1997].

Other glacial erosional features are uncommon in the ice-free areas examined. On Fildes and Barton Peninsulas, *John and Sugden* [1971] identified large-scale roches moutonnées, stoss and lee slopes, and striations. However, in my opinion, these features are not abundant. Smaller-scale roches moutonnées occur on Fildes Peninsula, particularly near the Río Foca site (Figure 2a), and in isolated locations on Barton Peninsula. These all are oriented roughly parallel to the present flow direction of Collins Ice Cap. Some skerries, particularly those off the northwest coast of Fildes Peninsula and adjacent Nelson Island, have whaleback forms that could be attributed to glacial erosion. However, these rocks are intermixed with numerous other shoals and stacks that are jagged in appearance and lack glacial smoothing. Farther west on Byers Peninsula, channels with overdeepened basins are cut into the upper platform and likely are the result of subglacial erosion. Other glacial features are rare on Byers Peninsula, except on Ray Promontory in the northwest where there are stoss and lee forms.

Depositional Landforms and Sediments

Pre-Holocene glacial deposits are thin and have a patchy distribution. On Fildes Peninsula, most of the surface is covered with highly weathered volcanic regolith.

However, thin, scattered patches of till are present, particularly between Gemel Peaks and Collins Ice Cap (Figure 2a). For example, at the Caleta Geologia site, ~10 cm of till overlies highly weathered volcanic rock and bedrock. The till has a silty-clayey matrix and contains boulders that stand as much as 40 cm above ground surface. Ten percent of these boulders are far-travelled granodiorite (likely derived from interior King George Island; *Smellie et al.*, 1984), whereas the remainder are local volcanic rocks. Another till patch occurs near Muschelbach and overlies sediments containing shells dated to 87 ± 14 kyr B.P. by the uranium-thorium method [*Mäusbacher*, 1992].

Barton Peninsula also lacks widespread glacial deposits. Much of the surface is covered by weathered bedrock, colluvium, and frost-heaved marine deposits. Till occurs primarily at sites close to present-day ice, particularly on the southeastern part of the peninsula. Other thin till patches, consisting of a scattering of debris over striated bedrock, exist to the west [*John and Sugden*, 1971]. Other than in low-elevation, presumably Holocene [*John and Sugden*, 1971], moraines adjacent to both Marian and Potter Coves, which contain numerous pieces of granodiorite, we did not find any erratics on Barton Peninsula.

Till sheets are also rare on Byers Peninsula, except near present-day ice. Erratics occur on some marine platforms, particularly those in the west. However, it is unclear whether these rocks were transported glacially, or if they were brought by icebergs when the platforms were forming. A scattering of boulders of the same lithology as and to the west of Chester Cone (Figure 2c) indicates that ice must have flowed from the east at some time in the past.

At present, there are no documented moraines of LGM or late-glacial age in the SSI. In general, moraines are rare and occur close to the present ice caps, outlet glaciers, and alpine glaciers. The youngest moraines are late Holocene in age, as determined by their cross-cutting relationships with raised beaches [*López-Martínez et al.*, 1992] or by lichenometry [*Birkenmajer*, 1979, 1997]. The older ridges are undated. All are within a few kilometers of present ice.

Submerged moraines are present offshore. Moraines in shallow water, such as those at Johnson's Dock [*López-Martínez et al.*, 1992] and Marian Cove, are assumed to be Holocene in age. Other, potentially older moraines have been recognized at 70 m depth in both South and False Bays on Livingston Island [*López-Martínez et al.*, 1992]. Using marine geophysical methods, *Griffiths and Anderson* [1989] located topographic highs near the heads of Admiralty and Maxwell Bays composed of sed-

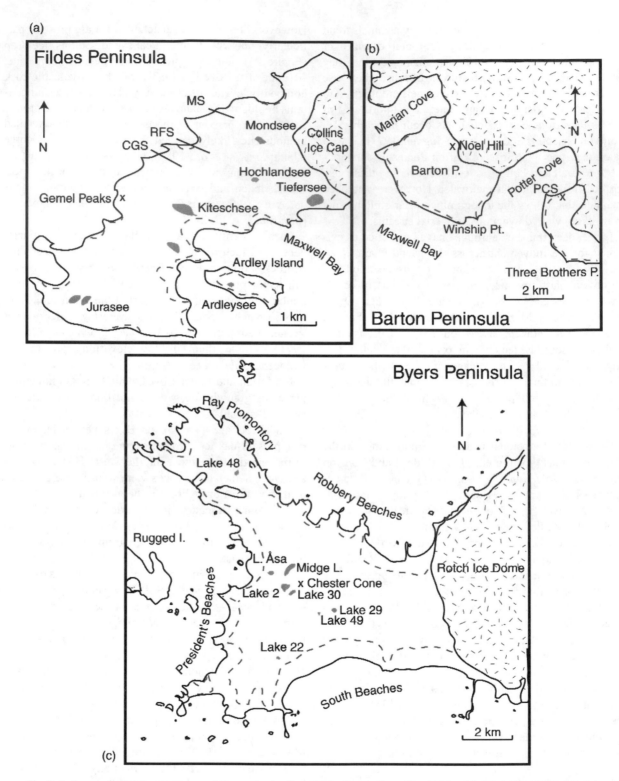

Fig. 2. Index map of Fildes, Barton, and Byers Peninsulas. PCS = Potter Cove Site, RFS = Río Foca Site, CGS = Caleta Geologia Site, MS = Muschelbach Shell Site. Dashed line shows the schematic representation of the uppermost Holocene raised beaches. Not all beach sites are shown.

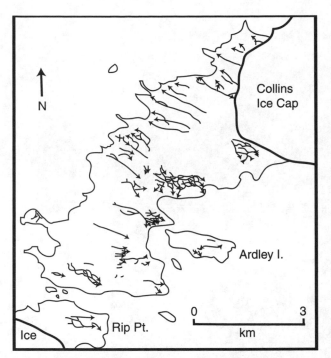

Fig. 3. Distribution of meltwater channels on Fildes Peninsula. Adapted from *John and Sugden* [1971].

iments that they attributed to a glacial origin. These moraine-like features are located in water depths of about 200–400 m [*Anderson*, pers. comm., 2002; *Yoon et al.*, 1997]. None of the submerged moraines is dated.

MARINE FEATURES

Platforms

The dominant landscape component of the SSI is a series of platforms that occur to as much as 275 m asl. These are thought to result from marine planation, although the uppermost may have formed subaerially [*John and Sugden*, 1971]. On Fildes Peninsula, platforms extend from 110 m below sea level to 155 m above sea level [*John and Sugden*, 1971]. These features are flat to gently undulating, seaward sloping, and separated by cliffs 20 to 60 m high. One of the best-developed platforms occurs on the northwest coast at 35–40 m elevation (Figure 4a). Here, it is as much as one kilometer wide and is laterally continuous for over six kilometers. Barton Peninsula also displays similar surfaces, the most prevalent of which are at 30-45 m, ~105 m, ~130 m, and ~190 m elevation. However, of the sites examined, Byers Peninsula is characterized the most by marine erosion. Widespread, well-preserved platforms, separated by

spectacular shore-parallel cliffs, occur from below sea level to at least ~100 m elevation (Figure 5). Remnants of a higher platform are at ~210–250 m elevation [*John and Sugden*, 1971]. The cliffs (10–60 m high) are remarkably well preserved, especially considering that they are cut in soft volcanic rock. A sea stack is associated with a cliff adjacent to the 85–100 m platform [*John and Sugden*, 1971].

Beaches and Beach Sediments

The lowest exposed platform is covered almost everywhere by a series of Holocene raised beaches to a maximum elevation of about 22 m (Figures 2, 4c). These Holocene beaches commonly are separated from the prominent 35 m-platform by a cliff (15–20 m high). Marine deposits also occur on higher platforms. For instance, the 35 m-platform displays widespread beach deposits, including well-rounded cobbles, and, in some cases, berm morphology (Figure 4b). Preserved beaches occur up to 66 m elevation in protected locations on Fildes Peninsula [*John and Sugden*, 1971]. On Barton Peninsula, a series of degraded berms is present near Winship Point (35–54 m elevation). There are higher, isolated beaches, such as a terrace at 138 m elevation near Marian Cove, which commonly are in association with marine-cut cliffs. On Byers Peninsula, degraded beaches and large areas of well-rounded beach cobbles are widespread to ~140 m elevation [*Arche et al.*, 1995]. The highest known beach in the SSI occurs at 275 m elevation on Noel Hill [Figure 2b; *John and Sugden*, 1971].

The elevation and age of the Holocene marine limit—key in determining the timing of deglaciation—is under debate. *John and Sugden* [1971] favored a limit of 54 m, based on their inference that beaches of this elevation have not been overrun by ice. Similarly, *Birkenmajer* [1995, 1997] suggested a limit of 45 m elevation at Three Sisters Point near King George Bay (Figure 1) and at 65 m elevation at Admiralty Bay. In contrast, others have placed the Holocene marine limit between 18 and 22 m elevation. *Mäusbacher* [1991] suggested that it had to be below 22 m elevation, because a lake at that elevation on Ardley Island contains only freshwater diatoms. Moreover, Jurasee, which has an age of 8700 ^{14}C yr B.P. for lacustrine sediments located ~25 cm above till, contains only freshwater material, thus constraining the Holocene marine limit to below 47 m, the elevation of the lake [*Mäusbacher et al.*, 1989; Figure 2a]. Here, I assume that Holocene raised beaches extend to a maximum of ~22 m elevation and that higher beaches are from previous interglacial periods. The reasons for

Fig. 4. Photographs of marine features in the South Shetland Islands: (a) 35-40 m raised platform, northwest Fildes Peninsula. Sea cliff is ~20 m high; (b) rounded beach cobbles on platform in figure (a). Largest clasts shown are ~15 cm diameter; (c) horizontal benches extending from sea level to halfway up Ardley Island (21 m elevation) are raised beaches; (d) west coast of Barton Peninsula showing modern and elevated sea stacks; (e) view south along Barton Peninsula to Potter Cove and Three Brothers Hill. Photo was taken from the 35-m platform, which is covered here by beach sediments.

choosing 22 m over ~60 m include 1) the fact that whereas beaches below 22 m occur almost everywhere, those above that elevation are more rare and discontinuous, 2) beaches above 22 m elevation are highly vegetated and degraded, 3) in many places, a prominent sea cliff separates the beaches below 22 m elevation from those above, and 4) Holocene marine material has yet to be found in lakes above 22 m elevation. Although the bulk of evidence points to a Holocene marine limit of 22 m elevation, one cannot at this time rule out the possibility that some isolated locations may have deglaciated earlier and have a higher local marine limit.

Chronologic Relationship Among Glacial, Marine, and Lacustrine Features

On Fildes Peninsula and elsewhere in the SSI, the subglacial meltwater channels are cut only into the upper marine platforms. Channels do not dissect platforms below 35 m elevation. In fact, some channels are truncated by the cliff that separates the ~20 m and ~35 m platforms [*John and Sugden*, 1971]. Therefore, the subglacial meltwater channels and the glaciation that they represent predate the formation of the lowest exposed platform. Similar to *John and Sugden* [1971], I attribute the channels and most large-scale glacial erosional forms to a pre-LGM glaciation.

Raised beaches show cross-cutting relationships with moraines in several locations and afford relative-age information. On Fildes Peninsula, beaches that occur to 18.5 m elevation extend up to and presumably beneath the Collins Ice Cap, suggesting that ice was less extensive than it is now when the beaches formed. Likewise, on Livingston Island, moraines rest on raised beaches [*López-Martínez et al.*, 1992], indicating ice readvance since the formation of the highest Holocene beaches. Many higher-elevation, older beaches in the SSI have been overrun, and thus predate glacial advance. For

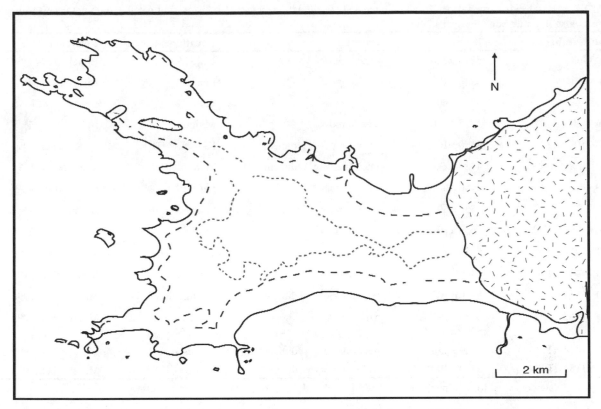

Fig. 5. Map of Byers Peninsula, showing locations of prominent marine platforms at 85-100 m elevation (central portion of peninsula enclosed in {....}) and at 30–50 m elevation{---}. A third platform is widespread adjacent to the present coastline. Modified from *John and Sugden* [1971].

example, the 275 m beach forms a well-preserved terrace on the south side of Noel Hill. However, striated beach cobbles occur in nearby till, becoming more common with increasing proximity to the beach [*John and Sugden*, 1971]. Ice must have advanced over the beach, but the exact timing of this expansion is unknown.

At Potter Cove, a stratigraphic section exposes till overlying raised marine sediments containing shells and seaweed (Figure 2b). Radiocarbon dates from this site indicate that the area was ice-free by 8370 ± 230 ^{14}C yr B.P. (Birm-48; all dates of marine materials quoted in the text are corrected for a marine-reservoir effect of 1300 years; *Björck et* al., 1991a; *Berkman and Forman*, 1996) with glacial readvance taking place sometime after 6383 ± 86 ^{14}C yr B.P. (Birm-23, *Sugden and John*, 1973).

The age of the highest-elevation Holocene beach (18–22 m) also affords a minimum age for deglaciation. Remains of two penguins from this beach (18 m elevation) on east-central Fildes Peninsula yielded corrected ages of 5260 ± 55 (HD-9426-9106) and 5350 ± 90 (HD 9425-9100)[14]C yr B.P. [*Barsch and Mäusbacher*, 1986].

Beaches of this elevation are widespread in the SSI. If one assumes that beaches of the same elevation are contemporaneous, then ice extent along the coast ~5400 ^{14}C yr B.P. must have been similar to or less than that of today.

Minimum ages for deglaciation, as well as inferences about former ice extent can be obtained from lake sediments. The oldest postglacial sediments are 4300 to 8700 ^{14}C on Fildes Peninsula [*Mäusbacher et al.*, 1989; *Mäusbacher*, 1991] and ~3000 to 5000 ^{14}C yr B.P. on Byers Peninsula (Table 1; *Björck et al.*, 1991b, 1993, 1995). Unfortunately, few cores penetrated to diamicton, which is necessary to determine the precise timing of deglaciation, as well as to document former ice extent. Moreover, radiocarbon dates of lacustrine material may require correction for a reservoir effect. This is due primarily to the introduction of old carbon from the bedrock, although guano contamination by marine mammals and birds can have an effect on some lakes [*Björck et al.*, 1991a]. *Björck et al.* [1991a] found that pure aquatic moss yields acceptable ages. However, mixed samples of aquatic moss and organic-rich sediments tend to produce incon-

TABLE 1. Basal radiocarbon dates for lakes in the SSI. These dates represent the lowest unit dated and do not necessarily indicate a close minimum age for deglaciation.

Lab Number	Date	Material	Location	Reference
Lu-3463	3210 ±220	aquatic moss	Lake 29	Björck et al., 1995
Lu-3463	3340 ±180	aquatic moss	Lake 22	Björck et al., 1995
Lu-3473	3730 ±400	aquatic moss	Lake 49	Björck et al.,1995
Ua-1220	3735 ±250	aquatic moss	Midge L.	Björck et al., 1991b
HD 11942-11732	4160 ±105	freshwater org. sediment	Kiteschsee (II)	Mäusbacher, 1991
—	4325 ± 90	freshwater org. sediment	Tiefersee (III)	Mäusbacher, 1991
Ua-3925	4480 ±60	aquatic moss	Lake 30	Björck et al., 1995
Ua-3923	4570 ±60	aquatic moss	Lake 48	Björck et al., 1995
Lu-3088	4600 ±100	aquatic moss	Lake Åsa	Björck et al., 1993
HD 11420-11161	5380 ±165	freshwater org. sediment	Tiefersee (I)	Mäusbacher, 1991
HD 11416-11147	5410 ±255	freshwater org. sediment	Ardleysee	Mäusbacher, 1991
HD 11417-11148	5410 ±185	freshwater org. sediment	Hochlandsee	Mäusbacher, 1991
HD 11162-10997	6950 ±195 5650 ± 195*	marine algae	Kiteschsee (I)	Mäusbacher, 1991
HD 11415-11133	7200 ±250	freshwater org. sediment	Mondsee	Mäusbacher, 1991
HD 11166-11024	8700 ±300	freshwater org. sediment	Jurasee	Mäusbacher, 1991

*Indicates date corrected for marine reservoir effect. Lu = Lund, Ua = Uppsala, HD = Heidelberg

sistent ages. In Table 1, those samples of specific materials (i.e., aquatic moss) should yield correct ages. Dates of samples of mixed origin (i.e., freshwater organic sediment) are less certain.

EXTENT OF THE LGM ICE CAP

What was the extent of the LGM ice cap in the SSI? Evidence is generally circumstantial. Few sites have any chronological control for deposits that might relate to the build up or demise of the LGM ice cap. Two of these sites occur on Livingston Island and consist of lake sediments overlying diamicton, presumably till. At Lake 2 on Byers Peninsula (Figure 2c), tephra layer AP14 lies above diamicton [Björck et al., 1995]. As this tephra has been dated elsewhere at 4700 yr B.P., Björck et al. [1995] suggested deglaciation of Lake 2 at ~4800 ^{14}C yr B.P. Likewise, moss dated at 4570 ± 60 ^{14}C yr B.P. overlies diamicton in Lake 48 [Björck et al., 1995]. Björck et al. [1995] proposed an age of ~4600 ^{14}C yr B.P. for deglaciation of this lake, because AP14 is absent. Both of these sites can be used to suggest that there was an expansion of ice and that the ice remained until mid-Holocene time. It is more difficult, however, to prove

that the expansion dates to the LGM and not to a later readvance, particularly as Lake 48 is in a highland region that even today supports ice.

A third site is on northwestern Fildes Peninsula and consists of till overlying shells dated to ~87 ka by the uranium-thorium method [Mäusbacher, 1991]. Although this site likely indicates ice advance over a Stage 5 deposit, caution must be exercised because uranium-thorium dates of shells are notoriously inaccurate [Kaufman et al., 1971, 1996]. Even if the dates are correct, this site does not prove expansion of an LGM-age ice sheet, only glacial advance at some time after 87 ka. Other sites with age control for till on Fildes Peninsula are at Tiefersee, Hochlandsee, Kiteschsee, Jurasee, and Mondsee (Figure 2a). Cores taken at all of these lakes reveal lacustrine sediments overlying diamicton [Mäusbacher, 1991]. However, it is difficult to determine whether the basal diamicton dates to the LGM or if it is the result of Holocene readvance. Of the five Fildes Peninsula sites mentioned, three are located close to present ice and could have been overrun during any Holocene advances. Only Jurasee and Kiteschsee, located about six and two kilometers, respectively, from the Collins Ice Cap, seem immune from this problem.

Indirect evidence in favor of expanded ice during the LGM comes from the widespread existence of Holocene raised beaches. These beaches have been thought to represent glacio-isostatic uplift that followed deglaciation. If so, then the age of the marine limit should yield the timing of the final unloading of ice. However, the presence of widespread, high-elevation marine platforms and beaches is worrisome, particularly given the tectonically active nature of the Scotia Arc region. One alternative hypothesis is that the presence of the raised beaches reflects tectonic uplift or a mix of tectonism and glacial isostasy. If so, then the existence and elevation of the beaches cannot be used as evidence of a large ice cap over the SSI. However, dates of the beaches would still yield minimum ages for deglaciation.

There is much evidence in favor of restricted ice at the LGM. First, the SSI display a predominantly *marine* landscape, not a glacial landscape. Other than the subglacial meltwater channels and troughs, glacial features are few and are confined primarily to alpine areas. Most predate the LGM or are Holocene in age. In contrast, marine deposits and erosional features are ubiquitous and well-preserved. The platforms are separated by steep cliffs that do not show evidence of glacial erosion. Even a sea stack is preserved in at least one location [*John and Sugden,* 1971]. The fact that these cliffs are formed primarily in soft, easily erodeable volcanic rock makes their preservation that more stunning. The existence of the dominant marine landscape does not disprove the hypothesis of expanded LGM ice, but it does indicate that if ice covered the presently ice-free areas it must have been thin, cold-based, and non-erosive.

With so few chronologically constrained data, reconstruction of LGM ice extent must still be considered speculative. One model is that ice caps over the islands expanded to the continental shelf edge [*Payne et al.,* 1989]. As an alternate working hypothesis, I suggest that the ice caps expanded only slightly, grounding at most a few kilometers offshore of their present positions (see dashed line in Figure 1). Although it cannot yet be proven, this hypothesis is consistent with the available geologic data outlined above. Key to this model is the assumption that submerged moraines in Maxwell, Admiralty, South, and False Bays date to the LGM. This may be a reasonable assumption given the fact that the moraines have not yet been masked by marine sedimentation. Glaciers in the SSI today generally terminate at sea level, probably because the sea undercuts the ice cliffs, as well as determines the flotation line. Lower eustatic sea level (-120 m; *Fairbanks,* 1989) at the LGM

would promote ice expansion, but isostatic depression due to the advancing ice would reduce the effect of this sea-level drop. If ice terminated at sea level at the LGM similar to at present, one would expect to find LGM moraines at water depths of less than 120 m. Increased sea ice (which would protect the ice cliff) and greater ice flux could have allowed the glaciers to flow into somewhat deeper water. One further requirement for this alternate hypothesis is that the ice over land must have been thin and primarily cold-based, in order to preserve the numerous pre-Holocene marine landforms and sediments.

DEGLACIATION

Deglaciation in the SSI commenced in or possibly before early Holocene time. At Potter Cove, it was well underway before 8300 ^{14}C yr B.P. (corrected). If the dates of organic lake sediment are correct, Jurasee at the southern tip of Fildes Peninsula was ice free before 8700 ^{14}C yr B.P. Both of these sites afford only minimum ages for the timing of ice retreat. Data from the adjacent sea floor in Maxwell Bay are scarce and inconclusive. *Li et al.* [2000] presented dates as old as 12,161 ^{14}C yr B.P. (corrected) of total organic carbon from ice-proximal sediments at the mouth of Marian Cove. However, with the presence of Tertiary organic material in the nearby bedrock and the very low organic carbon percentages (<0.5%), this age must be taken with caution. The oldest proposed age for deglaciation of Maxwell Bay, 17,000 ^{14}C yr B.P., is based on extrapolation of sedimentation rates [*Yoon et al.,* 1997].

The early Holocene or late Pleistocene deglaciation suggested by the Potter Cove and Jurasee data is in agreement with a date obtained by *Pudsey et al.* [1994] of about 11,000 ^{14}C yr B.P. for the transition to open marine sedimentation from glacial or glacial marine sedimentation on the continental shelf just south of the SSI. It should be noted that this date, similar to that from Jurasee, is of total organic carbon and hence may overestimate the true age. However, this age is in accord with dates for deglaciation of about 13,100 cal yr B.P. from the Palmer Deep (64° 51.71' S, 64° 12.47' W) derived from both total organic carbon and biogenic calcite [*Domack et al.,* 2001].

In contrast, other evidence seems to point to a later, mid-Holocene deglaciation. Most lake sediments on Byers and Fildes Peninsulas date only to about 4000–5000 ^{14}C yr B.P. However, many of these dates are minimum ages only. For example, on Fildes Peninsula, only at Tiefersee and Hochlandsee were radiocarbon samples taken close to the diamicton boundary. Even where radiocarbon dates directly overlie diamicton, it is in some cases unclear whether the diamicton is of LGM age or if it relates to more recent ice-cap readvance or to residual highland ice. The general pattern, however, obtained from lake sediments is that of mid-Holocene deglaciation.

How does one resolve these conflicting data? From the Potter Cove site, it is apparent that ice was at or behind its present position by early Holocene time. This would have been accompanied by drawdown of the interior ice cap, making it unlikely that extensive ice existed over Fildes Peninsula at that time. Ardley Island must have been ice free before 5500 ^{14}C yr B.P.—the age of the 18 m beach, because the marine limit there extends to 21 m elevation. Likewise, Collins Ice Cap must have been less extensive than at present ~5500 ^{14}C yr ago, because it now overrides beaches of that age. Therefore, on western King George Island, deglaciation commenced in early Holocene or late Pleistocene time with some parts of the ice cap being similar to or less extensive than at present by at least 8300 ^{14}C yr B.P.

The timing of deglaciation on Livingston Island is less certain. Raised beaches as high as 22 m elevation occur near South Bay and on western Byers Peninsula [Arche et al., 1995; Pallàs et al., 1995]. If one assumes that these beaches are of the same age as those on King George Island, then ice retreat must have occurred prior to 5500 ^{14}C yr B.P., the age of the 18-m beach. Inland, Lakes 2 and 48 contain sediments dating to about 4600–4800 ^{14}C yr B.P. that overlie diamicton and afford close minimum ages for deglaciation [Björck et al., 1995]. This would suggest that although the coastal regions, particularly in the west, were deglaciated before 5500 ^{14}C yr B.P., some interior areas may have preserved residual ice or were affected by Holocene readvance.

CONCLUSIONS

There is little direct proof of LGM ice expansion in the SSI. Indirect evidence from undated offshore moraines, diamictons in lake cores, and raised beaches is consistent with a hypothesis that the present ice caps expanded only slightly out onto the continental shelf. If correct, then the timing and amount of expansion may have been controlled to some extent by the eustatic drop in sea level, which reduced wave erosion on the glacier cliffs and changed the location at which ice would come afloat. Ice over the island chain must have remained thin and cold-based in order to preserve the widespread pre-LGM marine features. The timing of deglaciation is less cer-

tain, but was underway in some areas by at least early Holocene time, if not before.

Acknowledgments. Ethan Perry assisted ably in the field. John Evans provided logistical and moral support. We'd also like to thank Captain Warren Sanamo and the crew and support staff of the Laurence M. Gould, as well as Hamilton College and the National Science Foundation for supporting the conference that led to this paper, and three reviewers for their useful comments. This research was supported by the Office of Polar Programs of the National Science Foundation.

REFERENCES

Arche, A., J. López-Martínez, E. Serrano, and Martínez de Pisón, E., Marine landforms and deposits, in *Geomorphological Map of Byers Peninsula, Livingston Island*, edited by J. López-Martínez, M.R.A. Thomson, and J.W. Thomson, J.W., *BAS Geomap, 5A*, 35-42, 1995.

Barsch, D., and R. Mäusbacher, New data on the relief development of the South Shetland Islands, Antarctica, *Interdisciplinary Science Reviews, 11*, 211-219, 1986.

Berkman, P.A., and S.L. Forman, Pre-bomb radiocarbon and the reservoir correction for calcareous marine species in the Southern Ocean, *Geophys. Res. Lett., 23*, 633-636, 1996.

Birkenmajer, K., Quaternary geology at Arctowski Station, King George Island, South Shetland Islands (West Antarctica), *Studia Geologica Polonica, 110*, 91-104, 1997.

Birkenmajer, K, Glacier retreat and raised marine beaches at Three Sisters Point, King George I. (South Shetland Islands, West Antarctica), *Bull. Pol. Acad. Sci., 43*, 135-141, 1995.

Birkenmajer, K., Lichenometric dating of glacier retreat at Admiralty Bay, King George Island (South Shetland Islands, West Antarctica), *Bull. Pol. Acad. Sci., 27*, 77-85, 1979.

Björck, S., C. Hjort, Ó. Ingólfsson, R. Zale, and J. Ising, Holocene deglaciation chronology from lake sediments, in *Geomorphological Map of Byers Peninsula, Livingston Island*, edited by J. López-Martínez, M.R.A. Thomson, and J.W. Thomson, J.W., *BAS Geomap, 5A*, 49-51, 1995

Björck, S., H. Håkansson, S. Olsson, L. Barnekow, and J. Janssens, Paleoclimatic studies in South Shetland Islands, Antarctica, based on numerous stratigraphic variables in lake sediments, *J. of Paleolimnology, 8*, 233-272, 1993.

Björck, S., C. Hjort, Ó. Ingólfsson, and G. Skog, Radiocarbon dates from the Antarctic Peninsula region —problems and potential, *Quaternary Proceedings, 1*, 55-65, 1991a.

Björck, S., H. Håkansson, R. Zale, W. Karlén, and B.L. Jönsson, A late Holocene lake sediment sequence from Livingston Island, South Shetland Islands, with paleoclimatic implications, *Ant. Sci., 1*, 61-72, 1991b.

Blunier, T., J. Chappellaz, J. Schwander, A. Dällenbach, B. Stauffer, T.F. Stocker, D. Raynaud, J. Jouzel, H.B. Clausen, C.V. Hammer, and S.J. Johnsen, Asynchrony of Antarctic and Greenland climate change during the last glacial period, *Nature*, *394*, 739-743, 1998.

Blunier, T., and E. Brook, Timing of millennial-scale climate change in Antarctica and Greenland during the last glacial period, *Science*, *291*, 109-112, 2001.

Broecker, W.S., Paleocean circulation during the last deglaciation: A bipolar seesaw?, *Paleoceanography, 13,*119-121, 1998.

Clapperton, C.M, and D.E. Sugden, Holocene glacier fluctuations in South America and Antarctica, *Quaternary Science Reviews*, *7*, 185-198, 1988.

Curl, J.E., A glacial history of the South Shetland Islands, Antarctica, *Inst. of Polar Stud. Rep.*, 63, 1-129, 1980.

Domack, E., and S. Ishman, Oceanographic and physiographic controls on modern sedimentation within Antarctic fjords, *GSA Bulletin*, *105*, 1175-1189, 1993.

Domack, E., A. Leventer, R. Dunbar, F. Taylor, S. Brachfield, C. Sjunneskog, and ODP Leg 178 Scientific Party, Chronology of the Palmer Deep site, Antarctic Peninsula: a Holocene palaeoenvironmental reference for the circum-Antarctic, *The Holocene*, *11*, 1-9, 2001.

Fairbanks, R., A 17,000-yr glacio-eustatic sea level record: Influence of glacial melting rates on the Younger Dryas event and deep-ocean circulation, *Nature*, *342*, 637-642, 1989.

Griffiths, T.W., and J.B. Anderson, Climatic control of sedimentation in bays and fjords of the northern Antarctic Peninsula, *Marine Geology*, *85*, 181-204, 1989.

John, B.S., and D.E. Sugden, D.E., Raised marine features and phases of glaciation in the South Shetland Islands, *BAS Bull.*, *24*, 45-111, 1971.

Kaufman, A., W.S. Broecker, T.-L. Ku, and D.L. Thurber, The status of U-series methods of mollusk dating, *Geochimica et Cosmochimica Acta*, *35*,1155-1183, 1971.

Kaufman, A., B. Ghaleb, J.F.Wehmiller, and C. Hillaire-Marcel, Uranium concentration and isotope ratio profiles within *Mercenaria* shells: Geochronological implications, *Geochimica et Cosmochimica Acta*, *60*, 3735-3746, 1996.

Lamb, H.H., Circulation of the atmosphere, in *Antarctic Research*, edited by R.E. Priestley, R.J. Adie, and G.de Q. Robin, pp. 265-277, Butterworth and Company, London, 1964.

Li, B., H.I. Yoon, and B.K. Park, Foraminiferal assemblages and CaCO$_3$ dissolution since the last deglaciation in the Maxwell Bay, King George Island, Antarctica, *Marine Geology*, *169*, 239-257, 2000.

Loewe, F., Precipitation and evaporation in the Antarctic, in *Meteorology of the Antarctic*, edited by M.P. van Rooy, pp. 71-90, Weather Bureau, Department of Transportation, Pretoria, 1957.

López-Martínez, J., E. Martínez de Pisón, and A. Arche, Geomorphology of Hurd Peninsula, Livingston Island, South Shetland Islands, in *Recent Progress in Antarctic Earth Science*, edited by Y.Yoshida, pp. 751-756, Terrapub, Tokyo, 1992.

Mäusbacher, R., Distribution and stratigraphy of raised interglacial marine sediments on King George Island, South Shetlands, Antarctica, *Zeit. für Geomor. Supp.*, *86*, 113-123, 1992.

Mäusbacher, R., Die jungquartäre Relief- und Klimageschichte im Bereich der Fildeshalbinsel Süd-Shetland-Inseln, Antarktis. Ph.D. Diss., Geogr. Inst. der Univ. Heidelberg, 1991.

Mäusbacher, R., J. Müller, and R. Schmidt, Evolution of postglacial sedimentation in Antarctic lakes (King George Island), *Zeit. für Geomorph.*, *33(2)*, 219-234, 1989.

Pallàs, R., J.M. Vilaplana, and F. Sábat, Geomorphological and neotectonic features of Hurd Peninsula, Livingston Island, South Shetland Islands, *Ant. Sci.*, *7*, 395-406, 1995.

Payne, J.A., D.E Sugden, and C.M. Clapperton, Modeling the growth and decay of the Antarctic Peninsula ice sheet, *Quaternary Research, 31*, 119-134, 1989.

Pudsey, C.J., P.F. Barker, and R.D. Larter, Ice sheet retreat from the Antarctic Peninsula shelf, *Continental Shelf Research*, *14*, 1647-1675, 1994.

Smellie, J.L., R.J. Pankhurst, M.R.A. Thomson, and R.E.S. Davies, The geology of the South Shetland Islands: VI. Stratigraphy, geochemistry and evolution, *BAS Sci. Rep.*, *87*, 1-85, 1984.

Sowers, T., and M. Bender, Climate records covering the last deglaciation, *Science*, *269*, 210-214, 1995.

Steig, E.J., D.L. Morse, E.D.Waddington, M. Stuiver, P.M. Grootes, P.A. Mayewski, M.S. Twickler, and S.I. Whitlow, Wisconsinan and Holocene climate history from an ice core at Taylor Dome, western Ross Embayment, Antarctica, *Geog. Ann.*, *82A*, 213-235, 2000.

Sugden, D.E., and B.S. John, The ages of glacier fluctuations in the South Shetland Islands, Antarctica, in *Palaeoecology of Africa, the surrounding islands, and Antarctica*, vol. 8, edited by E.M. van Zinderen Bakker, pp. 139-159, Balkema, Cape Town, 1973.

Yoon, H.I., M.W. Han, B.K. Park, J.K. Oh, and S.K. Chang, Glaciomarine sedimentation and palaeo-glacial setting of Maxwell Bay and its tributary embayment, Marian Cove, South Shetland Islands, Antarctica, *Marine Geology*, *140*, 265-282, 1997.

Brenda Hall, Institute for Quaternary and Climate Studies, 311 Bryand Global Sciences Center, University of Maine, Orono, Maine 04469

ICE CORE PALEOCLIMATE HISTORIES FROM THE ANTARCTIC PENINSULA: WHERE DO WE GO FROM HERE?

Ellen Mosley-Thompson

Byrd Polar Research Center, Department of Geography, The Ohio State University, Columbus, Ohio

Lonnie G. Thompson

Byrd Polar Research Center, Department of Geological Sciences, The Ohio State University, Columbus, Ohio

It is essential to determine whether the strong 20th century warming in the Antarctic Peninsula (AP) reflects, in part, a response to anthropogenically driven, globally averaged warming or if it is consistent with past climate variability in the region. The necessary time perspective may be reconstructed from chemical and physical properties preserved in the regional ice cover and ocean sediments. Only three multi-century climate histories derived from ice cores in the AP region have been annually dated with good precision (± 2 years per century). The longest record contains only 1200 years and the three histories do not provide a coherent picture of 20th century climate variability. The highest elevation core, Dyer Plateau, reveals a 20th century increase in accumulation and isotopic enrichment (warming) and indicates that the high ice plateau extending southward to the West Antarctic Ice Sheet is strategically situated to capture large-scale climate variability in the region. Such histories are critical to unraveling the role of various forcing mechanisms and discerning the leads and lags in the system. Ocean-atmosphere connections between Antarctica and the tropics have gained considerable attention as Antarctic Intermediate Water (AAIW) links conditions along the Antarctic margin with those in both the tropical Pacific and the North Atlantic. Current knowledge of past climate conditions in the Antarctic Peninsula is limited. High resolution proxy histories from ice cores drilled at carefully selected sites along the AP spine offer tremendous opportunities to examine past climate variability and place regional climate changes within a much broader geographical context.

INTRODUCTION

Chemical constituents and physical properties preserved within the permanent ice cover along the spine of the Antarctic Peninsula (AP) contain a record of climatic and environmental changes over tens to thousands of years. Ice cores drilled at higher accumulation sites that are carefully selected to minimize disturbance of the ice strata by flow at depth will provide land-based proxy records to complement the climate histories emerging from high deposition rate sediment cores along the Peninsula margin [*Domack et al.*, this volume]. An ice core drilling project currently underway on Berkner Island is expected to produce a long history, possibly back into the Last Glacial Stage (LGS). However, at present the longest annually dated ice core-based history for the western side of the Ronne-Filchner Ice Shelf (RFIS) is the Dyer Plateau record that extends back to AD 1504 (all dates henceforth are AD). Here we review the status of ice-core derived climate records in the Peninsula region with emphasis upon those offering multi-century histories, summarize what we have learned from these records, discuss what new insights may be gleaned from the ice, and emphasize the need to extract longer, high resolution ice core-based climate histories.

MODERN CONTEXT

Temperature records for Antarctica are sparse and short with few extending prior to the International Geophysical Year (1957–58). This is particularly true for the continental interior. The longest and most dense network of meteorological records is in the Antarctic Peninsula region where the temperature record at Orcadas (South Orkney Islands) extends to 1903. *King et al.* [this volume] review the surface temperature records in the Peninsula that extend to the late 1940s and the upper air measurements that began in 1956. Their analyses demonstrate marked differences between the temperature trends in the AP and the rest of the continent (East and West Antarctica). *Jones et al.* [1993] also noted that temperature variations in the AP region are poorly correlated with those on the main part of the continent and concluded that extending the Antarctic temperature record by using the longer temperature histories from the Peninsula would be inappropriate. Within the AP region there are strong interannual differences among the station temperatures [*Peel*, 1992; *Limbert*, 1974] that reflect differential responses to forcing by sea ice extent (proximity to open water) and temporal variations in atmospheric and oceanic circulation patterns. Nearly two decades ago *Schwerdtfeger* [1984] highlighted the strong climatological differences between the east and west sides of the Peninsula's mountain range that extends over ~16°of latitude (~80–64°S) and acts as a natural barrier to the circum-Antarctic current, the prevailing westerly winds, and the katabatic barrier winds. With the advent of routine satellite-borne observations, installation of automatic weather stations (AWS), and computer models to synthesize these data, the spatial variability of climate in the Antarctic Peninsula is now better documented [*Simmonds*, this volume].

Near-surface air temperatures have warmed markedly in the AP region [*Marshall et al.*, 2002; *Morris and Vaughan*, this volume]. *King et al.* [this volume] report

that over the past 50 years summer air temperatures have risen about ~1.0°C, increasing ~0.2°C per decade. Temperatures in the lower- and mid-troposphere have warmed at a rate that is consistent with Southern Hemisphere trends. However, they note the most striking observation is the marked difference in the strength of the surface warming trend in summer and winter along the west side of the Peninsula. Here winter temperatures near the surface are warming at a rate of 0.1°C annually, making this the region of strongest warming in the Southern Hemisphere [*King et al.*, this volume]. They note that this strong warming is confined to the lower layers of the atmosphere. *Smith et al.* [1999] also observed this seasonal difference and suggested a strong oceanic role in the forcing. Observations reveal that the winter trend of +0.11°C per year at Faraday/Vernadsky Station is among the largest in the Southern Hemisphere. In 2001 the AP (Palmer Land) was one of Earth's three warmest regions [*Hansen et al.*, 2002].

This well-documented warming in the AP is of significant concern in light of the climatological and ecological sensitivity of the region. The warming in the last 50 years has been accompanied by other marked physical and biological changes. These include the distribution and persistence of coastal fast ice and a reduction in the extent and integrity of many of the ice shelves [*Vaughan and Doake*, 1996; *Vaughan et al.*, 2001; *Scambos et al.*, this volume; *Skvarca and De Angelis*, this volume], the persistence of lake ice [*Quayle et al.*, this volume], and changes in the range and size of sea bird populations [*Emslie et al.*, 1998]. The region is fragile from both a physical and biological perspective. A vast volume of land-based ice and floating ice shelves exists very close to the –9°C isotherm, the upper limit for viability of ice shelves [*Morris and Vaughan*, this volume], that appears to be creeping southward. This raises concerns about the potential melting and/or enhanced calving of the land-based ice that could contribute significantly to global sea level rise. The ecological systems in the region are complex and highly dependent on the physical and chemical cycles that will undoubtedly be affected by warming temperatures, reduced ice cover in the near-shore environment and increased fluxes of fresh water to the sea [*Convey*, this volume; *Emslie et al.*, 1998; *Smith et al.*, 1999].

A number of pressing questions emerge: Why are temperatures in the AP increasing so strongly while the rest of the Antarctic continent shows little change? What processes are responsible for the strong warming in winter along the west side of the Peninsula? Is this recent warming anthropogenically forced and if so, how much? Is the recent multi-decadal warming in the AP just one more in a sequence of naturally forced oscillations within the complex, coupled ocean-atmosphere system?

Progress has been made on the first question. *Thompson and Solomon* [2002] present an analysis of radiosonde temperatures and geopotential heights since 1979 that suggest a long-term trend toward a stronger circumpolar flow that warms the AP and southern tip of South America, but isolates and thus cools eastern Antarctica and the high polar plateau. The question remains whether this trend that is linked to the strength of the Southern Annual Mode is influenced by anthropogenically produced trace gases and/or stratospheric ozone depletion [*Simmonds*, this volume]. Whether the AP warming is anthropogenically forced or part of a longer-term oscillatory feature can only be addressed from the much longer temporal perspective available from proxy (indirect) climate histories preserved in stratified deposits such as marine sediments and ice cores.

This volume contains a number of papers presenting marine-based histories and to complement these, this paper reviews the existing ice core-based climate histories. The discussion below highlights the dearth of multi-century ice-core records in the AP, their temporal limitations, and argues strongly for an immediate and sharply focused effort to recover cores from those sites offering the potential for both long and high temporal resolution records. The discussion in the following sections highlights the need for rigorous development of a modern context for the climatological interpretation of the proxy indicators preserved within the accumulating snow. Implementing an array of AWS that include an acoustic depth gauge and sensors at multiple levels at drilling sites would reveal the timing of snowfall (when the record is produced) and allow calculations of surface fluxes for energy balance considerations. In addition, a geophysical survey is needed to guide site selection and thereby ensure the longest and least disturbed records possible. Currently, it is unknown whether the ice fields along the spine of the AP are gaining or losing mass, but such an evaluation is essential in light of the strong regional warming and the recent loss of extensive areas of shelf ice. If the plateau ice fields are threatened by the current warming, then attaining the paleoclimate histories preserved therein would assume a new level of urgency.

EXISTING ICE CORE HISTORIES FROM THE PENINSULA REGION

In the past two decades scores of shallow depth (< 40 meters) cores have been collected in the Antarctic Peninsula region by researchers from various international institutions. The major efforts have been by British, Argentine and German scientists and no attempt is made here to compile a list of all these cores because most of them are short (in both length and time represented). Further, it would be virtually impossible to use many of the existing cores to compile a 'representative' ice core-based climate history because of the differences in their chemical and physical analyses, their temporal resolution and coverage, and the quality of their time scales. In fact, due to the diversity of the climatological regime in the region [*Jones et al.*, 1993; *King et al.*, this volume], a single composite would not be appropriate. The strong climatological differences between the east and west sides of the Peninsula, as well as between the east and west sides of the Weddell Sea sector, argue that such an integration on annual and decadal scales would diminish important details of the regional climate histories. However, a large-scale composite would be quite useful on centennial to millennial scales.

A search of the literature reveals six cores from the AP that extend to earlier centuries. The locations of these core sites are shown in Figure 1. *Peel* [1992, his figure 28.6] synthesizes the results from three sites: James Ross Island (JRI), Site T340, and Dolleman Island (DI). Thus, we review them only briefly and chronologically. The ice core drilled in 1979 on JRI (Fig. 1) extended back to ~1850 and provided the first ice core-based climate study in the AP [*Aristarain et al.*, 1987]. Their comparison of the hydrogen isotopic (deuterium) records with surface air temperatures measured since 1956 at nearby stations (Esperanza and Faraday) suggested that the climate on James Ross Island is influenced by the climate regimes of both the eastern and western sides of the Peninsula, but the western climate regime dominates. Their filtered isotopic record (Fig. 2c), calibrated with AP temperature records (1953–1980), suggests a cooling of ~2°C for the northern part of the Antarctic Peninsula over this 27-year span and a general cooling trend over most of the 20th century. This isotopic record is inconsistent with the known air temperature trends in the AP. Their observation calls into question the use of isotopic records for paleoclimate reconstruction. As discussed later, the isotope-temperature relationship must be critically evaluated as part of future paleoclimatic research endeavors in the AP.

In 1989 a 100-meter core was drilled at Site T340 (~78°60'S; 55°W; Fig. 1) on the RFIS by the German Antarctic Research Program [*Graf et al.*, 1988]. Here accumulation is estimated to be ~155 mm w.e. (water equivalent) and the core was dated using seasonal $\delta^{18}O$ variations preserved throughout most of the core. The

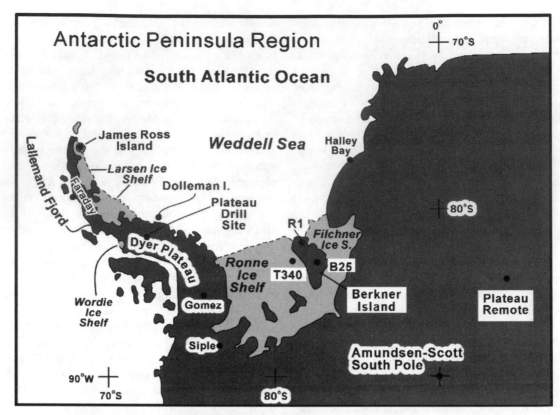

Fig. 1. The locations of the Antarctic Peninsula ice core sites discussed in the text are shown.

quality of the $\delta^{18}O$ signal was compromised by partial melting in the upper 89 meters during storage so that of the 479 annual $\delta^{18}O$ layers identified, about 80 (~16%) were expressed only as small maxima on shoulders of larger peaks. Five meters of core (in the upper part) were not recovered and extrapolation from surrounding sections resulted in the addition of 41 years so that the record extends back to 1460. The northward flow of the ice means that the ice (originally deposited as snow) did not accumulate at a single location, but over time as the ice progressed northward along its flowline. Thus, *Graf et al.* [1988] attempted to correct the $\delta^{18}O$ record downcore (back in time) for increasing continentality. The T340 $\delta^{18}O$ record, smoothed with an unspecified filtering function, suggests a slow cooling from the mid-nineteenth century toward the present and no recent warming, inconsistent with instrumental observations.

In 1986 British Antarctic Survey (BAS) scientists recovered three cores, the deepest to 133-m, on Dolleman Island (Fig. 1) where annual accumulation is roughly 420 mm w.e. *Peel et al.* [1988] reconstructed the DI climate ($\delta^{18}O$) record that extends back to 1795 (Fig. 2b). In 1981 two shorter (30.5 and 83-m) cores were collected by BAS scientists at the Gomez Nunatak (record

not shown) where the very high accumulation rate of 880 mm w.e. resulted in a much shorter record extending back only to 1942. In brief, *Peel et al.* [1988] note that from 1938 to 1986, the $\delta^{18}O$ records reflect most of the pronounced interannual variations observed in AP surface temperatures and from 1960 to 1980 the enrichment of $\delta^{18}O$ is consistent with the overall warming in the AP. However, from the middle of the nineteenth century to the 1960s the Dolleman $\delta^{18}O$ history also shows a broad cooling trend toward the present similar to that at T340 and JRI. Thus, from 1902 to 1960 the DI isotopic record does not show an isotopic enrichment (warming) contemporaneous with the warming trend evident in the Orcadas surface temperature record [*Peel et al.*, 1996, their Figure 28.6] and throughout the Peninsula region [*Jones et al.*, 1993].

In 1988 a cooperative glaciological-climatological ice core project was initiated on the Dyer Plateau (70°40'16"S; 64°52'30"W; 2002 masl; Fig. 1) where the mean annual temperature is about -21°C. In 1988–89 the first geophysical observations were made [*Raymond et al., 1996*] and two cores were drilled to 108-m depth at a site 6 km west of the ice divide. In 1989–90 two cores (233.8 m and 235.2 m) were drilled one meter apart at the

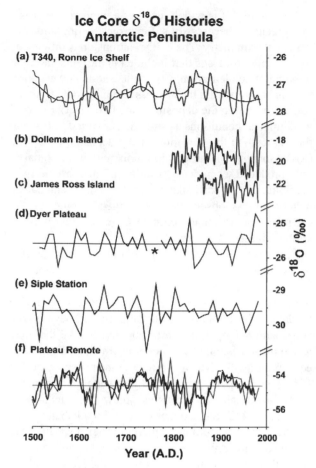

Ice Core δ¹⁸O Histories
Antarctic Peninsula

Fig. 2. The isotopic records from six of the cores discussed in the text are shown: (a) The record from T340 on the Ronne Ice Shelf is from *Graf et al.* [1988]; (b) The Dolleman Island record is from *Peel* [1992]; (c) The James Ross Island core is from *Aristarain et al.* [1986]; (d) The decadal averages of δ¹⁸O from the Dyer Plateau are from Thompson *et al.* [1994]; (e) The decadal averages of δ¹⁸O from the Siple Station record are from *Mosley-Thompson* [1992]; and (f) The decadal averages of δ¹⁸O, along with the 20-year un-weighted running mean (darker line) are from *Mosley-Thompson* [1996].

crest in the ice divide. The 1989 cores were analyzed only to 181-meters because the quality of the core recovered below that depth was very poor. The upper 181 meters provided an annually dated record extending back to 1505 [*Thompson et al.*, 1994]. Figure 3(a) shows δ¹⁸O and sulfate (SO_4^{2-}) concentrations for each sample in the upper 40-m of the core to illustrate the seasonal variations used to date the cores. The beta radioactivity profile reveals the 1963–64 time-stratigraphic marker associated with thermonuclear testing [*Crozaz*, 1969] that confirms the annual character of chemical signals used for dating. Figure 3(b) shows δ¹⁸O and sulfate (SO_4^{2-}) concentra-

tions for each sample in the lowest 10 meters of the 181-m core to confirm that the seasonal features are still excellently preserved. The time scale was reconfirmed by the identification of the major known volcanic events in the last 500 years [Fig. 4 in *Dai et al.*, 1995].

In 1995 a joint BAS-AWI (Alfred Wegner Institute) project recovered two cores (151 and 181-m), one from each of the two main domes of Berkner Island (Fig. 1). *Mulvaney et al.* [2002] present the chemical and isotopic records that extend back ~700 years for the R1 core (northern dome) and ~1200 years for the B25 core (southern dome). At present these are the longest, best dated cores in the AP region and the borehole temperatures suggest a warming of 0.5 to 1.0 °C in the recent past. The shape of borehole temperature records can be modeled to extract past temperature changes if past accumulation rates, the ice flow pattern, and geothermal heat flux are known or can be approximated [*Dahl-Jensen et al.*, 1998]. The Berkner cores were analyzed in very high temporal resolution and dated using seasonal variations in electrical conductivity (ECM) supplemented by chemical records in parts of the core where ECM was vague. The ECM signal does identify some of the largest known volcanic events (e.g., Tambora), but *Mulvaney et al.* [2002] report that in general the ECM does not give as clear a volcanic history as sites higher on the polar plateau because the volcanic signal is masked by the large seasonal ECM signal. The relationship between the isotopically inferred temperatures and those calculated from sea level temperatures using the dry adiabatic lapse rate suggest that Berkner Island experiences a persistent surface inversion. Their 1200-year δ¹⁸O history oscillates around a fairly stable mean, lacks much variability, and does not record a 20th century warming. The discrepancy between the δ¹⁸O-inferred temperature history and that from borehole measurements warrants future investigation. The chemical signals retain their annual character to the bottom of the core which bodes well for their preservation deeper in the ice dome. Currently a project is underway to recover an ice core to bedrock from the southern dome where ice is ~1000 m thick, accumulation is high (~1300 mm w.e.), the underlying bedrock is flat, the winds are low, and basal temperatures are expected to be about –12°C. *Mulvaney et al.* [2002] conclude that this is an excellent site from which to capture the long-term climate history of the region, possibly extending back to the Last Glacial Maximum (LGM) and resolving whether Berkner Island ice cap emerged from the overlying West Antarctic Ice Sheet to become an independent dome or whether the ice dome was deposited *in situ*.

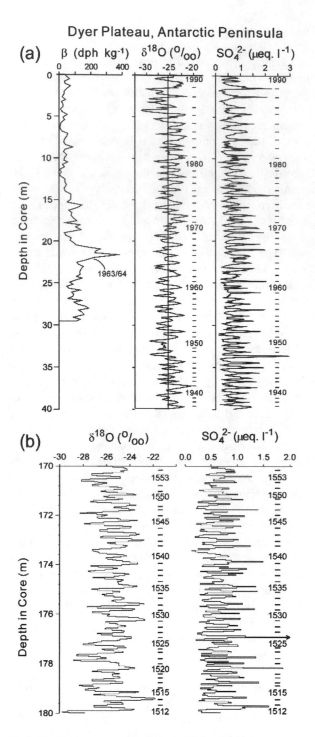

Fig. 3a,b. The seasonal variations in δ¹⁸O and sulfate concentration used to date the entire DP core are shown in (a) for the upper 40 meters. The major beta radioactivity peak in 1963/64 confirms the seasonality of the chemical signals. (b) The excellent preservation of the annual variations in δ¹⁸O and sulfate concentrations is shown for the lowest 10-m (170 to 180 m) of the 181-m core.

Although δ¹⁸O is widely used as a temperature proxy, interpretations necessarily suffer from the limitations of isotope thermometry. The δ¹⁸O relationship is influenced by multiple processes that include (1) air temperature at the time of condensation; (2) condensation and evaporation that occur within the airmass along its route from the source (ocean) to the deposition site (ice sheet); (3) local glaciological conditions as snow is converted to firn and then to ice; (4) the elevation of the deposition site; and (5) seasonality in the precipitation regime (a signal is only produced when snow is falling). The seasonality of precipitation is particularly critical when comparing meteorological observations averaged throughout the calendar year to an ice core δ¹⁸O record that may be strongly biased to a specific season.

Peel and Clausen [1982] conducted an extensive investigation of the temperature-isotope relationship using 10-meter temperatures and δ¹⁸O analyses of 10-meter cores at 25 sites located on or in close proximity to the Peninsula (for sites see their Figure 1). They report a systematic δ-T relationship throughout the AP that they were able to describe by a model in which an air mass in equilibrium with SMOW (standard mean ocean water) at 10°C (i.e., a mixing ratio of 7.8 g of water per kg of air) cools isobarically as it moves poleward, reaching the coast at ~ –4°C. Subsequent cooling by adiabatic uplift results in further isotopic depletion. Their investigation, designed to identify potential drill sites, led them to conclude that the best sites would be on the ice rises along the east coast (such as Dolleman Island) and along the spine of the Peninsula at sites south of 73°S.

A more recent comparison of the existing temperature and isotopic records in the Peninsula region [*Jones et al.*, 1993] highlights the lack of a consistent picture of climate variability in the AP as pieced together from the available ice core records, observations made at permanent stations, and records taken by early expeditions [*Jones et al.*, 1990]. In general, the δ¹⁸O records (interpreted as a proxy for temperature and subject to the limitations discussed above) are at odds with the limited meteorological observations over the last century. While the AP has experienced a century-long warming, the ice core evidence is mixed. The James Ross record suggests that the warmest temperatures occurred in the mid-nineteenth century with subsequent cooling. This broad scale trend is also evident in the cores from Dolleman Island and T340, although the isotopic trend on DI reversed in 1960 to become more consistent with observed air temperatures.

These core records suggest an increase in net annual accumulation (more precipitation would likely reflect warmer air temperatures [*Jones et al.*, 1993]) throughout the 20th century and this observation is also at odds with the isotopic records. This discussion highlights the perplexing nature of the ice core evidence of climate variability in the Peninsula. Clearly, more progress must be made in understanding the linkages among the various climate regimes in the Peninsula, the relationship of the Peninsula climate to that of the rest of the Antarctic Continent, and the utility of ice cores from the Peninsula to provide representative multi-millennial histories of climate variability in the region. Certainly expecting five cores, from widely diverse climatological settings, to provide a coherent picture for a region with complex meteorological and oceanographic forcing is not realistic. In the concluding section of this paper we return to this point along with recommendations for future research.

As with any ice cores, those from the AP reflect local, regional and larger-scale processes, but regional and larger-scale processes are of greater climatological interest. Excluding the Gomez core that extends back only to 1942, the Dyer Plateau cores provide the only paleohistory from the ice plateau that blankets the spine of the Peninsula. To expand our spatial perspective on climate variability along the axis of the Peninsula, two other ice core histories are included in this discussion. A 302-meter core was drilled in 1985 in West Antarctica, at the base of the Peninsula at Siple Station (75°55'S; 84°15'W; 1054 msal; Fig. 1). The Siple core provides a 550-year proxy climate history [*Mosley-Thompson et al.*, 1991; *Mosley-Thompson*, 1992] that covers nearly the same time interval as the Dyer cores. To expand the spatial view even further, we have included the most recent 500 years of a 4000-year record from a 200-meter core drilled in 1986 at Plateau Remote (84°S; 43°E; 3330 masl; Fig. 1) in East Antarctica [*Mosley-Thompson*, 1996].

Figure 2 illustrates all the existing multi-century records from the Peninsula region, except the new 1200-year history from Berkner Island [*Mulvaney et al.*, 2002]. The top three records, T340, DI and JRI are reproduced as they were originally presented and were smoothed using different filters as discussed in their respective references. The Dyer and Siple records are shown as simple decadal averages. The PR δ18O record is plotted as decadal averages with a 20-year un-weighted running mean (darker line in Fig. 2f). Unlike the DP and SS cores in which each annual value was the average of 6 to 10 samples, the low accumulation (~40 mm w.e.) at PR resulted in only one or two samples per year such that a single sample might reflect only winter or summer pre-

cipitation or even parts of two different years when accumulation was unusually low. The decadal averages of annual accumulation and dust flux for the DP, SS, and PR cores are shown in Figure 4a and 4b, respectively.

The δ18O record from DP (Fig. 2d) varies consistently around the long-term mean (1505–1989) until ~1840 after which conditions apparently cooled until 1920. From 1920 to 1940 the average δ18O became progressively more enriched and since 1940 the decadal averages have remained above the long-term mean. Thus, unlike the JRI, DI and T340 records, the DP site recorded much of the 20th century warming evident in the meteorological records. Moving southward to Siple Station, the δ18O record shows little variation about the mean (1500–1986) and no 20th century warming. At the PR site on the East Antarctic Plateau δ18O is more variable, in part reflecting the low resolution sampling interval discussed above. The Plateau Remote (PR) record contains some longer-term (~century scale) oscillations with a brief (~3 decades), but strong cooling in the early 17th century. Conditions remain at or above the long-term mean from 1660 to 1780 after which a gradual cooling trend persists until 1870 after which conditions warm rapidly, peaking at the turn of the 20th century. Since that time the δ18O record indicates a cooling trend to the present. The PR δ18O record, like those from South Pole, does not show 20th century 18O enrichment (warming) [*Mosley-Thompson*, unpublished data]. Similarly, the recently published isotopic record from Berkner Island [*Mulvaney et al.*, 2002] also does not show a 20th century warming. Thus, with the exception of the Dyer Plateau and Dolleman Island records, the ice core isotope histories in the AP do not record a 20th century warming. Unlike the Dyer Plateau where warming begins around 1920, the warming on DI begins four decades later in the 1960s. In light of the precision with which these core are dated (Fig. 3a, b), dating differences cannot account for this offset.

Peel [1992] noted that the Dolleman Island cores show a 20th century increase in accumulation that is counterintuitive if conditions are not warming. Figure 4a illustrates the decadal averages of the 500-year records of net annual accumulation from DP, SS, and PR. The PR accumulation record should be viewed cautiously given the dating limitations as noted above. Since 1900 accumulation has been consistently above the long-term mean at Dyer and at Siple it has been above the mean in all decades but one. Both accumulation records were corrected for compression with depth [*Thompson et al.*, 1994; *Mosley-Thompson et al.*, 1993, respectively], although the effect is small in the upper layers of the ice

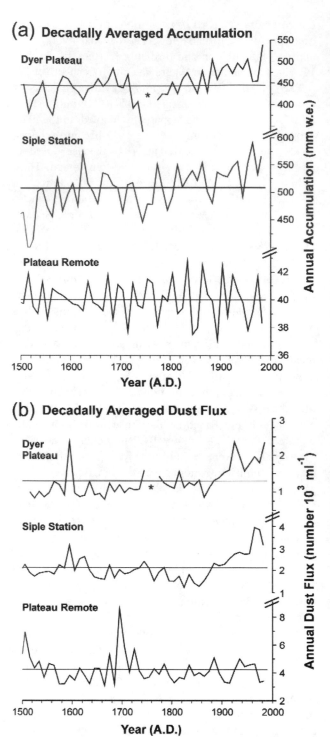

Fig. 4a,b. The decadal averages of net annual accumulation are shown in (a) and the annual dust flux is shown in (b) for the Dyer Plateau core [*Thompson et al.*, 1994], the Siple Station core [*Mosley-Thompson*, 1992], and the Plateau Remote core [*Mosley-Thompson*, 1996].

plateau. The accumulation increase at DP is also consistent with geophysical measurements that suggest a gradual increase in accumulation over the last 200 years from 460 mm ice equivalent to 540 mm ice equivalent [*Raymond et al.*, 1996]. The PR site shows no trend in accumulation.

To summarize, there is strong evidence for a regional increase in 20th century accumulation (i.e., Dolleman, Dyer, Siple), but proxy-based ($\delta^{18}O$) evidence of warming over the entire 20th century is evident only in the Dyer Plateau records. Clearly, meteorological evidence from the Antarctic Peninsula confirms the 20th century warming in both the near surface and mid-troposphere. So why have most of the ice core records failed to capture this larger-scale atmospheric signal? *Jones et al.* [1993], *Peel* [1992], *Peel and Clausen* [1982] and *Peel et al.* [1996] have addressed this issue. *Jones et al.* [1993] note that correlations (r) between temperature and both $\delta^{18}O$ and δD are on the order of –0.5 ($R^2 = 0.25$), leaving 75% of the variance unexplained. Other factors influencing $\delta^{18}O$ include temporal changes in moisture sources for the different sites, proximity to the sea ice edge and hence the differential influence of sea ice cover, seasonal differences in delivery of snowfall to the site, and glaciological controls on the preservation of the $\delta^{18}O$signal after snow deposition. The few ice cores in hand, coupled with potential dating errors and the lack of analysis of both $\delta^{18}O$ and δD for the same samples, limits resolution of these differences. *Peel et al.* [1996] suggest that the isotopic signals at sites strongly exposed to the weather systems in the Weddell Sea may not be good recorders of regional air temperature trends. For example, the $\delta^{18}O$ records suggesting mid-19th century warmth (Fig. 2a, b, c) may reflect the strong influence of changes in sea ice cover. *Peel et al.* [1996] note that these cores contain elevated concentrations of methane sulfonate (MSA$^-$) and reduced deuterium excess, d=δD-8($\delta^{18}O$), suggesting proximity of the sites to the ice edge (open water) or the influence of intermittent polyna. They conclude, and we agree, that cores from the higher altitude sites along the ice-covered spine of the AP are more likely to record the regional temperature variations than the lower elevation sites that are more strongly influenced by changes in the local sea ice distribution and persistent surface inversions.

Limited data suggest that on millennial time scales the climate in the Antarctic Peninsula may be more tightly coupled to that on the East Antarctic Plateau. The $\delta^{18}O$ record from the Plateau Remote core extends back 4000 years with nearly annual resolution (Fig. 5d). As discussed in more detail by *Mosley-Thompson* [1996] the long-term isotopic trend shows a cooling over the last four millennia with three broad oscillations of roughly 1200 years. The 4000-year record is much too short to determine whether the oscillations are periodic. The PR $\delta^{18}O$ record suggests warmer conditions from 4000 to 2500 ka, broadly contemporaneous with more open water in Lallemand Fjord (Fig. 1) inferred from elevated total organic carbon (TOC) in ocean sediments (Fig. 5c) [*Shevenell et al.*, 1996], with a mid-neoglacial cooling (~2.5 ka BP) on South Georgia Island (Fig. 5b) [*Clapperton et al.*, 1989], and with warmer conditions inferred from more depleted $\delta^{18}O$ in foraminiferal tests in the high resolution sediment record from the Bermuda Rise [Fig. 5a; *Keigwin*, 1996]. Twelve high to medium resolution sediment cores recently collected on the west side of the AP provide more details about the Holocene climate [*Domack et al.*, this volume]. Their records reflect warmer conditions dominating from the Early to the Middle Holocene with the onset of neoglacial cooling at ~ 3.2 ka. The PR $\delta^{18}O$ record (Fig. 5d) is broadly consistent (within dating uncertainties) with their marine record. The average $\delta^{18}O$ from 3.8 to 2.5 ka BP is −52.11‰ and from 2.5 ka BP to AD 1986 it is −54.12‰, nearly 2‰ more depleted (cooler). *Domack et al.* [this volume] report their cores contain a Medieval Warm Period (1.15 ka to 0.7 ka), a Little Ice Age signal (0.7 ka to ~0.15 ka) and 200-year oscillations in the regional climate/oceanographic conditions. At present there are no high-resolution ice core histories from the Antarctic Peninsula longer than 1200 years so that land-based evidence to support these oceanographic records does not exist. The 4000-year $\delta^{18}O$ record from Plateau Remote (Fig. 4d) provides a tantalizing hint of teleconnections from the East Antarctic Plateau through the Peninsula region and up through the Atlantic Basin as suggested by its similarity to the high resolution $\delta^{18}O$ record from the Bermuda Rise (BR). This linkage is not as speculative as it might appear upon first consideration.

A potential mechanism linking Antarctic proxy histories and those from lower latitudes (e.g., the BR record) is provided by the Antarctic Circumpolar Current (ACC) that constitutes the primary moisture source for snow that accumulates on the Antarctic ice fields. The ACC is a combination of recirculated waters from Pacific Ocean, Indian Ocean and modified North Atlantic Deep Water [*Shevenell and Kennett*, 2002]. *Keigwin* [1996] reports that changes in the production of North Atlantic Deep Water and in sea surface temperatures account for one-third and two-thirds, respectively, of the variability in the $\delta^{18}O$ content of *G. Ruber* at the BR site. *Hodell et al.* [2001] illustrate (their Fig. 5b) the similarity of their

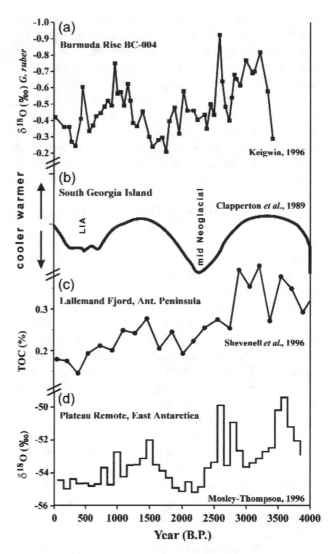

Fig. 5. Four different proxy records covering nearly the last four millennia are shown. These are (a) the isotopic abundance in G. Ruber tests in a high resolution sediment core from the Bermuda Rise [Keigwin, 1996]; (b) the climate history on South Georgia Island inferred from glacial geologic evidence [Clapperton et al., 1989]; (c) total organic carbon from a sediment core in the Lallemand Fjord indicates warmer conditions (more open water and higher TOC) and cooler conditions (more ice cover and lower TOC) [Shevenell et al., 1996]; and (d) the 100-year averages of $\delta^{18}O$ from an ice core at the Plateau Remote site (see Fig. 1) in East Antarctica [Mosley-Thompson, 1996].

diatom $\delta^{18}O$ record from a piston core at ~53°S with Keigwin's $\delta^{18}O$ record from BR. In fact, the near synchroneity of a large mid-Holocene climate change, ~5–6 ka, in both the South and North Atlantic regions [*Hodell et al.*, 2001], as well as in the tropics [*Thompson et al.*,

2002], support the hypothesis [*McIntyre and Molfino*, 1996] that such large-scale changes are initiated in the tropics and transmitted poleward by changes in surface winds that control the poleward advection of warm surface waters. The dynamics of these mechanisms by which climate changes may be propagated over long distances require further investigation as more similarities are reported among polar and low latitude proxy histories. Longer, higher resolution ice cores from the spine of the AP are needed to link climate changes in the AP with those at more distance sites.

CONCLUSION: WHERE DO WE GO FROM HERE?

At present, the number of multi-century ice core records from the AP–Weddell Sea region that are dated with good precision (± 2 years per century) is three: Dolleman Island, Dyer Plateau, and Berkner Island (in press). Dolleman and Dyer were drilled more than a decade ago and show similar accumulation histories, but different isotopic trends. The Berkner Island $\delta^{18}O$ record does not show 20th century warming or an increase in accumulation [*Mulvaney et al.*, 2002]. If the 20th century warming in the Peninsula, currently one of the strongest signals on the planet, is to be placed within a longer-term perspective, more well-dated ice cores extending further back in time must be drilled from sites that are sensitive to large-scale regional climate variations. The cores must be analyzed in the greatest possible temporal resolution for numerous chemical and physical constituents, but most particularly for both $\delta^{18}O$ and δD [*Jones et al.*, 1993].

Data from existing cores suggest that the best sites for capturing large-scale climate variability will be on the ice fields that blanket the spine of the AP. Here the higher annual accumulation and strong seasonal variations in a number of chemical constituents, coupled with known volcanic horizons, will allow very accurate dating back thousands of years. Importantly, these records will also provide the critical link between the histories from existing Antarctic cores (Byrd Station, Dome C, Taylor Dome, and Siple Dome) as well as those emerging from more recent cores (Dome Fuji, EPICA-Dome C, EPICA-Droning Maude Land, and Law Dome) and the low latitude ice core histories from the Andes of South America (Sajama and Huascarán) and Kilimanjaro in Africa. At present the climate trends in the Peninsula appear decoupled from those in the Antarctic interior, but only contemporaneous ice core-based climate histories can reveal whether this relationship has persisted over thousands of

years or is a late Holocene phenomenon. Moreover, the 20th century warming trend observed in the AP is clearly recorded in the tropical and subtropical ice core records collected around the Pacific Basin [*Thompson*, 2000; *Thompson et al.*, 2003].

The ocean-atmosphere connections between Antarctica and the Tropics are gaining considerable attention. Antarctic Intermediate Water (AAIW) provides an intimate link between conditions along the Antarctic margin and those in both the tropical Pacific and the North Atlantic. Thus, understanding the history of tropical and north Atlantic climate variability also requires knowledge of the conditions along the Antarctic coastal margin where the AAIW forms. *Pierrehumbert* [2000] notes that wind-driven upwelling in the Southern Ocean around Antarctica affects both the volume of AAIW that spreads northward and the depth of the tropical thermocline. Thus, a small shift in the circum-Antarctic wind stress pattern has the potential to modulate the climatic behavior of the tropical Pacific. Further, the rejection of brine by sea ice formation is critical to the formation of the dense water that gives rise to the Antarctic Bottom Water. Thus, *Pierrehumbert* [2000] concludes that the changes along the Antarctic ice margin are likely to be felt in the tropics and from there these effects will ripple to the rest of the planet. However, the hemispheric linkages are complex and issues of leads and lags abound [*Steig and Alley, in press*]. Millennial scale climate variability is well preserved in the inland cores from Vostok, Byrd Station, and the new EPICA core at Dome C but the major cold events are not in phase with those in the Greenland cores (*Blunier et al.*, 1998) leading to the suggestion of a bi-polar seesaw [*Broecker*, 1998]. However, methane concentrations from the Law Dome ice core [*Morgan et al.*, 2002] place the initiation of the Antarctic Cold Reversal before the Bölling transition (~14.5 ka BP in GRIP). These results support other evidence [*Steig et al.*, 1998; *Jouzel et al.*, 2001] that argue against a bi-polar seesaw. For example, Taylor Dome has a deglaciation pattern similar to that in Greenland [*Steig et al.*, 1998; 2000], but concerns remain about the timing of the deglaciation in that core [*Mulvaney et al.*, 2000]. To further complicate matters, the Siple Dome climate history (also from the Ross Embayment) looks more like those from East Antarctica (Dome C and Vostok). Thus, the timing of both the deglaciation and the major warm/cold events between the hemispheres, or even within Antarctica, has yet to be unraveled. New high resolution ice core records from carefully selected sites on the ice field extending along the spine of the Antarctic Peninsula, coupled with the long core currently being drilled on Berkner Island, will contribute new insight to the leads and lags in the global ocean-atmosphere system.

It is time to consider seriously a project to acquire a suite of cores along the spine of the Peninsula. To ensure long and interpretable records, the program must include site selection using radar profiles of the basal topography and internal reflection layers. South of the 1989/90 DP drill site toward Siple Station the ice field becomes thicker and wider, thus decreasing the percent of the ice core that will be disturbed by basal conditions and complex flow. As mentioned previously, the contemporary relationships among annual and seasonal air temperatures, the timing of precipitation, and the isotopic composition of the accumulated and preserved snowfall must be quantitatively evaluated. *In situ* monitoring at selected drill sites by installation of an AWS with an acoustic depth gauge system, coupled with pit sampling and isotopic modeling efforts [e.g., *Werner and Heimann*, 2002] must be included if the chemical signals in the ice cores are to be correctly interpreted [*Jouzel*, 1999]. The ice cores could be recovered using Ohio State's suite of portable, light-weight electro-mechanical and thermal-alcohol drills that recovered good to excellent quality core to a depth of 460 meters on the Bona-Churchill col in southeastern Alaska in 2002. The depth capability of these drills can be extended to 1000 meters with modest modification. The complete drill system [*Zagorodnov et al.*, 2002] including the camp and personnel can be moved from site to site by two twin otters as long as the fuel, ethanol, and core boxes are positioned by other flights.

The key questions that must be addressed with new cores from the ice field along the spine of the Antarctic Peninsula include:

1) Is the apparent decoupling of the climate in the Peninsula and on the polar plateau a 20th century phenomenon (possibly anthropogenic) or does this relationship persist over many millennia?

2) New sediment cores indicate warmer conditions in the Mid to Early Holocene in the AP, but what is the land-based evidence for Early Holocene warmth and how does it compare to the 20th century warming?

3) Can the ice core chemistry shed light on the history of the disintegration of ice shelves in the region? Sediment coring is planned to reconstruct the history of ice shelf growth and decay, but contemporaneous ice core proxy records will be needed to explore the climate forcing. Is the current diminishment of the ice shelves climatically driven, glaciologically driven or both? Long histories from new cores will be essential to unravel the controlling processes.

4) What is the history of local, regional and global volcanism recorded in these cores? This requires comparing excess sulfate histories from AP cores to similar histories from other parts of Antarctica as well as from Greenland.

5) Does the ice field along the spine of the Peninsula contain glacial stage ice? If so, what is the nature and timing of the deglaciation and how does it compare with the transition as recorded in the Ross Sea Sector, Greenland, or South America? This is critical for unraveling the leads and lags in the climate system as it transitions from glacial to interglacial conditions and from stadial to interstadial.

6) Was a large dust event at 4.2 ka BP globally distributed? At present there is great interest in this event that appears to have blanketed much of the tropics during a three century long drought that may have been responsible for societal disruptions in the Middle East [i.e., the beginning of the First Dark Age]. The timing of this dust event is at issue. However, the dust concentrations in the Dyer Plateau core are very low so that if this dust is present it should be easy to detect. As this dust event is prominent in the Andean ice fields of Peru it is reasonable to expect that it might be recorded in ice fields of the AP. If so, carefully sited cores could provide a firm date for this dust event that may have disrupted the course of human history [*Thompson et al.*, 2002].

As the U.S. and its scientific partners formulate their long-range research plans for the Antarctic, it is clear that the ice fields along the spine of the Antarctic Peninsula and the proxy records preserved therein should figure heavily into the program.

Acknowledgments. The contributions of the scientists from BAS, OSU, and the University of Washington in the acquisition of the Dyer Plateau cores and the interpretation of the results were invaluable. We thank the British Antarctic Survey for the logistical support that made the Dyer Plateau drilling project possible and particularly David Peel and Rob Mulvaney for their infectious enthusiasm for both the field work and the science. We are indebted to Jihong Cole-Dai, Mary Davis and Ping Nan-Lin for the ion chemistry, dust and $\delta^{18}O$ analyses, respectively. The acquisition and analysis of the OSU ice cores presented in this paper were supported over the years by the National Science Foundations' Office of Polar Programs. We thank Phil Jones, David Peel and one anonymous reviewer for constructive comments that significantly improved the quality of the paper. This is contribution number 1271 of the Byrd Polar Research Center.

REFERENCES

Aristarain, A. J., J. Jouzel, and M. Pourchet, Past Antarctic Peninsula climate, *Clim. Change, 8*, 69-89, 1986.

Blunier, T., J. Chappellaz, J. Schwander, A. Dällenbach, B. Stauffer, T. F. Stocker, D. Raynaud, J. Jouzel, H. B. Clausen, C. U. Hammer, and S. J. Johnsen, Asynchrony of Antarctic and Greenland climate change during the last glacial period, *Nature, 394*, 739-743, 1998.

Broecker, W., Paleocean circulation during the last glaciation: A bipolar seesaw?, *Paleoceanography, 13,* 119-121, 1998.

Clapperton, C. M., D. E. Sugden, J. Birnie, and M. J. Wilson, Late-glacial and Holocene glacier fluctuations and environmental change on South Georgia, Southern Ocean. *Quat. Res., 31*, 210-228, 1989.

Convey, P., Antarctic Peninsula climate change: signals from terrestrial biology, this volume.

Crozaz, G., Fission products in antarctic snow. An additional reference level in January 1965, *Earth Planet. Sci. Lett., 6*, 6-8, 1969.

Dahl-Jensen, D., K. Mosegaard, N. Gundestrup, G. D. Clow, S. J. Johnsen, A. W. Hansen, and N. Balling, Past temperatures directly from the Greenland Ice Sheet, *Science, 282*, 268-271, 1998.

Dai, J. C., L. G. Thompson, and E. Mosley-Thompson, A 485 year record of atmospheric chloride, nitrate and sulfate: results of chemical analysis of ice cores from Dyer Plateau, Antarctic Peninsula, *Ann. Glaciol., 21*, 182-188, 1995.

Domack, E. W., A. Leventer, J. Ring, E. Williams, D. Carlson, W. Wright, and G. Burr, Marine sediment record of natural environmental variability and recent warming in the Antarctic Peninsula, this volume.

Emslie, S. D., W. Fraser, R. C. Smith, and W. Walker, Abandoned penguin colonies and environmental change in the Palmer Station area, Anvers Island, Antarctic Peninsula, *Antarctic Science, 10*, 257-268, 1998.

Graf, W., H. Moser, H. Oerter, O. Reinwarth, and W. Stichler, Accumulation and ice-core studies on Filchner-Ronne Ice Shelf, Antarctica, *Ann. Glaciol., 11*, 23-31, 1988.

Hansen, J., R. Ruedy, M. Sato, and K. Lo, Global warming continues, *Science, 295*, 275, 2002.

Hodell, D. A., S. L. Kanfoush, A. Shemesh, X. Crosta, C. D. Charles, and T. P. Guilderson, Abrupt cooling of antarctic surface waters and sea ice expansion in the South Atlantic sector of the Southern Ocean at 5000 cal yr B.P., *Quat. Res., 56*, 191-198, 2001.

Jones, P. D., Antarctic temperatures over the present century - A study of the early expedition record, *J. Clim., 3*, 1193-1203, 1990.

Jones, P. D., R. Marsh, T. M. L. Wigley, and D. A. Peel, Decadal timescale links between Antarctic Peninsula ice core oxygen-18, deuterium and temperature, *The Holocene, 3*, 14-26, 1993.

Jouzel, J., Calibrating the isotopic paleothermometer, *Science, 286*, 910-911, 1999.

Keigwin, L. D., The Little Ice Age and Medieval Warm Period in the Sargasso Sea, *Science, 274*, 1503-1508, 1996.

King, J. C., J. Turner, G. J. Marshall, W. M. Connolley, and T. A. Lachlan-Cope, Antarctic Peninsula climate variability and its causes as revealed by instrumental records, this volume.

Limbert, D. W. S., Variations in the mean annual temperatures for the Antarctic Peninsula, *Polar Record, 17*, 303-306, 1974.

Marshall, G. J., V. Lagun, and T. A. Lachlan-Cope, Changes in Antarctic Peninsula tropospheric temperatures from 1956-99; a synthesis of observations and reanalysis data. *Int. J. Climat., 22*, 291-310, 2002.

McIntyre, A., and B. Molfino, Forcing of Atlantic equatorial and subpolar millennial cycles by precession, *Science, 274*, 1867-1870, 1996.

Morgan, V., M. Delmotte, T. van Ommen, J. Jouzel, J. Chappellaz, S. Woon, V. Masson-Delmotte, and D. Raynaud, Relative timing of deglacial climate events in Antarctica and Greenland, *Science, 297*, 1862-1864, 2002.

Morris, E. M., and D. G. Vaughan, Spatial and temporal variation of surface temperature on the Antarctic Peninsula and the limit of viability of ice shelves, this volume.

Mosley-Thompson, E., Paleoenvironmental conditions in Antarctica since A.D. 1500: Ice core evidence, in *Climate since A.D. 1500*, edited by R. S. Bradley and P. D. Jones, pp. 572-591, Routledge, London, 1992.

Mosley-Thompson, E., Holocene climate changes recorded in an East Antarctica ice core, in *Climatic Variations and Forcing Mechanisms of the last 2000 Years*, edited by P. D. Jones, R. S. Bradley, and J. Jouzel, pp, 263-279, Springer-Verlag, Berlin, 1996.

Mosley-Thompson, E., J. Dai, L. G. Thompson, P. M. Grootes, J. K. Arbogast, and J. F. Paskievitch, Glaciological studies at Siple Station (Antarctica): potential ice-core paleoclimatic record, *J. Glaciol., 37*, 11-22, 1991.

Mosley-Thompson, E., L. G. Thompson, J. Dai, M. Davis, and P.-N. Lin, Climate of the last 500 years: High resolution ice core records. *Quat. Sci. Rev., 12*, 419-430, 1993.

Mulvaney, R., R. Röthlisberger, E. W. Wolff, S. Sommer, J. Schwander, M. A. Hutterli, and J. Jouzel, The transition from the last glacial period in inland and near-coastal Antarctica, *Geophys. Res. Lett., 27*, 2673-2676, 2000.

Mulvaney, R., H. Oerter, D. A. Peel, W. Graf, C. Arrowsmith, E. C. Pasteur, B. Knight, G. C. Littot, and W. D. Miners, 1000-year ice core records from Berkner Island, Antarctica, *Ann. Glaciol., 35*, in press.

Peel, D. A., Ice core evidence from the Antarctic Peninsula region, in *Climate since A.D. 1500*, edited by R. S. Bradley and P. D. Jones, pp. 549-571, Routledge, London, 1992.

Peel, D. A., R. Mulvaney, and B. M Davison, Stable-isotope / air temperature relationships in ice cores from Dolleman Island and the Palmer Land Plateau, Antarctic Peninsula, *Ann. Glaciol., 10*, 130-136, 1988.

Peel, D. A., R. Mulvaney, E. C. Pasteur, and C. Chenery, Climate changes in the Atlantic sector of Antarctica over the past 500 years from ice-core and other evidence, in *Climatic variations and forcing mechanisms of the last 2000 years*, edited by P.D. Jones, R.S. Bradley, and J. Jouzel, pp. 243-262, Springer-Verlag, Berlin, 1996.

Peel, D. W., and H. B. Clausen, Oxygen isotope and total Beta radioactivity measurements on 10m ice cores from the Antarctic Peninsula, *J. of Glaciol., 28*, 43-55, 1982.

Pierrehumbert, R. T., Climate change and the tropical Pacific: The sleeping dragon awakes, *Proc. Nat. Acad. Sci., 96*, 1355-1358, 2000.

Quayle, W. C., L. S. Peck, J. C. Ellis, Evans, H. C. Butler, H. J. Peat, and P. Convey, Ecological responses of maritime antarctic lakes to regional climate change, this volume.

Raymond, C., B. Weertman, L. Thompson, E. Mosley-Thompson, D. Peel, and R. Mulvaney, Geometry, motion and mass balance of Dyer Plateau, Antarctica, *J. Glaciol., 42*, 510-518, 1996.

Scambos, T., C. Hulbe, and M. Fahnestock, Climate-induced ice shelf disintegration in Antarctica, this volume.

Schwerdtfeger, W., *Weather and Climate in the Antarctic Peninsula*, Developments in Atmospheric Science, 15, Elsevier, Amsterdam, 1984.

Shevenell, A. E., and J. P. Kennett, Antarctic Holocene climate change: A benthic foraminiferal stable isotope record from the Palmer Deep, *Paleoceanography, 17(2)*, 8000, doi:10.1029/2000PA000596, 2002.

Shevenell, A. E., E. W. Domack, and G. M. Kernan, Record of Holocene paleoclimate change along the Antarctic Peninsula: Evidence from glacial marine sediments, Lallemand Fjord. *Pap. Proc. R. Soc. Tas., 130(2)*, 55-64, 1996.

Simmonds, I., Regional and large-scale influences on Antarctic Peninsula climate, this volume.

Skvarca, P., and H. De Angelis, Impact assessment of climatic warming on glaciers and ice shelves on northeastern Antarctic Peninsula, this volume.

Smith, R., D. Ainley, K. Baker, E. Domack, S. Emslie, B. Fraser, J. Kennett, A. Leventer, E. Mosley-Thompson, S. Stammerjohn, and M. Vernet, Marine ecosystem sensitivity to historical climate change: Antarctic Peninsula, *Bioscience, 49*, 395-404, 1999.

Steig, E. J., and R. B. Alley, Phase relationships between Antarctic and Greenland climate records, *Ann. Glaciol., 35*, 451-456.

Steig, E. J., E. J. Brook, J. W. C. White, C. M. Sucher, M. L. Bender, S. J. Lehman, D. L. Morse, E. D. Waddington, and G. D. Clow, Synchronous climate changes in Antarctica and the North Atlantic, *Science, 282*, 92-95, 1998.

Steig, E. J., D. L. Morse, E. D. Waddington, M. Stuiver, P. M. Grootes, P. A. Mayewski, M. S. Twickler, and S. I. Whitlow, Wisconsinan climate history from an ice core at Taylor Dome, Western Ross Embayment, Antarctica, *Geogr. Ann. A, 82A*, 212-235, 2000.

Thompson, D. W. J., and S. Solomon, Interpretation of recent Southern Hemisphere climate changes, *Science, 296*, 895-899, 2002.

Thompson, L. G., Ice core evidence for climate change in the Tropics: implications for our future, *Quat. Sci. Rev., 19*, 19-35, 2000.

Thompson, L. G., D. A. Peel, E. Mosley-Thompson, R. Mulvaney, J. Dai, P. N. Lin, M. E. Davis, and C. F. Raymond, Climate since AD 1510 on Dyer Plateau, Antarctic Peninsula: evidence for recent climate change, *Ann. Glaciol., 20*, 420- 426, 1994.

Thompson, L. G., E. Mosley-Thompson, M. E. Davis, K. A. Henderson, H. H. Brecher, V. S. Zagorodnov, T. A. Mashiotta, P-N. Lin, V. N. Mikhalenko, D. R. Hardy, and J. Beer, Kilimanjaro ice core records: Evidence of Holocene climate change in tropical Africa, *Science, 298*, 589-593, 2002.

Thompson, L. G., E. Mosley-Thompson, M. E. Davis, P.-N. Lin, K. Henderson, and T. A. Mashiotta, Tropical glacier and ice evidence of climate change on annual to millennial time scales, *Clim. Change*, 2003, in press.

Vaughan, D. G., and C. S. M. Doake, Recent atmospheric warming and the retreat of ice shelves on the Antarctic Peninsula, *Nature, 379*, 328-330, 1996.

Vaughan, D. G., G. J. Marshall, W. M. Connolley, J. C. King, and R. Mulvaney, Climate change - Devil in the detail, *Science, 293*, 1777-1779, 2001.

Werner, M., and M. Heimann, Modeling interannual variability of water isotopes in Greenland and Antarctica, *J. Geophys. Res. (Atmospheres), 107(D1)*, 4001, doi:10.1029/2001JD900253, 2002.

Zagorodnov, V. S., L. G. Thompson, E. Mosley-Thompson, and J. J. Kelley, Performance of intermediate depth portable ice core drilling systems on polar and temperature glaciers, *Mem. Natl. Inst. Polar Res., Spec. Issue 56*, 67-81, 2002.

Ellen Mosley-Thompson, Byrd Polar Research Center, 108 Scott Hall, 1090 Carmack Road, The Ohio State University, Columbus, OH 43210; thompson.4@osu.edu.

Lonnie G. Thompson, Byrd Polar Research Center, 108 Scott Hall, 1090 Carmack Road, The Ohio State University, Columbus, OH 43210; thompson.3@osu.edu.

Ecological Responses

PALMER LONG-TERM ECOLOGICAL RESEARCH ON THE ANTARCTIC MARINE ECOSYSTEM

Raymond C. Smith[1], William R. Fraser[2], Sharon E. Stammerjohn[3], and Maria Vernet[4]

Long-term studies in the western Antarctic Peninsula (WAP) region, which is the location of the Palmer LTER, provide the opportunity to observe how climate-driven variability in the physical environment is related to changes in the marine ecosystem. During the past 50 years the WAP region has experienced a statistically significant warming trend. Associated with this warming trend, during the last two decades, sea ice extent has trended down and the sea ice season has shortened. Ecosystem response to these trends is becoming evident at all trophic levels but is most clearly seen in a shift in population size and distribution of penguin species with different affinities to sea ice. Our results show that these trends in penguin populations are in accord with climate change predictions.

1. INTRODUCTION

The Palmer Long-Term Ecological Research (Palmer LTER) program seeks to understand the structure and function of the Antarctic marine ecosystem in the context of physical forcing (atmospheric, oceanic and sea ice) on seasonal to millennial time-scales (http://www.icess.ucsb.edu/lter/lter.html). The western Antarctic Peninsula (WAP) region, the location of the Palmer LTER (Figure 1), is proving to be an exceptional area to study ecological response to climate variability (*Smith et al.,* 1995, 1999; *Ross et al.,* 1996). Mounting evidence (*McCarthy et al.,* 2001) suggests that the Earth is experiencing human-induced climate variability, and air temperatures from the last half-century confirm the rapidity of warming in the WAP area (*Sansom,* 1989; *Weatherly et al.,* 1991; *King,* 1994; *Stark,* 1994; *Smith et al.,* 1996; *King and Harangozo,* 1998; *Marshall and King,* 1998; *van den Broeke,* 1998a, 1998b; *Smith and Stammerjohn,* 2001). Consistent with this warming, the sea ice season

is shorter and the maritime system of the northern WAP is expanding southward and replacing the continental, polar system of the southern WAP. Ecosystem response to this latitudinal climate change is becoming increasingly evident at all trophic levels (*Smith et al.,* 1998; *Smith and Stammerjohn,* 2001; *Quetin et al.,* 1996). However, this change is most clearly seen in a shift in population size and distribution of penguin species with different affinities to sea ice (*Fraser et al.,* 1992; *Fraser and Trivelpiece,* 1996; *Smith et al.,* 1999, 2003). The Palmer LTER seeks to understand the full ecological response to and implications of this climate-related change in physical forcing that is presently occurring in the WAP region.

There is now widespread recognition of numerous natural and anthropogenic forced phenomena, such as El Nino-Southern Oscillation (ENSO), atmospheric ozone depletion, green-house gas accumulation, global deforestation, and global warming, that can potentially cause significant ecological change, but that require long-term studies to resolve (*McCarthy et al.,* 2001). The NSF-funded LTER program consists of a community of 24 ecological sites—referred to as the LTER Network (http://lternet.edu)—that study ecological systems with a long-term perspective. The LTER Network maintains a comprehensive set of core measurements made across individual sites and project-specific field experiments that test ecosystem-specific hypotheses to climate

[1]ICESS, University of California, Santa Barbara, California
[2]Polar Oceans Research Group, Sheridan, Montana
[3]Lamont-Doherty Earth Observatory of Columbia University, Palisades, New York
[4]Scripps Institution of Oceanography, University of California, San Diego, California

Copyright 2003 by the American Geophysical Union
10.1029/079ARS11

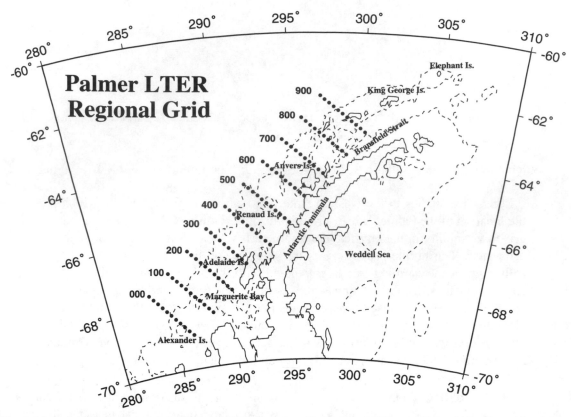

Fig. 1. Map of the western Antarctic Peninsula (WAP) region. The Palmer LTER sampling grid is denoted by the dots and each dot is a sampling station (perpendicular to the peninsula, the stations are 20 km apart, along the peninsula the sampling lines are 100 km apart). Palmer Station (64° 46' S, 64° 03' W) is located on the southern end of Anvers Island and the Faraday/Vernadsky (formerly British, now Ukrainian) Station is roughly 30 nautical miles south of Palmer Station.

change (*Callahan*, 1984; *Franklin et al.*, 1990). Within this context of ecosystem variability on inter-annual, decadal and longer time scales the Palmer LTER conducts research on the Antarctic marine ecosystem. In the following section we review some Palmer LTER findings of relevance to the theme of Antarctic Peninsula climate variability on seasonal to multi-decadal time-scales. A summary is given in the last section.

2. TEMPERATURE AND SEA ICE

Paleoclimate records are consistent in showing that the WAP region has moved from a relatively cold regime between approximately 2700 BP and 100 BP to a relatively warm regime during the past century (*Mosley-Thompson*, 1992; *Peel*, 1992; *Domack et al.*, 1993; *Thompson et al.*, 1994; *Dai et al.*, 1995; *Domack and McClennen*, 1996; *Leventer et al*, 1996; *Smith et al.*, 1999; and references in this volume). Instrument records, principally British Antarctic Survey (BAS) air temperature records for stations along the WAP, show a dramat-

ic warming trend during the past half century (*Sansom*, 1989; *Weatherly et al.*, 1991; *King*, 1994; *Stark*, 1994; *Smith et al.*, 1996; *King and Harangozo*, 1998; *Marshall and King*, 1998; *van den Broeke*, 1998a, 1998b; *Smith and Stammerjohn*, 2001). The warming trend in Faraday/Vernadsky air temperatures is strongest in midwinter months and peaks in June at 0.11 C° y⁻¹, representing a 6° C increase in June temperatures over the 51 year record. Satellite observations of sea ice extent have been available for the past two decades and, during this period a statistically significant anti-correlation between air temperatures and sea ice extent has been observed for this region (*Weatherly et al.*, 1991; *King*, 1994; *Smith et al.*, 1996). During the period of satellite observations sea ice extent in the WAP has trended down and the sea ice season has shortened (*Stammerjohn and Smith*, 1996; *Smith and Stammerjohn*, 2001). These observations are summarized in Figures 2 and 3.

Figure 2a shows the Faraday/Vernadsky annual average air temperatures from 1951 to 2001 (N=51). Note, we no longer use the 'full' record (N=57) since the earli-

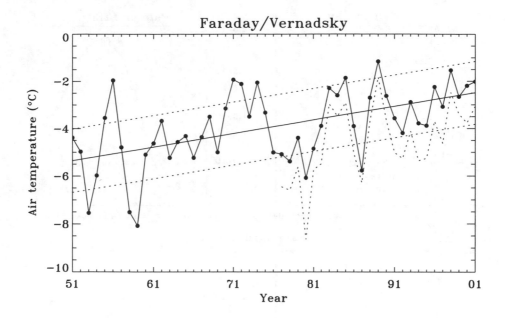

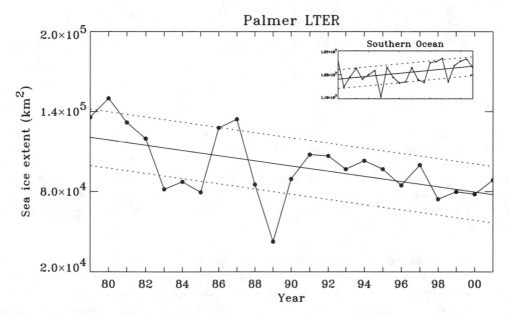

Fig. 2. (a) Faraday/Vernadsky (65° 15' S, 64° 15' W) annual average air temperatures, 1951-2001 (N=51). The solid line is the least-squares regression line with a gradient of 0.057° C a⁻¹, and the dotted lines indicate ± 1 s.d. from this line. A linear regression model shows the warming trend over this period to be significant at greater than the 99% confidence level. (b) Mean annual sea ice extent for the Southern Ocean (insert) and the Palmer LTER region. See *Stammerjohn and Smith* (1997) for details on the satellite data used. Faraday/Vernadsky air temperature data provided by the British Antarctic Survey, and sea ice data provided by the National Snow and Ice Data Center.

er data is now deemed less reliable. The solid line represents the least-squares regression line. After accounting for serial correlation present in this 51 year record (for method, see *Smith et al.,* 1996), the trend is statistically significant at a >99% confidence level. The dotted line indicates the ± 1 standard deviation (s.d.) from the regression line and has been used as a designator for defining "high" (above one s.d.) or "low" (below one s.d.) temperature years. The record from Rothera station (further south on the WAP) shows a strong temporal

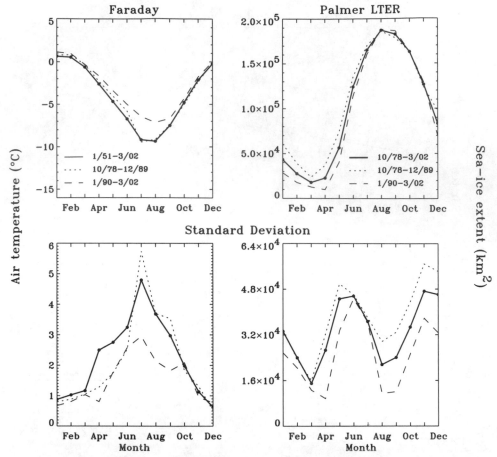

Fig. 3. Annual curves of monthly mean (a) Faraday/Vernadsky air temperature and (b) Palmer LTER sea-ice extent, shown for three different periods: the full record (bold line with solid dots) that includes 1/51-3/02 for (a) and 10/78-3/02 for (b); the period covering the first half of the sea-ice record (dotted line) that includes 10/78-12/89; and the period covering the second half of the sea-ice record (dashed line) that includes 1/90-3/02. Standard deviations of the monthly mean (c) Faraday/Vernadsky air temperature and (d) Palmer LTER sea-ice extent, shown for the same three periods. Faraday/Vernadsky air temperature data provided by the British Antarctic Survey, and sea ice data provided by the National Snow and Ice Data Center.

coherence (*King*, 1994; *Smith et al.*, 1996) with the mean annual temperatures averaging a few degrees cooler than Faraday/Vernadsky but displaying similar trends. This evidence suggests that there is a north/south temperature gradient along the WAP and that observed trends are coherent throughout the region.

Associated with the increasing trend in annual WAP air temperature is a decreasing trend in annual sea ice extent (Figure 2b) from 1979 to 2001 for the Palmer LTER region (see *Stammerjohn and Smith* (1996) for details on the satellite data used). The satellite sea-ice record is too short and the variability too high for the decreasing trend to be statistically robust, however the strong anticorrelation between WAP air temperature and sea-ice poses the possibility that sea-ice extent has been decreasing over

the last 50 years in association with the increase in WAP air temperature. Seasonal analysis also shows that the decreasing sea-ice trend is due mostly to decreasing sea ice during spring, summer, and fall, in conjunction with the observed decrease in the length of the sea-ice covered period (*Smith and Stammerjohn*, 2001). In contrast to the decreasing trend in WAP sea-ice extent, the inset in Figure 2b shows that sea-ice extent for the whole Southern Ocean is trending upward. *Thompson and Solomon* (2002) provide an interpretation of recent Southern Hemisphere climate change leading to the observed spatial variability.

The seasonal variability of air temperature and sea ice change during the past two decades is illustrated in Figure 3 (see *Smith and Stammerjohn*, 2001 for methods). Figures 3a and 3b show monthly mean air temperature and sea-ice extent for the following periods: the full instrument record (solid line for air temperature, 1/51-3/02, and sea-ice, 10/78-3/02); roughly the 1980s, i.e., the first half of the sea-ice record (dotted line for 10/78-12/89); and roughly the 1990s, i.e., the second half of the sea-ice record (dashed line for 1/90-3/02). Figures 3c and 3d show the standard deviations of monthly mean surface air temperatures and sea ice for the same periods plotted in Figures 3a and 3b, respectively.

Visual inspection of Figure 2a shows that the last two decades (1980s and 1990s) were warmer than the previous several decades. Figure 3a shows that the largest temperature changes between the 1980s and 1990s have occurred in winter (June-August) in contrast to relatively less change in late spring and summer (November-March). Figure 3c illustrates that in general (i.e., for the entire time period) there is significantly less variation in air temperatures during summer, when sea ice-free conditions and maritime conditions prevail, as compared to late fall and winter (May through September), when both sea-ice conditions and polar continental influences are highly variable. However, the winter variability in the 1990s is less than that for the 1980s, indicating that warmer air temperatures in winter are becoming more the norm.

The seasonal variability of sea ice (Fig. 3b) is inversely related to the seasonal cycle of air temperature (Fig. 3a) except that the summer sea ice extent minimum lags the summer air temperature maximum by two to three months. Most noteworthy, however, is that the 1990s winter warming and increased variability (shown in air temperature) is not associated with a seasonal anticorrelated response in sea-ice extent. There is no clear decrease in winter sea-ice extent between the 1980s and 1990s, but instead there has been a detectable decrease in summer-early fall (January to May) sea-ice extent. Figure 3d illustrates that in general there is increased variability during periods of sea-ice advance (May-July) and retreat (October-January). However, as seen for air temperature, variability has decreased from the 1980s to the 1990s indicating that later sea-ice advances and earlier sea-ice retreats are also becoming more the norm.

The overall timing and magnitude of the annual cycle of sea ice advance and retreat is influenced not only by long-term trends but also by El Niño-Southern Oscillation (ENSO) time scales of variability. Air temperature and sea ice extent for various regions of the Southern Ocean show some correlation with the Southern Oscillation Index (SOI, determined by the standardized sea-level pressure difference between Tahiti and Darwin, Australia). Several authors have suggested

(*Carleton,* 1988; *Mo and White,* 1985; *van Loon and Shea,* 1985; *van Loon and Shea,* 1987; *Smith et al.,* 1996; *White and Peterson,* 1996; *White et al.,* 1998) that these relationships support the idea of linkages between sea ice, cyclonic activity and global teleconnections. Recently, strong statistical associations between ENSO variability and Southern Ocean climate have been demonstrated (*Yuan and Martinson,* 2000, 2001; *Kwok and Comiso,* 2002). Martinson and co-workers, studying opposing anomalies in surface air and sea surface temperatures, sea-level pressure and sea ice extent between the southeastern Pacific and southwestern Atlantic sectors, describe the Antarctic Dipole (ADP) as an out-of-phase relationship which is coherent with ENSO variability. Recent work (*Liu et al.,* 2003) also discuss possible teleconnection mechanisms whereby the variability in the ADP is associated with ENSO-related variability in the regional mean meridional atmospheric circulation. Since the ENSO time scale of variability is second only to seasonal variability in driving worldwide weather patterns, an understanding of these teleconnected spatial patterns of variability are essential to an understanding of climate variability, sea ice forcing and ecological response within the WAP. Also, *Thompson and Solomon* (2002) have presented evidence that trends in the Southern Hemisphere annular mode, a large-scale pattern of variability characterized by fluctuations in the strength of the circumpolar vortex, have contributed to the observed warming over the WAP.

While the mechanistic processes that link WAP temperature and sea ice trends continue to be researched, the role of the mean position of the atmospheric circumpolar low-pressure trough (CPT) continues to be viewed as a possible linking mechanism. The Antarctic Peninsula is the only area in the Southern Ocean where the CPT crosses land. The seasonal cycle displayed in temperature, pressure, wind, and precipitation (*Schwerdtfeger,* 1984; *van Loon,* 1967) is linked to both increased cyclonic activity and a southward shift (of approximately 10° of latitude) of the CPT during spring and autumn. The relative position of the CPT influences not only the semiannual cycle of atmospheric variables but also the timing and distribution of sea ice and other oceanographic variables. Van Loon suggested that this seasonal temperature cycle is associated with enhanced meridional flow from middle to high latitudes during spring/fall. Indeed, more recent work by *Meehl* (1991) confirms that transient eddy heat flux likely contributes to this seasonal cycle in the Antarctic coastal zone.

King and co-workers (*King,* 1994; *King and Harangozo,* 1998; *Marshall and King,* 1998) also show a strong correlation between surface air temperature and meridional sea-level pressure indices calculated for the WAP area. Their results demonstrate that increased boundary-layer winds, flowing from the northwest sector toward the WAP, are associated with increased cyclonic activity and warm air advection from lower latitudes. The increase in surface temperatures associated with the increase in northerly winds consequently produces an environment with more maritime (warm and moist) characteristics, as opposed to the continental environment (cold and dry) that would result due to the effects of southerly winds and colder temperatures. *Stammerjohn et al.* (2003) discuss in detail the response of sea ice extent and drift dynamics to synoptic forcing in the WAP region and present a conceptual model of ice-SAO-ENSO linkages that may explain both seasonal and long-term variability. They suggest such a conceptual model may provide a linkage between synoptic-scale systems and a mechanism for long-term climate variability. Regardless of the mechanisms, both the modern instrument record and the paleohistory of the WAP suggest competitive interactions between two distinct latitudinal climatic zones—maritime to the north and continental to the south. The observed statistically significant warming trend over the past century is modulated by tropical-polar teleconnections with ENSO-like variability and the mechanisms behind these changes is an area of active and future research.

3. PHYTOPLANKTON AND SEA BIRDS

The physical forcing of the WAP marine ecosystem is undergoing change in conjunction with the seasonal variability and trends in air temperature and sea-ice as discussed above. As noted, these observations are indicative of a climate change where continental influences (cold and dry) are giving way to increasing maritime influences (warm and moist) along the WAP latitudinal climate gradient. The Palmer Station area in particular, located approximately two thirds of the way up the peninsula on the south edge of Anvers Island, has been within the region of increased maritime influence for at least the last several decades.

Ecological response to this environmental change can be manifest, or not, dependent upon circumstances. For example, key species sensitive to changes in sea-ice can amplify this change by causing a cascading response within the entire ecosystem (*Smith et al.*, 2003). One notable amplification mechanism occurs because the temperature threshold for an ice-to-water phase change may create a pronounced non-linear ecosystem response

to what is a relatively small temperature shift (*Welch et al.*, 2003). From the perspective of a sea ice-dominated ecosystem, the later advance and earlier retreat of sea ice in the 1990s (as compared with the 1980s) translates into a shorter sea ice season (the average period for sea ice coverage at Palmer Station) by roughly two weeks. A shift to a more oceanic marine ecosystem reduces the seasonality and geographical extent of the sea ice habitat and consequently will influence all trophic levels of the WAP sea ice dominated marine ecosystem.

The Palmer LTER is accumulating evidence that links the timing and magnitude of sea ice advance and retreat to the seasonal progression and life history patterns of phytoplankton, krill and sea birds (*Fraser et al.*, 1992; *Quetin et al.*, 1996; *Fraser and Trivelpiece*, 1996; *Smith et al.*, 1998, 1999, 2002; *Fraser and Hofmann*, 2003; *Quetin and Ross*, 2001; *Vernet et al.*, personal communication) as well as key biogeochemical processes (*Karl et al.*, 1996; *Carrillo and Karl*, 1999). Marine ecosystems are tightly coupled to physical forcing so we can anticipate that, as the temperature and sea ice trends described above continue, there will be changes in the abundance and distribution of key species in response to climate variability. In the following, we focus on indicators at the base (primary productivity) and at the top (seabird abundance) of the food web, respectively.

Within the context of climate variability, ecological response, and the interpretation of paleoclimatic records, it is important to recognize that the interannual variability in primary productivity in the WAP area is relatively large. Figure 4 (updated from *Smith et al.*, 2001) shows the annual primary production (gC m^{-2} yr^{-1}) determined near Palmer Station for each growing season from 1991/92 to 2001/02 using three different methods as described in the figure legend. The comparison shows that there is good agreement between the three methods (making use of surface and satellite observations as well as results from modeled primary productivity). There is also wide spatial variability as can be seen in Plate 1 which shows the Chl-a concentration (mg chl-a m^{-3}) averaged within the euphotic zone of the Palmer LTER grid for each January cruise from 1993 to 2002. The surface observations shown in Plate 1, and corresponding ocean color satellite (*Smith et al.*, 2001) data (not shown), illustrate both the seasonal and interannual variability as well as the spatial variability of phytoplankton production in the WAP region. Mechanisms underlying this observed variability within the WAP region have been previously discussed (*Smith et al.*, 1996, 2001) and include factors that can control growth rates (temperature, light, and nutrients) and/or by those that control the

Annual Primary Production at Palmer

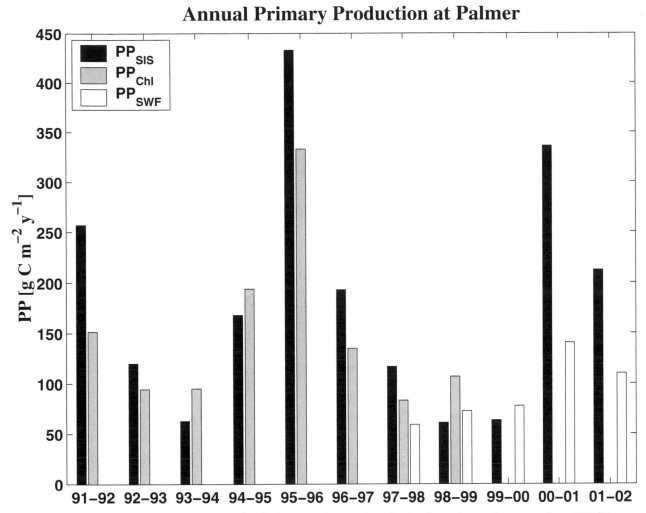

Fig. 4. Annual primary production (gC m^{-2} yr^{-1}) determined near Palmer Station for each growing season from 1991/92 to 2000/02. First, annual estimates have been made from integrated near-weekly surface sampling over the growing season from November to March (152 days) at Palmer Station (PP$_{SIS}$). Second, annual production estimated based on Palmer Station chl-a measurements using our production model and integrated over the same growing season (PP$_{CHL}$). Third, estimates from modeled primary production based on average monthly chl-a retrieved from SeaWiFS ocean color data (for the years since the satellite has been in orbit) (PP$_{SWF}$).

accumulation rate of cells in the euphotic zone and hence population growth (grazing, water column stability, and sinking). This high variability at the base of the food chain influences organisms at all trophic levels.

Our observations show that this high variability is characteristic of several biogeochemical provinces (coastal shelf zone, seasonal sea ice zone, permanently open ocean zone, polar front zone) within the larger WAP region (*Treguer and Jacques*, 1992; *Smith et al.,* 1998). Given this high variability, detecting trends in primary productivity is problematic and emphasizes the need for long-term records to properly resolve mechanisms that may permit predictive estimates of this vari-

ability. In spite of this variability, results from nearly a decade of both in situ and satellite observations show links between sea ice variability and primary production as well as some persistent spatial patterns (*Smith et al.,* 1998, 2001; *Dierssen et al.,* 2000). The challenge now is to understand how the combined influence of a long-term warming trend and ENSO-related variability is modifying the WAP ecosystem.

While a positive correlation between primary production and sea ice has been observed and is consistent with the Palmer LTER's original hypothesis, we find that the seasonal timing of sea ice advance and retreat and other factors, including the recently observed increase in gla-

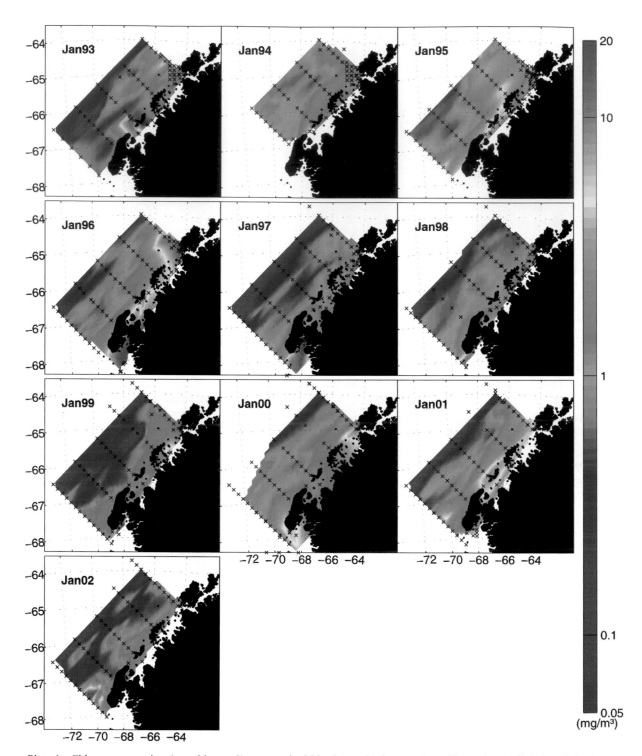

Plate 1. Chl-a concentration (mg chl-a m-3) averaged within the euphotic zone (roughly to the 1% light level) in the Palmer LTER grid for each January cruise from 1993 to 2002.

cial melt water input, are also important. Through several lines of evidence collected in conjunction with the Palmer LTER, *Dierssen et al.* (2002) have postulated that the freshening and warming of coastal surface waters over summer months are influenced not solely by sea ice melt, as suggested by the literature, but also by an influx of glacial melt water. Dierssen and co-workers suggest that glacial melt water in turn can significantly influence both the marine ecosystem and sea ice dynamics. They find that a stabilizing surface lens (top 50 meters of the water column) of low salinity melt water persists in recent years well past the late spring period that is influenced by sea ice-melt. This low salinity surface lens can extend up to 100 km from the coast and is associated with higher levels of primary productivity.

Currently the influence of glacial melt water on the WAP ecosystem is being subjected to more rigorous analysis. However, this example demonstrates the conflicting influences of a warming trend on primary production. The earlier departure of sea-ice would mean an earlier occurrence of a stabilizing fresh water lens (caused by sea-ice melt) that by late spring could be destroyed by wind-mixing caused by storms. However, the earlier departure of sea-ice also is associated with warmer summers and increased glacial-melt run-off, providing another fresh water source that could maintain or reestablish a stabilizing fresh water lens. Dierssen and co-workers also note that glacial melt-water may provide a replenishment of macro (and micro) nutrients in the later summer and early fall when the nutrient pool may be diminished from earlier blooms. Finally, a shallow fresh water lens will freeze more easily the following fall, thus possibly influencing the timing and location of sea ice advance in the WAP region. Understanding the mechanisms behind these conflicting influences is an important goal toward understanding the high temporal and spatial variability of primary productivity and the overall ecological response to the observed climate variability.

The factors controlling phytoplankton composition in the waters west of the Antarctic Peninsula, in addition to phytoplankton primary production, are related to water column stability, mainly driven by melt water, and expected to be affected by decreases in sea ice and/or increases in glacier melt water (*Dierssen et al.,* 2002). These changes, in turn, are expected to affect abundance and distribution of the main grazer, *Euphausia superba*. Phytoplankton assemblages in the WAP have varying proportions of diatoms, cryptomonads, prymnesiophytes, prasinophytes, chrysophytes, dinoflagellates and unknown flagellates (< 5 micrometers in diameter)

(*Garibotti et al.,* 2003). *Moline et al.* (2000) have proposed that glacier melt water is instrumental in promoting cryptomonad growth in the area near Anvers Island while *Walsh et al.* (2001) proposed krill grazing as the mechanism which shifts the system from diatom- to cryptomonad-dominated communities, a conclusion supported by *Garibotti et al.* (personal communication) based on krill grazing on phytoplankton total biomass. Furthermore, *Prezelin et al.* (2000) have proposed that upwelling promotes diatom growth, and identify the area around Anvers Island and in particular the area offshore of Renaud Island as regions where upwelling should sustain diatom growth. These authors suggested that the lower than expected diatom concentration in the water column during sampling was due to the time of sampling being too early in development of the seasonal phytoplankton succession. An alternative hypothesis would be that krill grazing in this region of high krill biomass control bloom dynamics and keeps the standing stock of diatoms low, as seen in other areas such as the North Pacific. However, the ten-fold interannual variability in krill biomass, in part due to episodic recruitment (*Quetin and Ross* 2002), means that krill are not always in high enough abundance to affect either bloom dynamics (*Ross et al.* 1998) or phytoplankton composition. If the various phytoplankton assemblies vary in space and time in response to climatic change, krill distributions may shift to match regions of optimal foraging. Alternatively, if krill remain in specific regions, and the phytoplankton assemblies shifts to one of lower than optimum food quality, the overall productivity of the krill population may decrease (*R. Ross and L. Quetin*, personal communication).

Garibotti and co-workers (personal communication) have also documented considerable variability in phytoplankton community composition, cell abundance and biomass alongshore in the inner shelf regions of the WAP. This variability is attributed not only to the selective removal of diatoms by krill but also appears to display different stages of seasonal succession as a consequence of the north to south retreat of sea ice. Thus, for phytoplankton within the WAP region, we can expect climate variability to influence changes not only in primary production but also on phytoplankton composition due to direct effects of ice and water-column physics or as consequence of direct and indirect effects of grazer's and their distribution and efficiency in controlling phytoplankton. Processes controlling the spatial and temporal patterns and variability of phytoplankton abundance (Plate 1) within the WAP region continue to be an active area of research by the Palmer LTER.

The ecosystem modifications most clearly manifest in the context of the WAP warming trend are the changes in upper level predator populations (*Fraser et al.,* 1992; *Fraser and Patterson,* 1997; *Smith et al.,* 1999, 2001; *Fraser and Hofmann,* 2003). Figure 5 shows the changes in Adelie and chinstrap penguin populations near Palmer Station during the past two decades and in gentoo penguins populations since founder colonies became established in the area during the early 1990s. Fraser and co-workers (*Fraser et al.,* 1992) hypothesized that a decrease in the number of cold years with heavy winter sea ice due to climate warming produced habitat conditions more suitable for the ice-intolerant (chinstrap and gentoo) penguins, as opposed to the ice-dependent (Adelie) penguins. The trends shown in Figures 2, 3 and 5 are consistent with this hypothesis.

Furthermore, the causal mechanisms suggested by this hypothesis have been implicated as key factors affecting penguin demography over a range of spatial and temporal scales in both paleoecological and demographic studies (*Taylor et al.,* 1990; *Baroni and Ormbelli,* 1991, 1994; *Denton et al.,* 1999; *Emslie,* 1995; *Emslie et al.,* 1998; *Fraser and Patterson,* 1997; *Smith et al.,* 1999, 2002; *Emslie,* this volume). Indeed, Emslie and co-workers (1998) have shown that the presence of chinstrap and gentoo penguins in the Palmer Station area is unprecedented in the 600-year fossil record, which is entirely dominated by Adelie penguin remains. This pattern stands in sharp contrast to trends evident 250 km north of the Palmer area where the relative dominance of Adelie and chinstrap penguins has changed cyclically in response to multi-century cooling and warming periods (*Emslie,* 1995, this volume). That chinstrap and gentoo penguins have invaded the Palmer Station region would seem to affirm the unusual nature of this 20th century WAP warming event.

Indicator species are invaluable for assessing ecological change. The founder colonies of chinstraps and gentoo species have increased dramatically while conversely Adelie penguins have decreased substantially in roughly just 25 years (Figure 5). Penguins integrate the influence of climate change across relatively wide spatial and temporal ranges, and the shifts in abundance and distribution of these species are consistent with expected habitat shifts based on the observed physical changes outlined above. In addition, Fraser and Hofmann (*Fraser and Hofmann,* 2003), who analyzed changes in the diets of Adelie penguins that span nearly a 30 year record, show there is a direct, causal relationship between variability in sea-ice cover, krill recruitment, krill abundance and predator foraging ecology. They suggest that the

variability inferred from the penguin diets might be related to changes in physical forcing associated with the Antarctic Circumpolar Wave (ACW) (*White and Peterson,* 1996). The periodicity of the ACW has been attributed to an ENSO-related teleconnection (e.g., *Gloersen and White,* 2001) and ENSO-related teleconnections with the WAP area have been suggested by others as discussed above.

4. SUMMARY

There is increasing evidence of the ecological impacts on natural systems of recent climate change (*Walther et al.,* 2002; *Parmesan and Yohe,* 2003; *Root et al.,* 2003). This evidence shows climate change is influencing a broad range of organisms in environments from the tropics to the poles. In the Climate Change 2001 report *McCarthy et al.* (2001) discuss the implications of climate change and describe two contrasting paradigms for the way ecosystems will respond to global change: ecosystem movement and ecosystem modification. The former is a gross simplification that assumes ecosystems will migrate relatively intact to a new location that is a closer analogue to their current climate and environment. Basic ecological knowledge suggests that this paradigm, while valuable for testing simple hypotheses, is unlikely to actually occur. The modification paradigm (*Kennedy,* 2002) assumes that, as climate and other environmental factors change with a consequent shifting of the coevolved synchrony of the food web, there will be a change in the abundance, distribution and dominance of key species. As the length of the sea ice season shortens (Figure 3) the seasonal timing of this sea ice dominated marine ecosystem can be expected to shift with corresponding changes in ecologically important events and life histories of key species (*Smith et al.,* 1995, Figure 4). Our results show that trends in penguin populations are in accord with climate change predictions.

The above data show that Adelies have declined in abundance in the Palmer Station area and suggest that the locus of their distributions will be forced further south along the WAP, while chinstrap and gentoo penguins emerge as the dominant top predators. The fossil record already supports such a scenario at more northern sites along the WAP, including evidence that squid and fish replaced krill as the dominant component in penguin diets as the climate warmed (*Emslie,* 1995; *Emslie et al.,* 1998, this volume). The Palmer LTER PI's have argued (*Ducklow et al.* personal communication) that these observations for upper level predators presage evidence

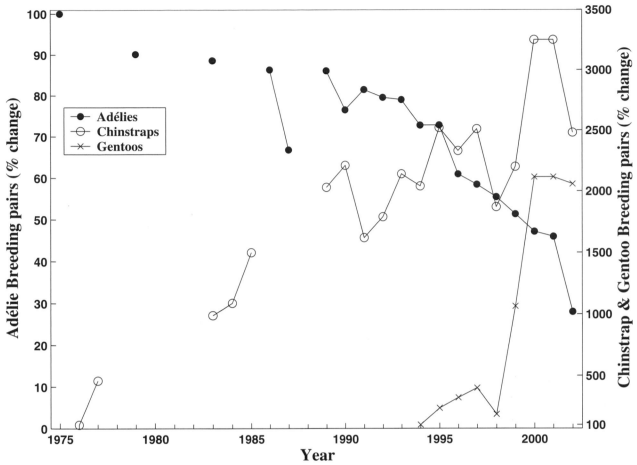

Fig. 5. Twenty six year trends in Adelie and chinstrap penguin populations at Arthur Harbor (Palmer Station) and for gentoo penguins since founder colonies became established in the early 1990's. Solid dots, Adelie penguins normalized to 100% in 1975 when the record began. The Adelie population of breeding pairs was 15,202 in 1975 and has declined to 7,161 in 2000. Chinstrap (open circles) and gentoo (plus signs) penguins are normalized to 100% in 1976 and 1994, respectively, when the founding colonies began. Chinstrap population of breeding pairs in 1976 was 10 and this founding population increased to 325 breeding pairs in 2000 while the gentoo population was 14 breeding pairs in 1994 and increased to 296 breeding pairs in 2000.

for ecosystem migration at lower trophic levels. A critical issue here is that the timing of change at different trophic and/or taxonomic groups may not be synchronous thus giving rise to profound ecological consequences. Future Palmer LTER studies are aimed at understanding these ecological consequences within the context of a changing environment.

The space/time variability in pigment biomass and primary production, discussed above, show that longer term observations will be needed to determine statistically significant trends. On the other hand, the interannual and ENSO-related variability in physical forcing and sea ice coverage provide an opportunity to "conduct" natural experiments by monitoring parameters and processes

during and after seasons/years of different physical forcing and sea ice coverage. A challenge is to understand how the combined influence of a long-term warming trend and ENSO-related variability is modifying the WAP ecosystem. Future Palmer LTER studies include studies to understand the mechanisms underlying these trends and the consequent spatial and temporal patterns in phytoplankton productivity. Climate variability along the Antarctic Peninsula offers a unique opportunity to study these changes within this Antarctic marine ecosystem.

Acknowledgments. This work was supported by National Science Foundation Grant OPP96-32763. This is Palmer LTER Contribution No. 232.

REFERENCES

Baroni, C., and G. Orombelli, Holocene raised beaches at Terra Nova Bay, Victoria Land, Antarctica, *Quaternary Research*, *36*, 157-177, 1991.

Baroni, C., and G. Orombelli, Abandoned penguin rookeries as Holocene paleoclimatic indicators in Antarctica, *Geology*, *22*, 23-26, 1994.

Callahan, J.T., Long-term ecological research, *BioScience*, *34* (6), 363-367, 1984.

Carleton, A.M., Sea ice-atmosphere signal of the Southern Oscillation in the Weddell Sea, Antarctica, *Journal of Climate*, *1* (2), 379-388, 1988.

Carrillo, C.C. and D.M. Karl, Dissolved inorganic carbon pool dynamics in northern Gerlache Strait, Antarctica, *Journal of Geophysical Research*, 104 (7), 15,873-15,884, 1999.

Dai, J.C., L.G. Thompson, and E. Mosley-Thompson, A 485-year record of atmospheric chloride, nitrate and sulfate: results of chemical analysis of ice cores from Dyer Plateau, Antarctic Peninsula, *Annals of Glaciology*, *21*, 182-188, 1995.

Denton, G.H., J.G. Bockheim, S.C. Wilson, and M. Stuiver, Late Wisconsin and early Holocene glacial history, inner Ross embayment, Antarctica, in *West Antarctic Ice Sheet Initiative*, edited by R.A. Bindschadler, pp. 55-86, NASA, Washington, DC, 1991.

Dierssen, H.M., M. Vernet, and R.C. Smith, Optimizing models for remotely estimating primary production in Antarctic coastal waters, *Antarctic Science*, *12* (1), 20-32, 2000.

Dierssen, H.M., R.C. Smith, and M. Vernet, Glacial meltwater dynamics in coastal waters West of the Antarctic Peninsula, *Proceedings of the National Academy of Science*, *99* (4), 1790-1795, 2002.

Domack, E.W., and C.E. McClennen, Accumulation of glacial marine sediments in fjords of the Antarctic Peninsula and their use as late Holocene paleoenvironmental indicators, in *Foundations for Ecological Research West of the Antarctic Peninsula*, edited by R.M. Ross, E.E. Hofmann, and L.B. Quetin, pp. 135-154, American Geophysical Union, Washington (DC), 1996.

Domack, E.W., T.A. Mashiotta, L.A. Burkley, and S.E. Ishman, 300-year cyclicity in organic matter preservation in Antarctic fjord sediments, in *The Antarctic Paleoenvironment: A perspective on Global Change (Part 2)*, edited by J.P. Kennett, and D.A. Warnke, pp. 265-272, American Geophysical Union, Washington (DC), 1993.

Emslie, S.D., Age and taphonomy of abandoned penguin rookeries in the Antarctic peninsula, *Polar Record*, *31* (179), 409-418, 1995.

Emslie, S.D., W.R. Fraser, R.C. Smith, and W.O. Walker, Abandoned penguin colonies and environmental change in the Palmer Station region, Anvers Island, Antarctic Peninsula, *Antarctic Science*, *10* (3), 255-266, 1998.

Franklin, J.F., C.S. Bledsoe, and J.T. Callahan, Contributions of the long-term ecological research program - an expanded network of scientists, sites, and programs can provide crucial comparative analyses, *BioScience*, *40* (7), 509-523, 1990.

Fraser, W.R., and E.E. Hofmann, Krill-sea ice interactions, part I: a predator's perspective on causal links between climate change, physical forcing and ecosystem response (accepted being revised), *Marine Ecology Progress Series*, 2003 (in press).

Fraser, W.R., and D.L. Patterson, Human disturbance and long-term changes in Adelie penguin populations: a natural experiment at Palmer Station, Antarctic Peninsula, in *Antarctic Communities: Species, Structure and Survival, Scientific Committee for Antarctic Research (SCAR), Sixth Biological Symposium*, edited by B. Battaglia, J. Valencia, and D.W.H. Walton, pp. 445-452, Cambridge University Press, New York, NY, 1997.

Fraser, W.R., and W.Z. Trivelpiece, Factors controlling the distribution of seabirds: winter-summer heterogeneity in the distribution of Adelie penguin populations, in *Foundations for Ecological Research West of the Antarctic Peninsula*, edited by R.M. Ross, E.E. Hofmann, and L.B. Quetin, pp. 257-272, American Geophysical Union, Washington, DC, 1996.

Fraser, W.R., W.Z. Trivelpiece, D.G. Ainley, and S.G. Trivelpiece, Increases in Antarctic penguin populations: reduced competition with whales or a loss of sea ice due to environmental warming?, *Polar Biology*, *11*, 525-531, 1992.

Garibotti, I.A., M. Vernet, W. Kozlowski, and M.E. Ferrario, Composition and biomass of phytoplankton assemblages in coastal Antarctic waters: a comparison of chemotaxonomy and microscopic analysis (In Press), in *Marine Ecology Progress Series*, 2002 (in press).

Glorsen, P. and W.B. White, Reestablishing the circumpolar wave in sea ice around Antarctica from one winter to the next, *Journal of Geophysical Research*, 106 (3), 4391-4395, 2001.

Karl, D.M., J.R. Christian, J.E. Dore, and R.M. Letelier, Microbiological oceanography in the region west of the Antarctic Peninsula: Microbial dynamics, nitrogen cycle and carbon flux, in *Foundations for Ecological Research West of the Antarctic Peninsula*, edited by R.M. Ross, E.E. Hofmann, and L.B. Quetin, pp. 303-332, American Geophysical Union, Washington, DC, 1996.

Kennedy, C., POTUS and the Fish, *Science*, *297* (5581), 477-477, 2002.

King, J.C., Recent climate variability in the vicinity of the Antarctic Peninsula, *International Journal of Climatology*, *14* (4), 357-369, 1994.

King, J.C., and S.A. Harangozo, Climate change in the western Antarctic Peninsula since 1945: observations and possible causes, *Annals of Glaciology*, *27*, 571-575, 1998.

Kwok, R., and J.C. Comiso, Southern ocean climate and sea ice anomalies associated with the southern oscillation, *Journal of Climate*, *15*, 487-501, 2002.

Leventer, A., E.W. Domack, S.E. Ishman, S. Brachfeld, C.E. McClennen, and P. Manley, Productivity cycles of 200-300 years in the Antarctic Peninsula region: understanding linkages among the sun, atmosphere, oceans, sea ice, and biota, *Geological Society of America Bulletin*, *108* (12), 1626-1644, 1996.

Liu, J., D. Rind, D.G. Martinson, and X. Yuan, Mechanistic study of the ENSO and southern high latitudes climate teleconnection, *Journal of Geophysical Research*, 2003 (in press).

Marshall, G.J., and J.C. King, Southern Hemisphere circulation anomalies associated with extreme Antarctic Peninsula winter temperatures, *Geophysical Research Letters*, *25* (13), 2437-2440, 1998.

McCarthy, J.J., O.F. Canziani, N.A. Leary, D.J. Dokken, and K.S. White, eds. *Climate Change 2001, Impacts, Adaptation, and Vulnerability*, 1032 pp., Intergovernmental Panel on Climate Change, Cambridge University Press, 2001.

Meehl, G.A., A reexamination of the mechanism of the semi-annual oscillation in the Southern Hemisphere, *Journal of Climate*, *4* (9), 911-926, 1991.

Mo, K.C., and G.H. White, Teleconnections in the Southern Hemisphere, *Monthly Weather Review*, *113* (1), 22-37, 1985.

Moline, M.A., H. Claustre, T.K. Frazer, J. Grzymski, O.M. Schofield, and M. Vernet, Changes in phytoplankton assemblages along the Antarctic Peninsula and potential implications for the Antarctic food web, in *Antarctic Ecosystems: Models for Wider Ecological Understanding*, edited by W. Davidson, C. Howard-Williams, and P. Broady, pp. 263-271, 2000.

Mosley-Thompson, E., Paleoenvironmental conditions in Antarctica since AD 1500: ice core evidence, in *Climate since A.D. 1500*, edited by R.S. Bradley, and P.D. Jones, pp. 572-591, Routledge, London, 1992.

Parmesan, C. and G. Yohe, A globally coherent fingerprint of climate change impacts across natural systems, *Nature*, 421, pp. 37-42, 2003.

Peel, D.A., Ice core evidence from the Antarctic Peninsula region, in *Climate since A.D. 1500*, edited by R.S. Bradley, and P.D. Jones, pp. 549-571, Routledge, London, 1992.

Prezelin, B.B., E.E. Hofmann, C. Mengelt, and J.M. Klinck, The linkage between upper circumpolar deep water (UCDW) and phytoplankton assemblages on the west Antarctic Peninsula Continental Shelf, *Journal of Marine Research*, *58* (2), 165-202, 2000.

Quetin, L.B., and R.M. Ross, Environmental variability and its impact on the reproductive cycle of Antarctic Krill, *American Zoologist*, *41* (1), 74-89, 2001.

Quetin, L.B., and R.M. Ross, Episodic recruitment in Antarctic krill, Euphausia superba, in the Palmer LTER study region (submitted), *Marine Ecology Progress Series*, 2002.

Quetin, L.B., R.M. Ross, T.K. Frazer, and K.L. Haberman, Factors affecting distribution and abundance of zooplankton, with an emphasis on Antarctic krill, Euphausia superba, in *Foundations for Ecological Research West of the Antarctic Peninsula*, edited by R.M. Ross, E.E. Hofmann, and L.B. Quetin, pp. 357-371, American Geophysical Union, Washington, DC, 1996.

Root, T. L., J.T. Price, K.R. Hall, S.H. Schneider, C. Rosenzweig and J. A. Pounds, Fingerprints of global warming on wild animals and plants, Nature, 421, pp. 57-60, 2003.

Ross, R.M., E.E. Hofmann, and L.B. Quetin, eds. *Foundations for Ecological Research West of the Antarctic Peninsula*, 448 pp., American Geophysical Union, Washington, DC, 1996.

Ross, R.M., L.B. Quetin, and K.L. Haberman, Interannual and seasonal variability in short-term grazing impact of Euphausia superba in nearshore and offshore waters west of the Antarctic Peninsula, *Journal of Marine Systems*, *17* (1-4), 261-273, 1998.

Sansom, J., Antarctic surface temperature time series, *Journal of Climate*, *2* (10), 1164-1172, 1989.

Schwerdtfeger, W., *Weather and Climate of the Antarctic*, 261 pp., Elsevier, New York, 1984.

Smith, R.C., and S.E. Stammerjohn, Variations of surface air temperature and sea ice extent in the Western Antarctic Peninsula (WAP) region, *Annals of Glaciology*, *33*, 493-500, 2001.

Smith, R.C., K.S. Baker, W.R. Fraser, E.E. Hofmann, D.M. Karl, J.M. Klinck, L.B. Quetin, B.B. Prezelin, R.M. Ross, W.Z. Trivelpiece, and M. Vernet, The Palmer LTER: A long-term ecological research program at Palmer Station, Antarctica, *Oceanography*, *8* (3), 77-86, 1995.

Smith, R.C., S.E. Stammerjohn, and K.S. Baker, Surface air temperature variations in the western Antarctic peninsula region, in *Foundations for Ecological Research West of the Antarctic Peninsula*, edited by R.M. Ross, E.E. Hofmann, and L.B. Quetin, pp. 105-121, American Geophysical Union, Washington, DC, 1996.

Smith, R.C., K.S. Baker, and M. Vernet, Seasonal and interannual variability of phytoplankton biomass west of the Antarctic Peninsula, *Journal of Marine Systems*, *17* (1-4), 229-243, 1998.

Smith, R.C., E.W. Domack, S.D. Emslie, W.R. Fraser, D.G. Ainley, K.S. Baker, J. Kennett, A. Leventer, E. Mosley-Thompson, S.E. Stammerjohn, and M. Vernet, Marine Ecosystems sensitivity to historical climate change: Antarctic Peninsula, *BioScience*, *49* (5), 393-404, 1999.

Smith, R.C., K.S. Baker, H.M. Dierssen, S.E. Stammerjohn, and M. Vernet, Variability of primary production in an Antarctic marine ecosystem as estimated using a multi-scale sampling strategy, *American Zoologist*, *41* (1), 40-56, 2001.

Smith, R.C., W.R. Fraser, and S.E. Stammerjohn, Climate variability and ecological response of the marine ecosystem in the western Antarctic Peninsula (WAP) region (In Press), in *Climate variability and ecosystem response at Long-Term Ecological Research (LTER) Sites*, edited by D. Greenland, D. Goodin, and R.C. Smith, Oxford Press, New York, 2003 (in press).

Stammerjohn, S.E., and R.C. Smith, Spatial and temporal variability of western Antarctic peninsula sea ice coverage, in *Foundations for Ecological Research West of the Antarctic Peninsula*, edited by R.M. Ross, E.E. Hofmann, and L.B. Quetin, pp. 81-104, American Geophysical Union, Washington, DC, 1996.

Stammerjohn, S.E. and R.C. Smith, Opposing Southern Ocean climate patterns as revealed by trends in se ice coverage, *Climatic Change*, 27 (4), 617-639, 1997.

Stammerjohn, S.E., M.R. Drinkwater, R.C. Smith, and X. Liu, Ice-atmosphere interactions during sea-ice advance and retreat in the western Antarctic Peninsula region, *Journal of Geophysical Research*, 2003 (in press).

Stark, P., Climatic warming in the central Antarctic Peninsula area, *Weather*, 49 (6), 215-220, 1994.

Taylor, R.H., P.R. Wilson, and B.W. Thomas, Status and trends of Adélie penguin populations in the Ross sea region, *Polar Record*, 26 (159), 293-304, 1990.

Thompson, D.W.J., and S. Solomon, Interpretation of recent southern hemisphere climate change, *Science*, *296*, 895-899, 2002.

Thompson, L.G., D.A. Peel, E. Mosley-Thompson, R. Mulvaney, J. Dai, P.N. Lin, M.E. Davis, and C.F. Raymond, Climate since AD 1510 on Dyer Plateau, Antarctic Peninsula: evidence for recent climate change, *Annals of Glaciology*, *20*, 420-426, 1994.

Treguer, P., and G. Jacques, Dynamics of nutrients and phytoplankton, and fluxes of carbon, nitrogen and silicon in the Antarctic Ocean, *Polar Biology*, *12*, 149-162, 1992.

van den Broeke, M.R., The semi-annual oscillation and Antarctic climate, part 1: influence on near surface temperatures (1957-79), *Antarctic Science*, *10* (2), 175-183, 1998a.

van den Broeke, M.R., The semiannual oscillation and Antarctic climate, part 2: recent changes, *Antarctic Science*, *10* (2), 184-191, 1998b.

van Loon, H., The half-yearly oscillations in middle and high southern latitudes and the coreless winter, *Journal of the Atmospheric Sciences*, *24* (5), 472-486, 1967.

van Loon, H., and D.J. Shea, The Southern Oscillation. Part IV: the precursors south of 15°S to the extremes of the oscillation, *Monthly Weather Review*, *113* (12), 2063-2074, 1985.

van Loon, H., and D.J. Shea, The Southern Oscillation. Part VI: anomalies of sea level pressure on the Southern Hemisphere and of Pacific sea surface temperature during the development of a warm event, *Monthly Weather Review*, *115* (2), 370-379, 1987.

Walsh, J.J., D.A. Dieterle, and J. Lenes, A numerical analysis of carbon dynamics of the Southern Ocean phytoplankton community: the roles of light and grazing in effecting both sequestration of atmospheeric CO2 and food availability to krill, in *Deep-Sea Research I*, pp. 1-48, 2001.

Walther, G., E. Post, P. Convey, A. Menzel, C. Parmesan, T.J.C. Beebee, J. Fromenthin, O. Hoegy-Guldberg and F. Bairlein, Ecological responses to recent climate change, *Nature*, 146, pp. 389-395, 2002.

Weatherly, J.W., J.E. Walsh, and H.J. Zwally, Antarctic sea ice variations and seasonal air temperature relationships, *Journal of Geophysical Research*, *96* (C8), 15,119-15,130, 1991.

Welch, K.A., W.B. Lyons, D.M. McKnight, P.T. Doran, A. Fountain, D. Wall, C. Jaros, T. Nylen, and C. Howard-Williams, Climate and hydrologic variations and implications for lake and stream ecological response in the McMurdo Dry Valleys, Antarctica, in *Climate variability and ecosystem response at Long-Term Ecological Research (LTER) Sites*, edited by D. Greenland, D. Goodin, and R.C. Smith, Oxford Press, New York, 2003 (in press).

White, W.B., and R.G. Peterson, An Antarctic circumpolar wave in surface pressure, wind, temperature and sea-ice extent, *Nature*, *380* (6576), 699-702, 1996.

White, W.B., S.-C. Chen, and R.G. Peterson, The Antarctic Circumpolar Wave: a beta effect in ocean-atmosphere coupling over the Southern Ocean, *Journal of Physical Oceanography*, *28* (12), 2345-2361, 1998.

Yuan, X., and D.G. Martinson, Antarctic sea ice extent variability and its global connectivity, *Journal of Climate*, *13* (10), 1697-1717, 2000.

Yuan, X., and D.G. Martinson, The Antarctic dipole and its predictability, *Geophysical Research Letters*, *28* (18), 3609-3612, 2001.

William R. Fraser, Polar Oceans Research Group, P.O. Box 368, Sheridan, MT 59749

Raymond C. Smith, ICESS & Dept. Geography, University of California, Santa Barbara, Santa Barbara, CA 93106

Sharon E. Stammerjohn, Lamont-Doherty Earth Observatory of Columbia University, Palisades, NY 10964

Maria Vernet, Scripps Institution of Oceanography, 2123 Sverdrup Hall, University of California, San Diego, CA 92093-0218

MARITIME ANTARCTIC CLIMATE CHANGE: SIGNALS FROM TERRESTRIAL BIOLOGY

Peter Convey

British Antarctic Survey, Cambridge, United Kingdom

The simplicity of maritime Antarctic terrestrial ecosystems, combined with rapid changes in several environmental variables, creates a natural laboratory probably unparalleled worldwide in which to study biological consequences of climate change. The Antarctic Peninsula and Scotia Arc provide a gradient from oceanic cool temperate to frigid continental desert conditions, giving a natural model of climate change predictions. Biota are limited by the twin environmental factors of low temperature and lack of water, while also facing changes in the timing of UV-B maxima, associated with the spring ozone hole. Biological changes consistent with predictions from climate amelioration are visible in the form of expansions in range and local population numbers amongst elements of the flora. Field manipulations demonstrate (i) potential for massive species and community responses to climate amelioration, (ii) the importance of existing soil propagule banks, and (iii) biochemical responses to changing radiation environments. Antarctic species possess considerable resistance/resilience and response flexibility to a range of environmental stresses. Wide environmental variability in Antarctic terrestrial habitats also means that predicted levels of change often fall well within the range already experienced. Thus, climate amelioration may generate positive responses from resident biota, at least while they remain protected through isolation from colonization by more effective competitors. Responses are likely to be subtle and multifactorial in origin, arising from changes in resource allocation and energy economics. The integration of subtle responses may lead to greater consequential impacts in communities and ecosystems.

INTRODUCTION

The Antarctic Terrestrial Environment

With a continental area greater than that of Australia or Western Europe, it is unsurprising that the terrestrial biota of Antarctica cannot be considered as a single entity. Three biogeographical zones are recognized in the Antarctic biological literature [*Smith, 1984; Longton, 1988*], defined on the basis of biological and climatological differences. These zones are generally referred to as the sub-, maritime and continental Antarctic. In detail, these zones are not equivalent to the geological separation of the continent into East and West Antarctica. The maritime Antarctic, which includes the west coast of the Antarctic Peninsula southwards to Alexander Island, and the Scotia Arc South Shetland, South Orkney and South Sandwich archipelagos, forms the focus of this paper. The east coast of the Antarctic Peninsula is included in the continental Antarctic biogeographic zone, but no biological information is available from the region other than a few opportunistically collected plant specimens.

The maritime Antarctic climate is strongly seasonal. The maritime influence is considerable, especially in the

10.1029/079ARS12

summer months, but is reduced during winter depending on the extent of seasonal sea ice formation. Mean air temperatures are negative for most of the year, only reaching 0–2°C for 1–4 months in summer. The fauna and flora are very restricted [*Convey, 2001a*]. Only two higher insects (chironomid midges (Diptera)) are present, one of which is a South American species with an Antarctic distribution limited to the South Shetland Islands. The terrestrial fauna is, otherwise, restricted to low diversity communities of soil arthropods (Acari and Collembola) and smaller meiofaunal invertebrates (Nematoda, Tardigrada, Rotifera). Similarly, higher plant diversity is very low, with only two species and, instead, a cryptogamic vegetation of bryophytes and lichens is dominant. Marine vertebrates may be locally abundant, using accessible terrestrial areas for breeding colonies and moult sites. These provide both considerable nutrient input and habitat disturbance at local scales.

Both plant and animal communities typically have much lower species richness and fewer trophic links and interactions than those of temperate, tropical and even Arctic latitudes. However, population densities of several Antarctic taxa (including mites, springtails, bryophytes and grass) may be comparable with those of lower latitude relatives [*Convey, 1996a*]. The same is true of some physiological variables when the measurement scale is appropriate to the length of an organism's active period or season, rather than considered on an annual timescale. The most extreme Antarctic environments include communities representing the first stages of colonization or succession, limited purely by the extreme physical conditions. These simple communities, and their analogues in the Arctic, are likely to be very sensitive to changes in climatic variables [*Callaghan and Jonasson, 1995; Freckman and Virginia, 1997*]. Some species and communities occur across exceptionally wide latitudinal and environmental gradients along the Antarctic Peninsula and Scotia Arc, giving a natural experiment modeling some climate change predictions. This gradient can be used to examine species and community biology across a wide range of natural conditions [*cf. Addo-Bediako et al., 2000; Chown and Clarke, 2000*].

The extreme geographical isolation of terrestrial habitats in the Antarctic provides a final advantage in studies of the consequences of climate change. The combination of physical isolation and recent emergence of terrestrial habitats from retreating ice is one important cause of low biodiversity, while increased colonization is itself a predicted outcome of climate amelioration. However, the geographical isolation will continue to restrict long-distance colonization, which may allow the direct consequences of change within existing communities to be separated from those relating to increased diversity through colonization.

Contemporary Climate Change

Three aspects of climate form the focus of studies of the biology of Antarctic terrestrial organisms in the context of climate change: temperature, water availability and radiation climate. A fourth factor, atmospheric CO_2 concentration, is clearly also of direct biological relevance, but has yet to be studied in the context of Antarctic biology and will not be considered further here [see *Oechel et al., 1997* and *Norby et al., 2001* for Arctic and wider commentaries].

Low thermal energy input is a key defining characteristic of Antarctic terrestrial habitats. Even so, several different features of temperature variation can be significant, including upper and lower thermal limits of organisms, diurnal and annual temperature ranges and means, and rates of change. The very low mean summer temperatures experienced by Antarctic Peninsula terrestrial habitats are particularly significant in the context of climate change, as they are near minimum threshold temperatures for many biological functions - a small temperature increment will have a considerably greater biological impact than one of the same magnitude in a less extreme environment.

Rapid trends of temperature increase are well documented from several stations along the Antarctic Peninsula and islands of the Scotia Arc [*Smith, 1990; Fowbert and Smith, 1994; King and Harangozo, 1998; Skvarca et al., 1998*; other papers in this volume], several of which have recorded increases in annual air temperatures of more than 1°C over the last 30–50 years. As described elsewhere in this volume, temperature trends in parts of the maritime Antarctic consist of a strong warming trend in the winter months, which is much reduced during summer [see also *King, 1994; King and Harangozo, 1998*]. In terms of potential biological impact, the relevance of mid-winter temperature increases could be questioned, as they may not lead to discernible consequences unless the temperatures achieved reach physiological or activity thresholds. However, winter increases will also lead to an effective shortening of the duration of the winter season (i.e. reducing the total time spent below biological thresholds), thereby providing a route to influence biology. Antarctic Peninsula temperature increase has been linked to a decrease in winter sea ice extent, which may further provide a link to El Niño Southern Oscillation (ENSO) events in the southern Pacific Ocean [*Cullather et al.,* 1996].

Arguably more important in the control of biological activity in Antarctic terrestrial habitats is the availability of liquid water [*Kennedy,* 1993; *Sømme,* 1995]. This is governed both directly by precipitation patterns (summer rain), and by thawing of seasonal snow banks and permanent glaciers. The presence of free water in terrestrial habitats is, therefore, often separated temporally and spatially from the precipitation event.

Although a general prediction of global circulation models is for changes in precipitation patterns, these predictions are spatially very coarse. Thus, increases in precipitation are predicted in the Antarctic coastal zone [*Budd and Simmonds,* 1991] and have been documented [*Turner et al.,* 1997]. However, climatic predictions and contemporary data do not exist at anything approaching a fine enough scale to be applied to terrestrial biological studies. It is important also to recognise that, at a local scale, precipitation (or melt) may decrease. Decreasing trends are also in evidence, particularly at some sub-Antarctic locations such as Marion I. and Îles Kerguelen [*Frenot et al.,* 1997; *Bergstrom and Chown,* 1999], and have also been reported from the South Orkney Is. [*Noon et al.,* 2001].

As free water is also derived from melting ice and snow, the contribution of current glacial retreat must also be considered. Recent ice shelf collapses on both east and west coasts of the Antarctic Peninsula [*Doake and Vaughan,* 1991; *Vaughan and Doake,* 1996; *Pudsey and Evans,* 2001], do not directly impact terrestrial biota other than by allowing maritime (i.e. moist) climatic influences closer to specific areas of coastline. Rapid glacial melting and retreat, in contrast, release water directly into terrestrial ecosystems as well as exposing new ground for colonization, and have been documented along the length of the Antarctic Peninsula, and Scotia Arc islands, including sub-Antarctic South Georgia [*Smith,* 1990; *Gordon and Timmis,* 1992; *Fowbert and Smith,* 1994, *Pugh and Davenport,* 1997; *Fox and Cooper,* 1998; *Vaughan et al.,* 2001]. In some cases, such as on Signy Island, the losses are considerable with a total of approximately 40% reduction in area over the last 50 years. At smaller sites or the microhabitat scale, previously 'permanent' areas of snow or ice may be lost completely.

The third aspect of climate change expected to influence terrestrial biota is that of radiation climate. Here,

the annual formation of a spring ozone hole over the Antarctic, through concentration of anthropogenic pollutants during the south polar winter, has occurred only since the early 1980s [*Farman et al.,* 1985]. This allows increased penetration of shorter wavelength UV-B radiation to the Earth's surface, but leaves UV-A and photosynthetically active radiation (PAR) levels unaffected (hence the biologically important ratio of incident UV-B:PAR is altered [*Smith et al.,* 1992]). Levels of UV-B received during the period of maximum ozone loss are little different from the normal summer solstice maxima which biota already face. However, there are two important differences: these levels now occur for an additional period earlier in the season than previously [*Huiskes et al.,* 1999] and, hence, lower wavelengths penetrate to the Earth's surface than is normal in spring.

Finally, it is important not to consider these variables in isolation, as interactions between them are likely to be significant [*Convey,* 1996a]. The obvious link between temperature and availability of free water through melting has been mentioned already. Precipitation patterns will almost certainly be linked to patterns of cloud cover and wind, and hence to insolation (PAR and UV) and temperature in terrestrial habitats. These links will become especially important at the microclimatic scale.

RESPONSES TO CONTEMPORARY CHANGE

Changes in climate are not a new threat to the terrestrial biota of Antarctica. Throughout the Pleistocene, wide variations in climate and the extent of ice cover are known to have occurred, including periods when ice extent and thickness have been considerably less than is present now [*Sugden and Clapperton,* 1977; *Clapperton and Sugden,* 1982, 1988; *Lorius et al.,* 1985; *Smith,* 1990; *Pudsey and Evans,* 2001]. Despite a recognition that the magnitude of changes being experienced in parts of Antarctica is likely to lead to clear, identifiable, consequences in the continent's simple terrestrial ecosystems [*Roberts,* 1989; *Smith and Steenkamp,* 1990; *Voytek,* 1990], studies of these systems initially lagged behind much of the rest of the world. A predictive literature exists, encompassing both the importance of Antarctic terrestrial biology as a focus of climate change studies, and predictions of potential consequences of change [*e.g. Wynn-Williams,* 1994, 1996; *Kennedy,* 1995a; *Convey,* 1997a; *Walton et al.,* 1997; *Bergstrom & Chown* 1999]. These predictions are summarized in Table 1. At the ecosystem scale, the consequences of changes in two of the three major environmental variables (temperature, water) can lead to predictions of an increase in rate of colonization by species new to the Antarctic, leading to increased diversity, biomass, trophic complexity and habitat structure. The effects of the third variable (radiation), while subtle, are likely to be negative, at least in terms of requiring increased resource allocation to mitigation strategies. In the longer term, existing Antarctic species and communities may be lost through the effects of increased competition.

More recently, an increasing number of Antarctic terrestrial biological studies have either been aimed directly at testing predictions associated with climate change, or have generated data which can be applied in this context. Broadly, these studies fall into two classes, descriptive or observational studies that inevitably provide only correlations and circumstantial evidence, and studies that involve identifying the consequences of more or less realistic manipulations of natural habitats. Both classes of study can be usefully further separated into those working at the level of the whole organism or above, addressing ecological questions, and those addressing physiological or biochemical questions. The following sections, based on these simple categorizations, summarize existing evidence of the consequences of climate change on terrestrial ecosystems and biota of the Antarctic Peninsula and Scotia Arc.

Field Observations—Ecology and Biogeography

Life history strategies [after *Southwood,* 1977, 1988; *Greenslade,* 1983; *Grime* 1988] of Antarctic terrestrial biota are generally "adversity" selected, with features relating to stress (cold and drought) tolerance, being

particularly refined and well-represented [*Cannon and Block* 1988; *Block* 1990; *Sømme* 1995; *Convey* 1996a, 1997b, 2000]. Of great relevance to any study of response to climate change is the degree of responsive flexibility found in many life history and physiological characteristics. This underlying flexibility is a direct consequence of the rapid and unpredictable variation in environmental characteristics, which is typical of the Antarctic terrestrial environment even before any systematic changes in climate are considered.

Life history flexibility could allow changes in development rate and shortening of life cycle duration in response to climate warming or extension of the active season. The South Georgian diving beetle, *Langustis angusticollis*, provides an example of a species likely to show a rapid response to warming of its lake habitat [*Arnold and Convey,* 1998]. It possesses an environmentally cued diapause, and thermal energy input to its lake environment currently limits this species to a two-year life cycle. However, an increase of only 1°C would allow an annual life cycle. Such an increase is realistic on South Georgia, and has already been reported from Signy Island lakes [*Quayle et al.,* 2002]. As *L. angusticollis* is the top predator in the lake ecosystem, such a change in annual population dynamics will have a large impact on lake trophic interactions.

Visually striking and recent increases in population numbers and extents of the two native Antarctic flowering plants (*Deschampsia antarctica* and *Colobanthus quitensis*) at sites along the Antarctic Peninsula have been linked to regional climate warming and local snow recession by several authors [*Fowbert and Smith,* 1994; *Smith,* 1994; *Grobe et al.,* 1997]. It

TABLE 1. General predictions of the biological impacts of changes in the three major climatic variables in Antarctica [modified from *Convey, 2001b*].

Temperature	Water	UV-B radiation
Increase:	*Increase:*	– Alter resource allocation strategies, damage to cellular structures and processes (*e.g.* pigments involved in screening, quenching, avoidance, repair processes)
– Extend active season, increase development rates and reduce life cycle duration	– Extend active season (may be counteracted by thermal consequence of increased cloud/ decreased insolation)	
– Alter physiological resource allocation strategies (*e.g.* reduced stress tolerance)	– Increase local distribution and expose new ground for colonization	
– Trophic impacts through altered feeding preferences and population growth		– Impact on food chain through induced changes in morphology and palatability
	Decrease:	
– Alter distribution by expanding range limits (upper thermal limits may restrict distributions)	– Reduce active season	– Range limitation
	– Local extinction	– Local extinction
– Exotic colonisation		

should be noted that these increases in numbers and local population densities and extent do not equate to any significant increase in the species' ranges, the southern limits of which are defined simply by the lack of suitable ice free ground south of the Terra Firma Islands in southern Marguerite Bay. The most important source of new plants is likely to be through mature seed production, which now occurs in a much higher proportion of seasons than found only two decades earlier [*Convey*, 1996b; *cf. Edwards*, 1974]. These seeds may also remain dormant in soil propagule banks [*McGraw and Day*, 1997].

Bryophytes and microbiota are similarly well-suited to take advantage of improved environmental conditions, further helped by possessing a range of asexually and sexually produced propagules facilitating dispersal over both very short and very long distances. Rapid population increases have again been seen at sites in the maritime Antarctic [*Smith*, 1993, 2001; *Wynn-Williams*, 1996]. Bryophyte spores may also remain dormant in the soil propagule bank for many years [*Smith and Coupar*, 1986; *Smith*, 1987, 1993; *During*, 1997, 2001]. Cultures derived from Antarctic propagule banks are dominated by locally occurring species. However, aerobiological studies demonstrate that biological material is carried infrequently into the Antarctic Peninsula region from southern South America associated with particular storm circulation patterns [*Marshall*, 1996] although, as yet, bryophyte or lichen propagules of distant origin have not been found [*Marshall and Convey*, 1997]. This is the most plausible explanation for the few recent records of species new to sites in the region (*e.g.* the moss *Polytrichum piliferum* on Signy Island [*Convey and Smith*, 1993]), and for the presence of lower latitude species at geothermally active sites [*Smith* 1991; *Convey et al.*, 2000a].

Local increases in vegetation, associated with development of habitat complexity, leads to rapid colonization by existing terrestrial invertebrate communities, although this simple observation does not appear to have been quantified. There are no reported examples of alien invertebrates becoming established in the maritime Antarctic, other than an enchytraeid worm and dipteran accidentally introduced by human agent from sub-Antarctic South Georgia to Signy Island [*Block et al.*, 1984]. However, transplant experiments (no longer permitted under the regulations of the Antarctic Treaty) [*e.g. Edwards and Greene*, 1973; *Edwards*, 1980] and accidental introductions to sub-Antarctic islands demonstrate that a range of taxa can establish when the problems of long distance transport are overcome—for instance, more than 50% of the vascular plant flora of the Scotia Arc island of South Georgia is accounted for by introduced species [*Smith*, 1996; *Bergstrom and Chown* 1999]. In any specific example, it is not possible to prove a causal link between climate amelioration and colonization, even though an increasing frequency of such events is expected.

On the Scotia Arc, a range of invertebrate introductions to South Georgia are documented, including beetles [*Ernsting*, 1993], springtails [*Convey et al.*, 1999] and mites [*Pugh*, 1994]. The impact of these introductions does not stand out visually, but detailed studies have shown clear consequences on the trophic dynamics of existing communities [*Ernsting et al.*, 1995, 1999], or proposed displacement of native species [*Convey et al.*, 1999]. A final warning note is struck by the observation that dangers exist not only in the introduction of alien invertebrates, which are frequently reported from various Antarctic Peninsula research stations often associated with cargo shipments, but also through the accidental movement of true Antarctic species between or within Antarctic zones [*e.g. Convey et al.*, 2000b].

The significance of such examples is twofold. It is clear that many species have the potential to establish in the Antarctic and that opportunities are likely to occur more frequently under climate amelioration. It is also clear that humans, as vector, will assume an increasingly pivotal role in the colonization of Antarctica, especially given the increased accessibility of the continent to both scientific and recreational activities [*Pugh*, 1994; *Smith*, 1996].

The scientific potential of the environmental gradient extending along the Scotia Arc is yet to be exploited in testing the predictions of climate change. A single study of reproductive investment in the oribatid mite, *Alaskozetes antarcticus*, at latitudes between South Georgia and Marguerite Bay (c. 53–69°S) [*Convey*, 1998] provides a strong suggestion of variation in resource allocation strategy, finding greater investment in eggs in the milder sub-Antarctic site. This pattern is expected to apply widely, as the inherently very flexible physiological and life history strategies typical of Antarctic invertebrates will facilitate such changes in resource allocation patterns.

Field Observations—Physiology and Biochemistry

The practicalities of Antarctic fieldwork mean that few observational datasets exist relating to physiological or biochemical characteristics, and even fewer relate to studies with a long enough run of data to allow exploration of links with changing features of climate.

Springtails (Collembola) are among the most common and widely distributed arthropod groups in Antarctic terrestrial ecosystems. Unlike other insects, gas exchange in Collembola occurs across the cuticle (rather than using a tracheal system), which reduces the animal's ability to control body water content. Springtails are therefore vulnerable to changes in the water status of their microenvironment, which feature led to the proposal that body water content may be a proxy for microhabitat water status [Block and Harrisson, 1995; Block and Convey, 2001]. Thus, a significant upward trend in the body water content of the springtail Cryptopygus antarcticus at Signy Island between 1984 and 1987 was postulated to result from an interaction between increased insolation and wind speed resulting in greater local ice and snowmelt, and increased liquid water in the microhabitat [Block and Harrisson, 1995]. Analysis of the continuation of this dataset over a longer 11 year period to 1995 has identified other long-term patterns [Block and Convey, 2001] particularly (i) upward trends in water content during the late autumn and early spring, and (ii) evidence of a downward trend during part of summer. These field measures of an ecophysiological characteristic indicate a lengthening of the summer season, with earlier water availability in spring and later availability in autumn. The decrease in part of summer may indicate local exhaustion of water supply, and highlights the need to avoid the assumption that climate amelioration will automatically lead to reduced stress.

Newsham et al. [2002] provide the only examples of measurement of a direct biochemical response to increased UV-B radiation during episodes of ozone depletion in non-manipulated field populations of two bryophyte species. These species show patterns of protective pigment synthesis and loss that are most strongly correlated with their recent (natural) radiation exposure history, indicating a rapid and dynamic biochemical response to this environmental stress.

Manipulations—Ecology

Simple field manipulation techniques are an attractive means of mimicking predicted changes in environmental variables, particularly for longer term studies in remote and infrequently accessible sites [see Kennedy, 1995b,c]. Their use has become widespread in Antarctic studies [e.g. Wynn-Williams 1992, 1996; Day et al., 1999; Smith, 2001; Convey and Wynn-Williams, 2002; Convey, 2003]. Methodologies differ in detail, but all involve the use of chambers or screens placed over selected areas of habitat

to alter aspects of the thermal and radiation climates. With access to field support and power, more complex methodologies may include water or nutrient amendments, and radiation supplementation with lamps. While their practical advantages are clear, the environmental changes achieved by such methodologies are more complex and inter-related than often appreciated [Kennedy, 1995b,c], requiring considerable care in experimental design.

The use of greenhouse warming methodologies at sites along the Antarctic Peninsula and Scotia Arc has led to rapid and spectacular responses at the population and community level in microbial [Wynn-Williams, 1993, 1996], bryological and flowering plant [Smith, 1990] and invertebrate [Kennedy, 1994, 1996; Convey and Wynn-Williams, 2002; Convey, 2003] studies, with species variously achieving greater ground coverage, lusher growth, increased population densities and increased reproductive output. These studies also confirm the importance of propagule banks in community development. Microbial and nematode populations may show these responses over timescales of as short as 1–2 years.

The complexities in interpretation of greenhouse-generated data are illustrated by studies of the fauna of manipulated sites. In a study on Alexander Island (c. 72°S) in the southern maritime Antarctic [Convey and Wynn-Williams, 2002], nematode worms showed very rapid local increases of one to three orders of magnitude in population density over the first year of manipulation. However, this appeared to be a rapid response of a single microbivorous genus, Plectus, to the sudden increase in food supply. After three years of manipulation, community diversity under manipulations had increased through local colonization, while population densities had returned to a level only slightly increased relative to non-manipulated sites [Convey, 2003]. Differences in response patterns is a prediction of studies across more than one trophic level [Day, 2001; Searles et al., 2001], as the mechanisms underlying responses will be complex and involve a range of biotic and abiotic factors and interactions.

A multitrophic level study at Palmer Station, Anvers Island, demonstrated clear negative impacts of UV-B exposure of terrestrial plant communities on the associated arthropod populations [Convey et al., 2002], hypothesized to be linked with a change in diet quality rather than a direct impact of UV-B on individual animals. This study also confirmed the key role of water availability in terrestrial ecosystems, with amendments leading to significant increases in populations of several arthropods. Counter-intuitively, warming was identified as a negative influence on numbers of some species (although a similar result has been reported in a single Arctic study [*Coulson et al.*, 1996; *Webb et al.*, 1998]), and interpreted as being linked with increased drying stresses. There is a clear need for further multivariate climate manipulation studies to address consequences across multiple trophic levels [*Day*, 2001].

Greenhouse manipulations affecting temperature and UV radiation are known to lead to changes in growth form of vegetation [*Smith*, 1990, 2001; *Day et al.*, 1999, 2001; *Sullivan and Rozema*, 1999; *Ruhland and Day*, 2000], and hence subtle alterations in microclimates and three-dimensional habitat structure. The consequences of such changes have not been studied in the Antarctic.

Manipulations—Physiology and Biochemistry

The consequences for photosynthetic organisms of increased exposure to UV-B radiation experienced during the spring ozone hole has formed a particular focus of biological research in the Antarctic. The potential for direct UV-B impacts on soil fauna has not, so far, received attention, through a rather simplistic assumption that these organisms are shielded from exposure to radiation by their habitat, although the use of screening pigments has been demonstrated in the Antarctic freshwater copepod, *Boeckella poppei* [*Rocco et al.*, 2002].

That UV-B radiation has damaging effects on aspects of cell biochemistry is well-known, as are the range of responses used to mitigate the damage [*e.g. Vincent and Quesada*, 1994; *Wynn-Williams*, 1994; *Cockell and Knowland*, 1999]. Results of laboratory experiments that mimic environmental conditions tend to conform with these general expectations [*e.g. Quesada et al.*, 1995]. In contrast, field manipulations of natural (and more complex) systems which change UV-B levels using lamps indicate that negative effects on plants may be very limited [*Fiscus and Booker*, 1995, *Allen et al.*, 1998], a result which is consistent with the simple observation that Antarctic plant communities have apparently shown little or no visible damage to exposure to elevated UV-B under the ozone hole over the last two decades. As a cautionary note, it should also be emphasized that there is currently a lack of data relating to the possibility of genetic damage to organisms in Antarctic terrestrial habitats, although it is clear that the potential exists [*George et al.*, 2002].

Both amendment and screening methodologies have been used to study the effect of changes in UV radiation exposure of autotrophic groups (algae, cyanobacteria,

bryophytes, phanerogams, lichens) [*Wynn-Williams*, 1996; *Quesada et al.*, 1998; *Huiskes et al.*, 1999; *Montiel et al.*, 1999; *George et al.*, 2001]. Maximum levels of UV-B experienced during the spring ozone hole, assuming no protection from winter snow cover, are similar to the levels received normally at midsummer. Fully hydrated physiologically active organisms may be able to activate appropriate mechanisms whether exposed to UV-B during spring or summer. However, the effectiveness of some repair mechanisms depends on the UV-B:PAR ratio, which is much higher during spring ozone depletion than in midsummer, while sensitivity to UV-B may also vary with other environmental factors or developmental stage. UV enhancement studies of the flowering plants, *Deschampsia antarctica* and *Colobanthus quitensis*, demonstrated no reduction of photosynthetic parameters when exposed to a range of realistic radiation stresses [*Montiel et al.*, 1999; *Lud et al.*, 2001]. Separate UV screening studies of the same species have shown changes in concentrations of screening pigments, growth and production in response to UV-B exposure level [*Day et al.*, 1999; *Ruhland and Day*, 2000; *Xiong and Day*, 2001].

Manipulation studies involving bryophytes and cyanobacteria have so far not generated consistent patterns of response. The hypothesis that increased exposure to UV-B should lead to reduced photosynthetic yield or increases in protective pigments has been supported in some, but not all, studies of Antarctic cyanobacteria [*Quesada et al.*, 1998; *George et al.*, 2001] and bryophytes [*Montiel et al.*, 1999; *Huiskes et al.*, 1999]. However, *Newsham et al.* [2002], while demonstrating increased concentrations of UV-B screening pigments and carotenoids in field measurements of bryophytes, found no change in photosynthetic yield. Likewise, *Lud et al.* [2001] reported no change in yield in a lichen. *George et al.* [2002] provide the first quantification of the potential for UV-related DNA damage in Antarctic terrestrial microbiota, highlighting that levels of damage observed are considerably lower than those experienced in the Antarctic shallow marine environment, thought to be a consequence of much greater development of screening systems in terrestrial organisms.

Recent general overviews of the biochemical impact of exposure to UV-B [*e.g. Rozema*, 1999; *Paul*, 2001] confirm the expected changes in utilization of reaction pathways, particularly relating to pigment production patterns. These changes have an important ecological implication in that, implicitly, they require alterations in the resource allocation strategies (the energy economy) utilised by the study organisms. These responses may appear to be very subtle, with minor specific costs. However, they may accumulate to have far greater impacts elsewhere in the ecosystem [see, for instance, *Day*, 2001; *Johnson et al.*, 2002].

The pigments involved in UV quenching and screening often include polyphenolic subunits, which, separately, influence digestibility to herbivores and detritivores. An impact of altered food quality within terrestrial food webs exposed to UV-B has been postulated [*Convey et al.*, 2002] but not demonstrated directly in the Antarctic, while conflicting conclusions have been reported from decomposition studies at lower latitudes [*Newsham et al.*, 1999; *Paul et al.*, 1999; *Rozema et al.*, 1999].

HISTORICAL PERSPECTIVE

Antarctic terrestrial biology has, thus far, received little attention in the context of historical climate reconstruction. In part, as mentioned above, this is because the low altitude, coastal and vegetated, habitats typical of the Antarctic Peninsula are considered to be of very recent origin. Radiocarbon dating of the deepest moss peat banks found in the region, on Signy Island (South Orkney Islands) and Elephant Island (South Shetland Islands), give maximum ages of 5–6,000 years [*Fenton*, 1980; *Björk et al.*, 1991a]. This is an age consistent with regional glacial reconstructions, which currently suggest an absence of exposed coastal ground at glacial maxima. More recent fluctuations in ice extent have resulted in burying of some peat banks by ice and their subsequent re-exposure [*Fenton*, 1982], providing biological corroboration for the timing of these events.

Studies from a range of other biological disciplines have the potential to be used to give at least circumstantial evidence relevant to climate reconstruction. Stratigraphic records of past environmental changes can be found in the biological material preserved in lake sediment cores. This includes macrofossils, microfossils, and biochemical fossils. Palaeolimnological studies use these materials to infer aspects of climate, including periods of changing temperature and moisture balance, production, depth and sea level change at the time of sediment deposition. Often a number of these biological records are statistically combined to provide comprehensive reconstructions of regional climate history [*Björck et al.*, 1996; *Jones et al.* 2000], although accurate dating is vital [*Hodgson et al.* 2001a].

Of the microfossils, diatoms have been most successfully used to reconstruct past environmental changes [*Spaulding and McKnight*, 1999]. This is particularly the

case when regional datasets can be used to statistically relate modern species assemblages and environmental variables, hence allowing interpretation of fossil data [*Roberts and McMinn*, 1998; *Hodgson et al.*, 2001b]. Pollen microfossils have also been used to detect the changing influence of air masses originating from the temperate latitudes, while peaks in Chlorophytea and moss spores have been used to indicate favourable growing conditions in lakes associated with warm periods [*Björck et al.*, 1991b; *Jones et al.*, 2000]. Considerable progress has also been made in using fossil pigments in applications where the abundance, production, and composition of past phototrophic communities are important response variables [Leavitt and Hodgson, 2001]. Long after the morphological remains of algae and bacteria are lost due to various degradation processes, sedimentary carotenoids (carotenes and xanthophylls), and chlorophylls can be used to track past populations [*Leavitt et al.*, 2002; *Squier et al.*, 2002; *Walker et al.*, 2002].

Macrofossils from cores obtained on Signy Island also include identifiable remains of vertebrate hairs, freshwater crustaceans and terrestrial arthropods [*Hodgson and Johnston 1997*; *Hodgson et al.*, 1998], showing quantitative changes in abundance over time. Systematic changes in abundance of some of these taxa in sediments have the potential to provide proxy measures of variables such as lake productivity and water inflow from surrounding terrestrial habitats, and hence clues to climatic conditions. Similarly, ultra-violet (UV) radiation screening pigments present in cyanobacteria from lake sediments are being used in reconstructions of historical UV climate regimes [*Leavitt et al. 2002*], while plant pollen and moss spores also have the potential to be used in the same way [*Rozema et al.*, 2001, 2002]. It has been demonstrated that the concentrations of pigments in pollen and spores can be manipulated by experimental exposure of plants to UV radiation and, hence, proposed that extraction and analysis of pollen and spore material preserved in dated moss peat cores will provide a proxy measure of environmental UV exposure.

Notwithstanding the modelled extent and thickness of ice cover at Pleistocene maxima, recent studies of the mite faunas of both the Antarctic Peninsula and parts of continental Antarctica have concluded that elements of these represent the remnants of an ancient, pre-glaciation, regional fauna [*Marshall and Pugh 1996*; *Pugh and Convey*, 2000]. For the East Antarctic fauna, it can be argued that the species concerned are currently montane, and hence could have occupied more limited nunatak refugia even at glacial maxima. However, this hypothesis appears inappropriate for the Antarctic Peninsula fauna, given that the majority of species are restricted to low altitude coastal/maritime localities and that, currently, there is no known obligate montane fauna from this region. Separately, a recent analysis of the Antarctic nematode fauna [*Andrássy,* 1998] concluded that (a) there is no overlap at species level between maritime and continental Antarctic representatives and (b) that the large majority of species in both regions were endemic, not being found outside the Antarctic. The clear implication of these observations is that a coastal fauna has been able to survive in situ at currently unknown refuge sites on the Antarctic Peninsula throughout the glacial episodes, a factor which requires taking into account when preparing glacial reconstructions.

Finally, a small number of biological studies may also be used to provide suggestive evidence of the existence within Antarctica of conditions suitable for continuity of terrestrial life over much longer timescales. Thus, the endemic Continental Antarctic mite family Maudheimiidae whose species inhabit inland nunataks, is argued to be an ancient relict predating the final breakup of Gondwana [*Marshall and Pugh,* 1996; *Marshall and Coetzee,* 2000], and hence must have had access to suitable habitats throughout this period. Molecular biology also has the potential to quantify rates of divergence between related species and hence, with certain assumptions, provide evidence for the timing of speciation events.

CONCLUSIONS

The Antarctic Peninsula region and associated islands of the Scotia Arc are experiencing change in three environmental variables of major biological significance, these being (a) rapidly increasing temperatures, (b) changes (both increases and decreases) in liquid water availability through precipitation and snow/ice melt, (c) changes in radiation climate, both associated with (b) and directly as a consequence of increased UV-B penetration via the spring ozone hole. This combination of factors is unique worldwide.

In general, Antarctic terrestrial biota, which are well-adapted to their highly variable and stressful environment, possess appropriate features allowing them to control the consequences of changing environmental variables, as the predicted levels of change are small compared with the natural variability already experienced. In the absence of long range colonization by more effective competitors, resident biota are expected to be able to take advantage of reduction in environmental stresses associated with regional climate amelioration (although, locally, increases in

these stresses may also be experienced, while any impacts of UV-B exposure are expected to be negative, if subtle).

Some highly visible examples of biological responses to recent Antarctic Peninsula climate amelioration exist. These include rapid local expansion of populations of the two resident flowering plants, comparable increases in bryophyte populations, and rapid colonization by local plant and animal species of previously sparsely vegetated ground and that exposed by snow and ice recession. Field manipulation experiments generate comparable positive responses. These responses have, thus far, been limited to species already resident in the region, with no occurrences on the Antarctic Peninsula of the predicted colonization events by lower latitude species that are known to possess appropriate physiological and ecological capacities. Sub-Antarctic islands, including South Georgia on the Scotia Arc, provide many examples and warnings of the impact of both naturally colonizing and anthropogenically introduced species.

Many biological responses to changing environmental variables are likely to subtle and even, apparently, insignificant. This is particularly true of detailed physiological and biochemical responses to small changes in stress. However, these responses involve changes in resource allocation strategies within the organism, an important feature controlling life history strategies. Such small responses may integrate to give considerably greater impacts at the community or ecosystem levels.

Terrestrial biological studies related to climate change have been focused on the recent timescale, with little attention yet given to the potential for the use of biological signals in longer-term reconstruction of climate in the Antarctic Peninsula region. This chapter highlights several possible biological contributions, which may offer assistance in climatological, glacialogical and geological reconstructions.

Acknowledgments. I am very grateful to Andrew Clarke, Lloyd Peck, Kevin Newsham, Dominic Hodgson and David Barnes for stimulating discussions on this subject over several years and, along with two anonymous referees, for constructive comments on earlier manuscript versions. This paper contributes to the British Antarctic Survey's BIRESA (Biological Responses to Environmental Stress in Antarctica) project and to the SCAR RiSCC (Regional Sensitivity to Climate Change in Antarctica) Program.

REFERENCES

Addo-Bediako, A., Chown, S. L. and Gaston, K. J., Thermal tolerance, climatic variability and latitude, *Proceedings of the Royal Society of London, series B* 267: 739-745, 2000.

Allen, D. J., Nogués, S. and Baker, N. R., Ozone depletion and increased UV-B radiation: is there a real threat to photosynthesis? *Journal of Experimental Botany* 49: 1775-1788, 1998.

Andrássy, I., Nematodes in the Sixth Continent, *Journal of Nematode Morphology and Systematics* 1: 107-186, 1998.

Arnold, R. J. and Convey, P., The life history of the world's most southerly diving beetle, *Lancetes angusticollis* (Curtis) (Coleoptera: Dytiscidae), on sub-Antarctic South Georgia, *Polar Biology* 20: 153-160, 1998.

Bergstrom, D. M. and Chown, S. L., Life at the front: history, ecology and change on southern ocean islands, *Trends in Ecology and Evolution* 14: 472-476, 1999.

Björk, S., Malmer. N., Hjort, C., Sandgren, P., Ingólfsson, O., Wallén, B., Smith, R. I. L. and Jónsson, B. L., Stratigraphic and palaeoclimatic studies of a 5500-year-old moss bank on Elephant Island, Antarctica, *Arctic and Alpine Research* 23: 361-374, 1991a.

Björck, S., Häkansson H., Zale R., Karlen W. and Jonsson B.L. A late Holocene sediment sequence from Livingston Island, South Shetland Islands, with palaeoclimatic implications. *Antarctic Science* 3: 61-72, 1991b.

Björck, S., Olsson, S., Ellis-Evans, C., Häkansson, H., Humlum, O., and deLirio, J. M..Late Holocene palaeoclimatic records from lake sediments on James Ross Island, Antarctica. *Palaeogeography, Palaeoclimatology, Palaeoecology* 121: 195-220, 1996

Block, W., Cold tolerance of insects and other arthropods, *Philosophical Transactions of the Royal Socety of London, series B* 326: 613-633, 1990.

Block, W. and Convey, P., Seasonal and long-term variation in body water content of an Antarctic springtail - a response to climate change? *Polar Biology* 24: 764-770, 2001.

Block, W. and Harrisson, P. M., Collembolan water relations and environmental change in the maritime Antarctic, *Global Change Biology* 1: 347-359, 1995.

Block, W., Burn, A. J. and Richard, K. J., An insect introduction to the maritime Antarctic, *Biological Journal of the Linnean Society* 23: 33-39, 1984.

Budd, W. F. and Simmonds, I., The impact of global warming on the Antarctic mass balance and global sea level. In: G. Weller, C.L. Wilson and B.A.B. Severin (eds.) *Proceedings of the International Conference on the Role of Polar regions in Global Change*, pp. 489-494, Geophysics Inst., Univ of Alaska, Fairbanks, 1991.

Callaghan, T.V. and Jonasson, S., Arctic terrestrial ecosystems and environmental change, *Philosophical Transactions of the Royal Society of London, series A* 352: 259-276, 1995.

Cannon, R. J. C. and Block, W., Cold tolerance of microarthropods, *Biological Reviews* 63: 23-77, 1988.

Chown, S. L. & Clarke, A., Stress and the geographic distribution of marine and terrestrial animals. In: K. B. Storey & J. Storey (eds.) *Environmental Stressors and Gene Responses*. Elsevier, Amsterdam, 2000.

Clapperton, C.M. & Sugden, D. E., Late quaternary glacial history of George VI Sound area, West Antarctica. *Quaternary Research* 18: 243-267, 1982.

Clapperton, C.M. and Sugden, D. E., Holocene glacier fluctuations in South America and Antarctica. *Quaternary Science Reviews* 7: 185-198, 1988.

Cockell, C. S. and Knowland, J., Ultraviolet radiation screening compounds. *Biological Reviews* 74: 311-345, 1999.

Convey, P., The influence of environmental characteristics on life history attributes of Antarctic terrestrial biota. *Biological Reviews* 71: 191-225, 1996a.

Convey, P., Reproduction of Antarctic flowering plants. *Antarctic Science* 8: 127-134, 1996b.

Convey, P., Environmental change: possible consequences for life histories of Antarctic terrestrial biota, *Korean Journal of Polar Research* 8: 127-144, 1997a.

Convey, P., How are the life history strategies of Antarctic terrestrial invertebrates influenced by extreme environmental conditions? *Journal of Thermal Biology* 22: 429-440, 1997b.

Convey, P., Latitudinal variation in allocation to reproduction by the Antarctic oribatid mite, *Alaskozetes antarcticus*, *Applied Soil Ecology* 9: 93-99, 1998.

Convey, P., How does cold constrain life cycles of terrestrial plants and animals? *Cryo-Letters* 21: 73-82, 2000.

Convey, P., Antarctic Ecosystems. In: S. A. Levin (ed.) *Encyclopedia of Biodiversity*, vol. 1., Academic Press, San Diego, pp. 171-184, 2001a.

Convey, P., Terrestrial ecosystem response to climate changes in the Antarctic. In: *"Fingerprints" of climate change - adapted behaviour and shifting species ranges*, G.-R. Walther, C.A. Burga & P.J. Edwards (eds.), Kluwer, New York, pp 17-42, 2001b.

Convey, P., Soil faunal community response to environmental manipulation on Alexander Island, southern maritime Antarctic. In *Antarctic Biology in a Global Context*, eds. A. H. L. Huiskes, W. W. C. Gieskes, J. Rozema, R. M. L. Schorno, S. van der Vies and W. J. Wolff. Backhuys, Leiden, pp. 74-78, 2003.

Convey, P. and Smith, R. I. L., Investment in sexual reproduction by Antarctic mosses, *Oikos* 68: 293-302, 1993.

Convey, P. and Wynn-Williams, D. D., Antarctic soil nematode response to artificial environmental manipulation. *European Journal of Soil Biology*, 38: 255-259, 2002.

Convey, P., Greenslade, P., Arnold, R. J. and Block, W., Collembola of sub-Antarctic South Georgia, *Polar Biology* 22: 1-6, 1999.

Convey, P., Smith, R. I. L., Hodgson, D. A and Peat, H. J., The flora of the South Sandwich Islands, with particular reference to the influence of geothermal heating, *Journal of Biogeography* 27: 1279-1295, 2000a.

Convey, P., Smith, R. I. L., Peat, H. J. and Pugh, P. J. A., The terrestrial biota of Charcot Island, eastern Bellingshausen Sea, Antarctica an example of extreme isolation, *Antarctic Science* 12: 406-413, 2000b.

Convey, P., Pugh, P. J. A., Jackson, C., Murray, A. W., Ruhland, C. T., Xiong, F. S. and Day, T. A., Response of Antarctic terrestrial arthropods to multifactorial climate manipulation over a four year period, *Ecology*, 83: 3130-3140, 2002.

Coulson, S. J., Hodkinson, I. D., Webb, N. R., Block, W., Bale, J. S., Strathdee, A. T., Worland, M. R. and Wooley, C.,

Effects of experimental temperature elevation on high-arctic soil microarthropod populations, *Polar Biology* 16: 147-153, 1996.

Cullather, R. I., Bromwich, D. H. and van Woert, M. L., Inter-annual variations in Antarctic precipitation related to El Niño—Southern Oscillation, *Journal of Geophysical Research* 101: 19109-19118, 1996.

Day, T. A., Multiple trophic levels in UV-B assessments - completing the ecosystem, *New Phytologist* 152: 183-186, 2001.

Day, T. A., Ruhland, C.T. and Xiong, F., Influence of solar UV-B radiation on Antarctic terrestrial plants: results from a 4-year field study. *Journal of Photochemistry and Photobiology B: Biology* 62: 78-87, 2001.

Day, T. A., Ruhland, C. T., Grobe, C. W. and Xiong, F., Growth and reproduction of Antarctic vascular plants in response to warming and UV radiation reductions in the field. *Oecologia* 119: 24-35, 1999.

Doake, C. S. M. and Vaughan, D. G., Rapid disintegration of the Wordie Ice Shelf in response to atmospheric forcing, *Nature* 350: 328-330, 1991.

During, H. J., Bryophyte diaspore banks, *Advances in Bryology* 6: 103-134, 1997.

During, H. J., New frontiers in bryology and lichenology: diaspore banks, *Bryologist* 104: 92-97, 2001.

Edwards, J. A., Studies in *Colobanthus quitensis* (Kunth) Bartl. and *Deschampsia antarctica* Desv.: VI.: Reproductive performance on Signy Island, *British Antarctic Survey Bulletin* 39: 67-86, 1974.

Edwards, J. A., An experimental introduction of vascular plants from South Georgia to the maritime Antarctic, *British Antarctic Survey Bulletin* 49: 73-80, 1980.

Edwards, J. A. and Greene, D. M., The survival of Falklands Island transplants at South Georgia and Signy Island, South Orkney Islands, *British Antarctic Survey Bulletin* 33 & 34: 33-45, 1973.

Ernsting, G., Observations on life cycle and feeding ecology of two recently-introduced predatory beetle species at South Georgia, sub-Antarctic, *Polar Biology* 13: 423-428, 1993.

Ernsting, G., Block, W., MacAlister, H. and Todd, C., The invasion of the carnivorous carabid beetle *Trechisibus antarcticus* on South Georgia (sub-Antarctic) and its effect on the endemic herbivorous beetle *Hydromedion sparsutum*, *Oecologia* 103: 34-42, 1995.

Ernsting, G., Brandjes, G. J., Block, W. and Isaaks, J. A., Life-history consequences of predation for a subantarctic beetle: evaluating the contribution of direct and indirect effects, *Journal of Animal Ecology* 68: 741-752, 1999.

Farman, J. C., Gardiner, B. G. and Shanklin, J. D., Large losses of total ozone in Antarctica reveal seasonal ClO_x/NO_x interaction, *Nature* 315: 207-210, 1985.

Fenton, J. H. C., Vegetation re-exposed after burial by ice and its relationship to changing climate in the South Orkney Islands, *British Antarctic Survey Bulletin* 51: 247-255, 1980.

Fenton, J. H. C., The rate of peat accumulation in Antarctic moss banks, *Journal of Ecology* 68: 211-228, 1982.

Fiscus, E. L. and Booker, F. L., Is increased UV-B a threat to crop photosynthesis and productivity? *Photosynthesis Research* 43: 81-92, 1995.

Fowbert, J. A. and Smith, R. I. L., Rapid population increase in native vascular plants in the Argentine Islands, Antarctic Peninsula, *Arctic and Alpine Research* 26: 290-296, 1994.

Fox, A. J. and Cooper, A. P. R., Climate-change indicators from archival aerial photography of the Antarctic Peninsula, *Annals of Glaciology* 27: 636-642, 1998.

Freckman, D.W. and Virginia, R.A., Low-diversity Antarctic soil nematode communities: distribution and response to disturbance, *Ecology* 78: 363-369, 1997.

Frenot, Y., Gloaguen, J.-C. and Trehen, P., Climate change in Kerguelen Islands and colonization of recently deglaciated areas by *Poa kerguelensis* and *P. annua*. In: B. Battaglia, J. Valencia and D.W.H. Walton (eds.) *Antarctic Communities: Species, Structure and Survival*, Cambridge University Press, Cambridge, 1997.

George, A. L., Murray, A. W. and Montiel, P. O., Tolerance of Antarctic cyanobacterial mats to enhanced UV radiation. *FEMS Microbiology Ecology* 37: 91-101, 2001.

George, A. L., Peat, H. J. and Buma, A. G. J., Evaluation of DNA dosimetry to assess ozone-mediated variability of biologically harmful ultraviolet radiation in Antarctica. *Photochemistry and Photobiology* 76: 274-280, 2002.

Gordon, J. E. and Timmis, R. J., Glacier fluctuations on South Georgia during the 1970s and early 1980s, *Antarctic Science* 4: 215-226, 1992.

Greenslade, P. J. M., Adversity selection and the habitat templet. *American Naturalist* 122: 352-365, 1983.

Grime, J. P., The C-S-R model of primary plant strategies—origins, implications and tests. In: R. M. Anderson, B. D. Turner and L. R. Taylor (eds.) *Population Dynamics*, Blackwell, Oxford. pp. 123-139, 1988.

Grobe, C. W., Ruhland C. T. and Day T. A., A new population of *Colobanthus quitensis* near Arthur Harbor, Antarctica: correlating recruitment with warmer summer temperatures, *Arctic and Alpine Research* 29: 217-221, 1997.

Hodgson, D.A. and Johnston N.M. Inferring seal populations from lake sediments. *Nature* 387: 30-31, 1997.

Hodgson, D. A., Johnston, N. M., Caulkett, A. P. and Jones, V. J., Palaeolimnology of Antarctic fur seal *Arctocephalus gazella* populations and implications for Antarctic management, *Biological Conservation* 83: 145-154, 1998.

Hodgson, D. A., Noon, P. E., Vyverman, W., Bryant, C. L., Gore, D. B., Appleby, P., Gilmour, M., Verleyen, E., Sabbe, K., Jones, V. J., Ellis-Evans, J. C. and Wood, P. B., Were the Larsemann Hills ice-free through the last glacial maximum? *Antarctic Science* 13: 440-454, 2001.

Hodgson, D. A., Vyverman, W. and Sabbe, K., Limnology and biology of saline lakes in the Rauer Islands, eastern Antarctica, *Antarctic Science* 13: 255-270, 2001

Huiskes, A., Lud, D., Moerdijk-Poortvliet, T. and Rozema, J., Impact of UV-B radiation on Antarctic terrestrial vegetation. In: J. Rozema (ed.) *Stratospheric Ozone Depletion: the*

Effects of Enhanced UV-B Radiation on Terrestrial Ecosystems, Backuys, Leiden, 1999.

Johnson, D., Campbell, C. D., Lee, J. A., Callaghan, T. V. and Gwynne-Jones, D., Arctic microorganisms respond more to elevated UV-B radiation than CO_2, *Nature* 416: 82-83, 2002.

Jones, V.J., Hodgson D.A. and Chepstow-Lusty A. Palaeolimnological evidence for marked Holocene environmental changes on Signy Island, Antarctica. *The Holocene* 10: 43-60, 2002

Kennedy, A. D., Water as a limiting factor in the Antarctic terrestrial environment: a biogeographical synthesis, *Arctic and Alpine Research* 25: 308-315, 1993.

Kennedy, A. D., Simulated climate change: a field manipulation study of polar microarthropod community response to global warming, *Ecography* 17: 131-140, 1994.

Kennedy, A. D., Antarctic terrestrial ecosystem responses to global environmental change, *Annual Review of Ecology and Systematics* 26: 683-704, 1995a.

Kennedy, A. D., Temperature effects of passive greenhouse apparatus in high-latitude climate change experiments, *Functional Ecology* 9: 340-350, 1995b.

Kennedy, A. D., Simulated climate change: are passive greenhouses a valid microcosm for testing the biological effects of environmental perturbations? *Global Change Biology* 1: 29-42, 1995c.

Kennedy, A. D., Antarctic fellfield response to climate change: a tripartite synthesis of experimental data, *Oecologia* 107: 141-150, 1996.

King, J. C., Recent climate variability in the vicinity of the Antarctic Peninsula, *International Journal of Climatology* 14: 357-369, 1994.

King, J. C. and Harangozo, S. A., Climate change in the western Antarctic Peninsula since 1945: observations and possible causes, *Annals of Glaciology* 27: 571-575, 1998.

Leavitt, P.R, Hodgson, D.A. and Pienitz R. Past UV radiation environments and impacts on lakes. In: E.W. Helbling, and H. Zagarese (eds.) *UV Effects in Aquatic Organisms and Ecosystems*. Comprehensive Series in Photosciences, Vol. 2, Royal Society of Chemistry, 2002.

Longton, R. E., *Biology of Polar Bryophytes and Lichens*, Cambridge University Press, Cambridge, 1988.

Lorius, C., Jouzel, J., Ritz, C., Merlivat, L. and Barkov N. I., A 150,000-year climate record from Antarctic ice, *Nature* 316: 591-596, 1985.

Lud, D., Huiskes, A. H. L., Moerdijk, T. C. W. and Rozema, J., The effects of altered levels of UV-B radiation on an Antarctic grass and lichen, *Plant Ecology* 154: 87-99, 2001.

McGraw, J. B. and Day, T. A., Size and characteristics of a natural seed bank in Antarctica *Arctic and Alpine Research* 29: 213-216, 1997.

Marshall, D. J. and Coetzee, L., Historical biogeography and ecology of a continental Antarctic mite genus, *Maudheimia* (Acari, Oribatida): evidence for a Gondwanan origin and Pliocene-Pleistocene speciation, *Zoological Journal of the Linnean Society* 129: 111-128, 2000.

Marshall, D. J. and Pugh, P. J. A., Origin of the inland Acari of continental Antarctica, with particular reference to Dronning Maud Land, *Zoological Journal of the Linnean Society* 118: 101-118, 1996.

Marshall, W. A., Biological particles over Antarctica, *Nature* 383: 680, 1996.

Marshall, W. A. and Convey, P., Dispersal of moss propagules in the maritime Antarctic, *Polar Biology* 18: 376-383, 1997.

Montiel, P., Smith, A. and Keiller, D., Photosynthetic responses of selected Antarctic plants to solar radiation in the southern maritime Antarctic. *Polar Research* 18: 229-235, 1999.

Newsham, K. K., Greenslade, P. D., Kennedy, V. H. and McLeod, A. R., Elevated UV-B radiation incident on *Quercus robur* leaf canopies enhances decomposition of resulting leaf litter in soil, *Global Change Biology* 5: 403-409, 1999.

Newsham, K. K., Hodgson, D. A., Murray, A. W. A., Peat, H. J. and Smith, R. I. L., Response of two Antarctic bryophytes to stratospheric ozone depletion, *Global Change Biology*, 8:972-983, 2002.

Noon, P. E., Birks, H. J. B., Jones, V. J. and Ellis-Evans, J. C., Quantitative models for reconstructing catchment ice-extent using physical-chemical characteristics of lake sediments, *Journal of Paleolimnology* 25: 375-392, 2001.

Norby, R. J., Kobayashi, K. and Kimball, B. A., Rising CO_2—future ecosystems. *New Phytologist* 150: 215-221, 2001.

Oechel, W.C., Cook, A.C., Hastings, S.J. and Vourlitis, G.L., Effects of CO_2 and climate change on arctic ecosystems. In: S. J. Woodin and M. Marquiss (eds.) *Ecology of Arctic Environments*. Blackwell, Oxford, 1997.

Paul, N., Plant responses to UV-B: time to look beyond stratospheric ozone depletion? *New Phytologist* 150: 5-8, 2001.

Paul, N., Callaghan, T., Moody, S., Gwynne-Jones, D., Johanson, U. and C. Gehrke, C, UV-B impacts on decomposition and biogeochemical cycling. In: Rozema, J. (Ed.), *Stratospheric ozone depletion, the effects of enhanced UV-B radiation on terrestrial ecosystems*, Backhuys, Leiden, pp. 117-133, 1999.

Pudsey, C. J. and Evans, J., First survey of Antarctic sub-ice shelf sediments reveals mid-Holocene ice shelf retreat, *Geology* 29: 787-790, 2001.

Pugh, P. J. A., Non-indigenous Acari of Antarctica and the sub-Antarctic islands, *Zoological Journal of the Linnean Society* 110: 207-217, 1994.

Pugh, P. J. A. and Convey, P., Scotia Arc Acari: antiquity and origin, *Zoological Journal of the Linnean Society* 130: 309-328, 2000.

Pugh, P. J. A. and Davenport, J., Colonisation vs. disturbance: the effects of sustained ice-scouring on intertidal communities, *Journal of Experimental Marine Biology and Ecology* 210: 1-21, 1997.

Quayle, W. C., Peck, L. S., Peat, H., Ellis-Evans, J. C. and Harrigan, P. R., Extreme responses to climate change in Antarctic lakes, *Science* 295: 645, 2002.

Quesada, A., Mouget, J. L. and Vincent, W. F., Growth of Antarctic cyanobacteria under ultraviolet radiation—UVA counteracts UVB inhibition, *Journal of Phycology* 31: 242-248, 1995.

Quesada, A., Goff, L. and Karentz, D., Effects of natural UV radiation on Antarctic cyanobacterial mats, *Proceedings of the NIPR Symposium on Polar Biology* 11: 98-111, 1998.

Roberts, D. and McMinn A. A weighted-averaging regression and calibration model for inferring lake water salinity from fossil diatom assemblages in saline lakes of the Vestfold Hills: a new tool for interpreting Holocene lake histories in Antarctica. *Journal of Paleolimnology* 19: 99-113, 1998.

Roberts, L., Does the ozone hole threaten antarctic life? *Science* 244: 288-289, 1989.

Rocco, V. E., Oppezzo, O., Pizarro, R., Sommaruga, R., Ferraro, M and Zagarese, H. E., Ultraviolet damage and counteracting mechanisms in the freshwater copepod *Boeckella poppei* from the Antarctic Peninsula, *Limnology and Oceanography* 47: 829-836, 2002.

Rozema, J. (Ed.), *Stratospheric ozone depletion, the effects of enhanced UV-B radiation on terrestrial ecosystems*, Backhuys, Leiden, 1999.

Rozema, J., van Geel, B., Björn, L. O., Lean, J. and Madronich, S., Toward solving the UV puzzle, *Science* 296: 1621-1622, 2002.

Rozema, J., Kooi, B., Broekman, R. and Kuijper, L., Modelling direct (photodegradation) and indirect (litter quality) effects of enhanced UV-B on litter decomposition. In: Rozema, J. (Ed.), *Stratospheric ozone depletion, the effects of enhanced UV-B radiation on terrestrial ecosystems*, Backhuys, Leiden, pp. 135-156, 1999.

Rozema, J., Noordijk, A. S. J., Broekman, R. A., van Beem, A., Meijkamp, B. M., de Bakker, N. V., van de Staaij, J. W. M., Stroetenga, M., Bohncke, S. J. P., Konert, M., Kars, S., Peat, H. J., Smith, R. I. L. and Convey, P., (Poly)phenolic compounds in pollen and spores of Antarctic plants as indicators of solar UV-B: a new proxy for the reconstruction of past solar UV-B? *Plant Ecology* 154: 9-25, 2001.

Ruhland, C. T. and Day T. A., Effects of ultraviolet-B radiation on leaf elongation, production and phenylpropanoid concentrations of *Deschampsia antarctica* and *Colobanthus quitensis* in Antarctica. *Physiologia Plantarum* 109: 244-251, 2000.

Searles, P. S., Kropp, B. R., Flint, S. D. and Caldwell, M. M., Influence of solar UV-B radiation on peatland microbial communities of southern Argentina. *New Phytologist* 152: 213-221, 2001.

Skvarca, P., Rack W., Rott H. and Ibarzábal y Donángelo T., Evidence of recent climatic warming on the eastern Antarctic Peninsula. *Annals of Glaciology* 27: 628-632, 1998.

Smith, R. C., Prezelin, B. B., Baker, K. S., Bidigare, R. R., Boucher, N. P., Coley, T., Karentz, D., MacIntyre, S., Matlick, H. A., Menzies, D., Ondusek, M., Wan, Z. and Waters, K. J., Ozone depletion: ultraviolet radiation and phytoplankton biology in Antarctic waters, *Science* 255: 952-958, 1992.

Smith, R. I. L., Terrestrial Plant Biology of the Sub-Antarctic and Antarctic. In: R. M. Laws (ed.) *Antarctic Ecology*. Academic Press, London, 1984.

Smith, R. I. L., The bryophyte propagule bank of Antarctic fellfield soils. *Symp. Biol. Hungarica* 35: 233-245, 1987.

Smith, R. I. L., Signy Island as a paradigm of biological and environmental change in Antarctic terrestrial ecosystems. In: K.R. Kerry and G. Hempel (eds.) *Antarctic Ecosystems, Ecological Change and Conservation*. Springer-Verlag, Berlin, 1990.

Smith, R. I. L., Exotic sporomorpha as indicators of potential immigrant colonists in Antarctica, *Grana* 30: 313-324, 1991.

Smith, R. I. L., The role of bryophyte propagule banks in primary succession: case study of an Antarctic fellfield soil. In: J. Miles and D.W.H. Walton (eds.) *Primary succession on land*. Blackwell, Oxford, 1993.

Smith, R. I. L., Vascular plants as indicators of regional warming in Antarctica, *Oecologia* 99: 322-328, 1994.

Smith, R. I. L., Introduced plants in Antarctica: potential impacts and conservation issues, *Biological Conservation* 76: 135-146, 1996.

Smith, R. I. L., Plant colonization response to climate change in the Antarctic, *Folia Fac. Sci. Nat. Univ. Masarykianae Brunensis, Geographia* 25: 19-33, 2001.

Smith, R. I. L. and Coupar, A. M., The colonization potential of bryophyte propagules in Antarctic fellfield soils, *CNFRA* 58: 189-204, 1986.

Smith, V. R. and Steenkamp, M., 1990, Climatic change and its ecological implications at a sub-Antarctic island, *Oecologia* 85: 14-24, 1990.

Sømme, L., *Invertebrates in Hot and Cold Arid Environments*. Springer-Verlag, Berlin, 1995.

Southwood, T. R. E., 1977, Habitat, the templet for ecological strategies, *Journal of Animal Ecology* 46: 337-365, 1977.

Southwood, T. R. E., Tactics, strategies and templets, *Oikos* 52: 3-18, 1988.

Squier, A.H., Hodgson D.A. and Keely B.J. Sedimentary pigments as markers for environmental change in an Antarctic lake. *Organic Geochemistry*, 33: 1655-1665, 2002.

Sugden, D. E. and Clapperton, C. M., The maximum ice extent on island groups in the Scotia Sea, Antarctica, *Quaternary Research* 7: 268-282, 1977.

Sullivan, J. and Rozema, J., UV-B effects on terrestrial plant growth and photosynthesis. In: Rozema, J. (Ed.), *Stratospheric ozone depletion, the effects of enhanced UV-B radiation on terrestrial ecosystems*, Backhuys, Leiden, pp. 39-57, 1999.

Turner, J., Colwell, S.R. and Harangozo, S., Variability of precipitation over the coastal western Antarctic Peninsula from synoptic observations, *Journal of Geophysical Research* 102: 13999-14007, 1997.

Vaughan, D. G. and Doake, S. M., Recent atmospheric warming and retreat of ice shelves on the Antarctic Peninsula. *Nature* 379: 328-331, 1996.

Vaughan, D. G., Marshall, G. J., Connolley, W. C., King, J. C. and Mulvaney, R., Devil in the detail, *Science* 293: 1777-1779, 2001.

Vincent, W. F. and Quesada, A., Ultraviolet effects on cyanobacteria: implications for Antarctic microbial ecosystems, *Antarctic Research Series* 62: 111-124, 1994.

Voytek, M. A., Addressing the biological effects of decreased ozone on the Antarctic environment, *Ambio* 19: 52-61, 1990.

Walker, S., Squier A., Hodgson D.A. and Keely B.J. Origin and significance of 132-Hydroxychlorophyll derivatives in sediments. *Organic Geochemistry,* 33: 1667-1674, 2002.

Walton, D. W. H., Vincent, W. F., Timperley, M. H., Hawes, I. and Howard-Williams, C., Synthesis: polar deserts as indicators of change. In: W. B. Lyons, C. Howard-Williams and I. Hawes (eds.) *Ecosystem Processes in Antarctic Ice-Free Landscapes.* Balkema, Rotterdam, 1997.

Webb, N.R., Coulson, S. J., Hodkinson, I. D., Block, W., Bale, J. S. and Strathdee, A.T., The effects of experimental temperature elevation on populations of cryptostigmatic mites in high Arctic soils, *Pedobiologia* 42: 298-308, 1998.

Wynn-Williams, D. D., Plastic cloches for manipulating natural terrestrial environments. In: D. D. Wynn-Williams (ed.) *BIOTAS Manual of Methods for Antarctic Terrestrial and Freshwater Research,* Scientific Committee on Antarctic Research, Cambridge, 1992.

Wynn-Williams, D. D., Microbial processes and the initial stabilisation of fellfield soil. In: J. Miles and D. W. H. Walton (eds.) *Primary Succession on Land,* Blackwell, Oxford, 1993.

Wynn-Williams, D. D., Potential effects of ultraviolet radiation on Antarctic primary terrestrial colonizers: cyanobacteria, algae, and cryptogams, *Antarctic Research Series* 62: 243-257, 1994.

Wynn-Williams, D. D., Response of pioneer soil microalgal colonists to environmental change in Antarctica, *Microbial Ecology* 31: 177-188, 1996.

Xiong, F. S. and Day, T. A., Effect of solar ultraviolet-B radiation during springtime ozone depletion on photosynthesis and biomass production of Antarctic vascular plants, *Plant Physiology* 125: 738-751, 2001.

Peter Convey, British Antarctic Survey, Natural Environment Research Council, High Cross, Madingley Road, Cambridge CB3 0ET, United Kingdom.

ECOLOGICAL RESPONSES OF MARITIME ANTARCTIC LAKES TO REGIONAL CLIMATE CHANGE

Wendy C. Quayle[1], Peter Convey, Lloyd S. Peck, Cynan J. Ellis-Evans, Helen G. Butler[2], and Helen J. Peat

British Antarctic Survey, Cambridge, United Kingdom

Polar lakes act as key early detectors of global change effects. Duration and thickness of snow and ice cover determines albedo, underwater light availability, thermal and chemical regimes and mixing properties. However, quantitative data on responses by aquatic systems to regional-scale change remain scant. Here, by utilizing 48 year (1947–1995) air temperature and 33 year (1963–1996) lake environmental datasets obtained at Signy Island, South Orkney Islands, we report one of the fastest responses to regional climate change so far documented in the southern hemisphere. Between 1980–1995 lake water temperatures increased by 2–3 times the local air temperature rise. Autumn freeze and spring break up dates of lake ice have changed, extending the open water period by up to 4 weeks. Winter total extractable phytopigments, alkalinity and orthophosphate show 2- to 10-fold increases, all linked with an areal reduction of the island's ice cover of ~ 45% in total in the last 50 years. These results indicate that its geographical position allows Signy Island to be a sensitive indicator of rapid ecological change in Antarctic freshwater environments. Such rapid change will be widespread in polar regions if current global change predictions are correct.

INTRODUCTION

Uncertainties relating to regional scale climate predictions and a lack of understanding about critical processes prompted IPCC [Intergovernmental Panel on Climate Change] Working Group II [*Watson et al.,* 1995; *McCarthy et al.,* 2001] to propose that assessing the sensitivity and adaptability of systems to regional climate change is necessary to understand its potential impacts. Obtaining more quantitative information on regional-scale climate change and responses of systems to this change was a major recommendation promulgated by the Group [*Watson et al.,* 1995].

The Antarctic Peninsula region is experiencing rapid changes in climate, considerably greater than the global mean. Regional warming has been evident since around 1950 and the rate has accelerated since about 1980 [*Smith,* 1990]. Most notable is air temperature which is increasing at a rate of 0.06°C a[-1], [*Smith,* 1990, *Fowbert and Smith,* 1994; *King,* 1994; *King and Harangozo,* 1998; *Skvarca et al.* 1998], possibly driven by El Niño Southern Oscillation (ENSO) events in the southern Pacific Ocean [*Cullather et al.,* 1996]. Although the period and seasons of observations have been different between different study locations, Signy Island, South Orkney Islands (60°43'S 45°38'W), on the Scotia Arc is one of a number of sites that has experienced significant annual air temperature increases over the last 30–50 years. [*Smith,* 1990, *Frenot et al.,* 1997, *Bergstrom and Chown,* 1999, *Convey,* 2001]. In sharp contrast with a 14

[1]Current affiliation: CSIRO Land & Water, Griffith, Australia
[2]Current affiliation: NERC, Swindon, UK

year data record (1986–1999) from the McMurdo Dry Valleys which reflects a 0.7°C drop in air temperature per decade [*Doran et al., 2002*; but see also *Turner et al., 2002* and *Walsh et al., 2002*], summer air temperatures on Signy have increased by 1°C over the last 50 years. During this period two further major environmental perturbations have been occurring on the island. Firstly, it is estimated that there has been an areal reduction in perennial ice cover of approximately 45% [*Smith, 1990, Quayle et al., 2002*]. Secondly, terrestrial and freshwater ecosystems adjacent to the coast have been severely impacted by rapidly increasing numbers of fur seals (*Arctocephalus gazella*) [*Hawes, 1985, Smith, 1988, 1990, Ellis-Evans, 1990, Butler, 1999a*].

Other than the Antarctic Peninsula region, the most sensitive regions responding to climate change are found in parts of the Arctic, where temperatures have increased by *c.* 0.05–0.07°C a^{-1} between 1961 and 1990, and the Tibetan Plateau [0.03°C a^{-1}, *Liu and Chen, 2000*]. Arctic ecosystems are currently predicted to warm by 3–6°C over the next 50 years (0.06–0.12 °C a^{-1}) although the implications of this increase are not fully understood [*McDonald et al., 1996*]. These values compare with global mean temperature increases of 0.016°C a^{-1} over the last 2 decades [*Jones et al., 1999*], with recent warming greater over land (0.024°C a^{-1}) than the ocean (0.016°C a^{-1}) [*Stott et al., 2000*].

Climate change has the potential to affect many aspects of ecosystem function. However, there are large differences in the sensitivity of particular systems to different regional climate scenarios, as they are subjected to multiple climatic and non-climatic stresses on a range of scales, which also may have non-linear interactions.

Effects of climate change have been recognized in the decreasing duration of ice cover on lakes and rivers around the northern hemisphere for time series that are longer than 100 years [*Magnuson et al., 2000*]. Linear trends from 1846–1995 from Russia, Finland and Japan averaged a freeze date that was 5.8days/100 years later and a break up date 6.5 days/100 years earlier. The change in freeze and break up dates corresponded to ~1.2°C/100 (0.012°C a^{-1}) years and conversion of change in ice cover to change in temperature equates to 0.2°C per day change in phenological date. Ice phenology is a useful climate proxy since it integrates change over the summer season. The break up date and freeze date is most strongly correlated to the temperature 1–2 months prior to the event and the break up date is strongly influenced by the timing, magnitude and rate of spring run off [*Magnuson et al., 2000*].

Responses to climate change have been closely studied in Alaskan Arctic lakes [*Hobbie et al., 1999*] and boreal lakes in northern Ontario near the area of predicted maximum greenhouse effect [*Schindler et al., 1990*]. These latter lakes warmed by 1–2°C between 1969 and 1988 (0.05–0.11 °C a^{-1}) consistent with, or in some cases higher than, model predictions [*King and Harangozo, 1998, Schindler et al., 1990*]. In these northern hemisphere boreal systems, the effects are dependent on watershed processes such as patterns of water renewal and less significantly, occurrences of forest fires in catchment areas [*Schindler et al., 1996*]. The thawing of permafrost in Arctic tundra systems is a particularly important consequence of environmental warming, leading to lake enrichment by increased phosphorus inputs [*Hobbie et al., 1999*].

In contrast, the sensitivity to climate change of Antarctic lakes and certain High Arctic freshwater systems (e.g. Char Lake, *Schindler et al., 1974*), which lie within relatively barren fellfield, rock and ice catchments, is more related to their being close to the threshold of water phase changes. This applies critical limits on many physical environmental variables including temperature, ice extent, snow cover, underwater light availability and albedo [*Vincent, 1988*].

Lakes on Signy Island experience large seasonal variations in photosynthetically active radiation (PAR) caused by changes in seasonal incident radiation and winter ice and snow cover [*Heywood et al., 1980*]. Such changes cause substantial and often rapid variations in phytoplankton activity [*Ellis-Evans, 1990*], with consequent changes in water column chemistry [*Hawes, 1983, 1985*].

Here we report dramatic changes in physical and chemical water parameters in nine lakes on Signy Island, South Orkney Islands whose general characteristics are known [*Heywood, 1967, Heywood et al., 1980*]. Out of the total of 17 lakes on the Island, these 9, [namely Changing, Moss, Light, Spirogyra, Tranquil, Gneiss, Twisted, Emerald and Tioga Lakes, see Figure 1], were selected as having minimal disturbance from local seabird and seal populations, which could confound the identification of links with climate change processes.

LOCATION DESCRIPTION

Signy Island lies at the confluence of the ice-bound Weddell Sea and warmer Scotia Sea. Air mass interactions between the two areas govern its climate, while the influence of Weddell Sea pack ice is strong [*Heywood et*

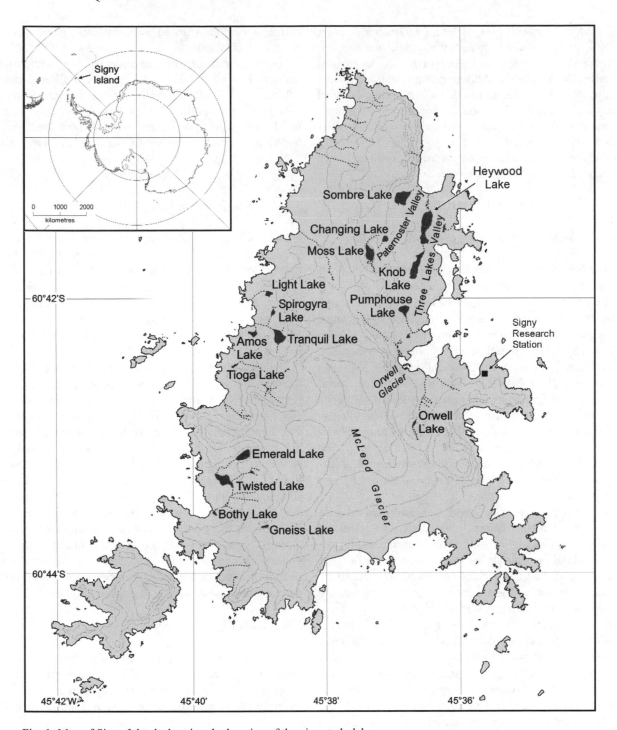

Fig. 1. Map of Signy Island, showing the location of the nine study lakes.

al., 1980, Murphy et al., 1995, Clarke and Leakey, 1996]. Its position, oceanic setting, small area and low altitude make it exceptionally sensitive to environmental change [*Smith*, 1990]. Although Signy Island is only about 20 km^2 in area, its 17 lakes cover a wide range of maritime Antarctic lacustrine environments. Altitudes range from 4.5–150 m above sea level and aspects differ markedly [*Heywood*, 1967, *Heywood et al.,* 1980]. Individual catchment areas are separated, small and often isolated [*Heywood et al.,* 1980].

METHODS

Meteorological and Limnological Data Collection

The British Antarctic Survey (BAS) recorded daily air temperatures on Signy Island between 1947 and 1995 (when year-round operations ceased at the Research Station) using standard methods, and since 1988 using a Synoptic Climatological Automatic Weather Station (SCAWS). Mean monthly values of air temperature (Stevenson's screen) were used in this study.

Between 1963 and 1996, BAS undertook year round studies of the freshwater lakes on Signy Island and, since the early 1970s, conducted a regular lake-monitoring program on the island. A number of chemical and physical variables were monitored in 16 lakes in the period 1980–1995. The data discussed here were obtained during winter from 9 lakes which were undisturbed by animals and birds.

The extent of lake ice cover was estimated on a % basis. The number of open water days was calculated between the loss of 100% ice cover in spring and formation of 100% ice cover in autumn. The ice break out date was taken as the first date any open water was observed, while ice formation was taken as the date after which the ice did not break up again for the rest of the winter. It was not practicable to make daily visits to all lakes during the critical freeze/melt periods, introducing a small amount of inaccuracy. However, we consider that the changes in the ice formation date and duration of open water period over the study period are too large and the increase too regular to be accounted for by estimation errors.

Water temperature was measured in the field using a YSI polarographic probe and thermistor. Prior to 1990, ammonium-N concentration was determined using the method of *Chaney and Marbach* [1962], orthophosphate following *Strickland and Parsons* [1972] and alkalinity according to *Mackereth* [1963]. From 1990, the analyses followed modifications to all these methods described by *Mackereth et al.* [1989]. Cross comparisons between old

and new methods showed no significant differences in the results obtained.

For measurements of phaeopigment concentrations aliquots [500 ml – 2 l] of water were filtered onto Whatman GF/C filters, which were then extracted with cold 95% methanol for 24 hours. Samples were acidified to 0.003 M with hydrochloric acid and measured spectrophotometrically according to *Marker et al.* [1980a,b].

RESULTS

Water Temperature and Ice Cover

Mean winter (July/August) water temperatures for all 9 lakes increased by 0.9°C between 1980 and 1995 (0.06°C a^{-1}). Each individual lake showed an increasing temperature trend over the study period, with increases ranging from 0.2 to 1.3°C. Trends in 6 of the 9 lakes were significant (linear regression, $p < 0.05$) [Figure 2a]. Local sea temperatures, however, did not change over this period [BAS, unpublished data].

The open water period was 34 days longer in 1994, and 63 days longer in 1993, than was observed in 1980 [Figure 3a]. The trend of the raw data indicated that autumn freeze occurred c. 10 days later and, more importantly, spring thaw c. 30 days earlier at the end compared to the start of the study [Figure 3b]. Lake ice phenology reflects a 0.5°C increase in mean summer air temperatures [0.033 °C a^{-1}; Figure 2b] but little or no increase in autumn or winter air temperatures at Signy Island during this study. However, we also recognize that interannual variability in these dates will be substantial, as ice formation and breakup are influenced by the timing of frontal systems arriving over the island and by highly localized factors such as wind speed and direction over each lake during spring and autumn [*Gardiner et al.,* 1998].

Extractable Phytopigments

Total extractable phytopigment concentrations increased significantly in 7 of the 9 lakes, from an overall mean of 1.4 µg l^{-1} in the early 1980s to 3.5 µg l^{-1} in the early 1990s, peaking at 6.8 µg l^{-1} in 1995 [Figure 4].

Alkalinity, Ammonium-N, and Orthophosphate

Mean winter alkalinity of all lakes increased significantly over the 15 years of study [linear regression, $p=0.05$] from approximately 230 to 350 µequiv l^{-1} [Figure 5a]. Mean winter orthophosphate concentration increased 5-fold from approximately 5 µg l^{-1} in the early

1980s to 25 µg l⁻¹ in the mid 1990s [Figure 5b]. Mean winter ammonium-N concentration also increased from 98 to 184 µg l⁻¹ over the course of the study. However, significant increases occurred in only 4 of the 9 lakes with the remaining 5 showing little or no change in concentration resulting in an overall non-significant trend in the mean data for all lakes [Figure 5c].

DISCUSSION

Water Temperature and Ice Cover

Polar lakes warm in winter through the processes of basal heat flow and release of latent heat of fusion during

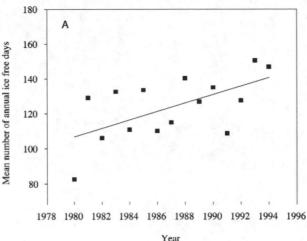

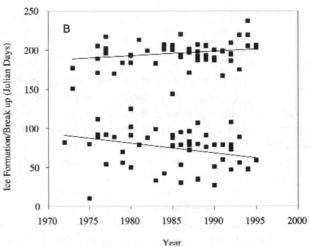

Fig. 3. Mean number of annual ice free days (mean across all lakes) [linear regression, $R^2 = 0.32$, $p < 0.05$]. b. trends of ice formation and break up dates.

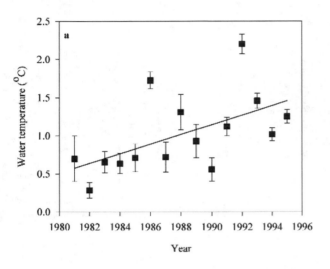

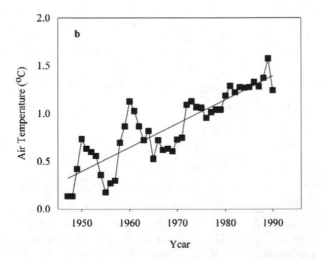

Fig. 2. a. Mean winter (July/August) lake water temperature [linear regression, $R^2 = 0.5671$, $p < 0.05$]; b. Five year running mean of summer [December to February] air temperature.

ice formation. Insulation of lake waters by ice and snow cover also dramatically reduces heat loss [*Ragotzkie*, 1978, *Vincent*, 1988]. Therefore, it is reasonable to hypothesize that the combination of reduction in lake ice cover duration and increase in exposed rock in lake catchments on Signy Island have, on balance, increased annual heat budgets of the lakes through two processes - [1] an extended period of incident summer radiation absorption by the water column and [2] increased absorption of incident radiation by lake bottom sediments and microbial mats and surrounding rock basins. The latter is stored as heat and released to the water column during winter when the lakes are ice covered.

Comparison of a regular photographic record of key ice retreat areas [*Quayle et al.*, 2002] corroborate previous estimates [*Smith*, 1990] that the perennial ice cover

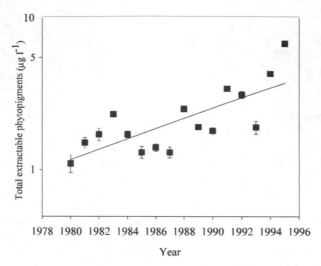

Fig. 4. Mean concentration of winter [July/August] extractable phytopigments over the course of the study [linear regression, $R^2 = 0.5735$, $p < 0.001$].

on Signy Island has reduced in area by roughly 45% since 1951. This high rate of deglaciation associated with climate change has affected lakes in several ways and amplified signals on the island through increased melting and loss by ablation in recent years [e.g *Noon et al., 2001, Smith, 1985, 1990 Block and Harrisson, 1995*].

The data indicate that the mean rate of lake water temperature increase is 3-4 times greater than global mean temperature increase [*Jones et al., 1999*] and 2–3 times the rate of local mean summer air temperature increase at Signy Island [Figure 2b] [*Skvarca et al., 1998*]. A key factor underlying water temperature increases may be the apparent reduction in the duration of lake snow and ice cover of approximately 30–40 days per annum. An analogous line of evidence has been drawn from terrestrial invertebrate ecophysiological studies on Signy Island. These indicate that the duration of the summer active season when liquid water is available in terrestrial habitats has increased recently, implying earlier thaw and/or later freezing events [*Block and Harrisson, 1995, Block and Convey, 2001*].

Lake ice formation and break up dates correlate most strongly with air temperatures during 1–2 months before the event [*Magnuson et al., 2000*]. However, the timing, magnitude and rate of spring runoff are also critical in controlling the phenology of lake ice break up. Recent changes in precipitation patterns on the island, with less snowfall in winter and a greater proportion of precipitation occurring as rain in summer, will also be influential [*Noon et al., 2001*]. On Signy Island, these factors are intimately linked to the characteristics of the Island's ice

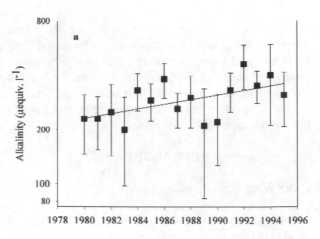

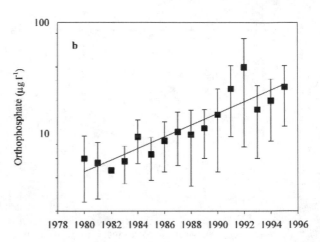

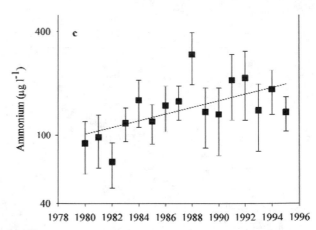

Fig. 5. Trends over the study period of a. alkalinity [linear regression, $R^2 = 0.333$, $p < 0.05$], b. orthophosphate [linear regression, $R^2 = 0.8146$, $p < 0.0001$] (all nine lakes increased significantly at $p < 0.05$), c. ammonium-N (four of the nine lakes increased significantly, leading to an insignificant but upward trend when all lakes were considered together). All error bars (SE) depict variation between lakes.

cap, which is currently very sensitive to fluctuations in summer air temperatures [*Smith,* 1990]. The ice phenology data equate to a change of approximately 0.02°C per day in the break up date, an order of magnitude more sensitive than the temperature change required to achieve the same effect in lakes and rivers of the northern hemisphere [*Magnuson et al.,* 2000].

Extractable Phytopigments

Accumulation of snow cover on the lakes from May to September reduces incident PAR penetration to the water column to < 1% of maximum summer values [*Hawes,* 1985]. During this period of the year, photosynthesis becomes undetectable, algal cells sediment out of the water column, chlorophyll-*a* concentrations decline to ~ 1 μg l^{-1} and the proportion of phaeopigments (chlorophyll breakdown products) increases [*Light et al.,* 1981, *Hawes,* 1985]. The efficiency of PAR exclusion from maritime Antarctic lakes by the large amount of snow cover present in July and August suggests that the increasing levels of pigments observed at this time of year over the study are unlikely to result from growth of phytoplankton populations. Rather, the increased concentrations are more likely derived from an enhanced standing crop of phytoplankton present in the water column at the end of each summer [*Hawes,* 1985]. This, coupled with the open water period extending further into autumn, will potentially provide a larger biomass contribution to the under-ice population.

There is also growing evidence that mixotrophy is present in Antarctic oligotrophic lakes [*Laybourn-Parry,* 1997, *Roberts and Laybourn-Parry,* 1999], while autotrophic flagellate species known to have mixotrophic capability are common in the Signy Island lakes [*Butler et al.,* 2000, Laybourn-Parry, pers. comm.]. These may contribute substantially to the observed increase in under-ice plankton populations.

Alkalinity, Ammonium-N, and Orthophosphate

Increased summer air temperatures at Signy Island [Figure 2b] and the extended period of open water for radiative energy inputs have caused changes through both deglaciation processes affecting the catchment areas and variations in in situ biogeochemical processes in the lakes. Deglaciation processes result in increasing quantities of allochthonous materials, including nutrients, micronutrients and base cations, being supplied to the lakes by meltwater streams [*Hawes,* 1983, *Smith,* 1985]. These inputs include not only soluble phases but also particulate organic and inorganic materials, which are subsequently biogeochemically transformed by microbially driven processes within the lake sediments [*Ellis-Evans,* 1985, *Caulkett and Ellis-Evans,* 1996]. Increases in alkalinity, orthophosphate and ammonium-N in the lake water column result from the interaction between increased autochthonous carbon, increased chemical imports from catchments, and reduction of iron, manganese, nitrate [nitrification and denitrification] and sulphate *via* microbial respiration, [*Schindler et al.,* 1990, *Dillon et al.,* 1997].

Thus, both autochthonous and allochthonous processes enrich the lakes, resulting in significant increases in mean winter alkalinity [Figure 5a] and 5-fold increases in mean winter orthophosphate concentrations over the study period [Figure 5b]. Identification of which source (autochthonous or allochthonous) is driving the alkalinity and nutrient increases observed in these lakes is difficult. In some lakes, it is possible that the observed increases are, in part, attributable to orthophosphate and base cations being transported into the lakes by meltwater streams, which have leached material from small bird colonies and small areas of seal impacted vegetation in the lake catchments. However, significant increases were also observed in lakes that have no animal colonies in their catchments. One alternative explanation for the considerable increases in orthophosphate concentrations is given by increases in catchment area, newly exposed to physical and chemical weathering. During the initial stages of the current deglaciation 40–50 years ago, meltwater streams would have drained largely over ice or over limited exposed tracts of frozen, weathered terrain with little opportunity for nutrient leaching. However, more recently, streams have been able to accumulate high nutrient concentrations (particularly orthophosphate) as they drain over greater extents of exposed fellfield soils [*Caulkett and Ellis-Evans,* 1997] and thawed ground. Chemical contributions are also obtained from water draining through plant and microbial communities that rapidly colonize newly exposed rock surfaces [*Smith,* 1985].

As has been found in Arctic systems [*Hobbie et al.,* 1999], the permafrost active layer depth has certainly increased at Signy Island since the early 1960s [M. Guglielmin and N. Cannone, pers. comm.], thus releasing bound inorganic and organic nutrients [*Caulkett and Ellis-Evans,* 1997]. A further contribution to the observed alkalinity and nutrient enrichment may arise through increased autochthonous activity resulting from the increased duration of the annual open water period and subsequent biogeochemical cycling in winter.

Our findings can, therefore, be explained by two factors, [1] an increase in snow and ice melt and permafrost active layer depth associated with deglaciation and [2] an increase of mineralization by lake microbes and anaerobic sediment release of nutrients in winter, associated with the increased autochthonous production enabled by the longer annual open water period.

Allochthonous organic and inorganic nutrients supplied by the spring melt to lakes on Signy Island support seasonal microbial blooms [*Hawes*, 1983]. However, additional nutrients enter the lakes in forms unavailable to spring and summer populations, such as organic residues resistant to rapid decomposition and orthophosphate ions covalently bonded to hydrated iron oxides [*Ellis-Evans and Lemon*, 1989]. The latter are only released under anaerobic conditions within winter lake sediments. In spring and summer, when water and surface sediments are fully aerobic, most bioavailable phosphorus is quickly taken up by microbes and phosphorus is the limiting nutrient in many of the lakes [*Hawes*, 1983]. In certain lakes on Signy Island, the concentration of sediment-derived orthophosphate available to the water column is reduced even further because a proportion co-precipitates with iron [which can occur in high concentrations] and sediments out of the water column [*Caulkett and Ellis-Evans, 1996*].

In winter, under complete ice and snow cover, oxygen concentrations decrease in all vertical lake profiles as the sediment/water interface is approached [*Ellis-Evans and Lemon*, 1989, *Butler*, 1999b], because of microbial respiration in near-surface sediments. Anaerobic processes therefore become increasingly significant. In the freshwater lakes studied here, sulphate is only present in low concentrations and, therefore, denitrifiers, manganese and iron reducing bacteria and methanogens, rather than sulphate reducing bacteria, are the primary organisms that remove oxidised anions (NO_3, MnO, Fe_2O_3, CO_2) and create the anoxic conditions essential for the release of iron, manganese, phosphorus and other base cations to lake bottom waters [*Dillon et al.*, 1997].

In all the lakes, sediment oxygen consumption under winter ice cover causes the oxic/anoxic boundary to migrate up towards the sediment/water interface [*Ellis-Evans*, 1985] and in some cases into the overlying water column [e.g. *Butler*, 1999a]. A large proportion of the phosphate mobilized is ultimately trapped at the oxic/anoxic boundary by co-precipitation with ferric oxyhydroxides. However, there is evidence that a small proportion is released to overlying waters [*Ellis-Evans and Lemon*, 1989]. This proportion is likely to increase as the abundance of organic and inorganic materials

delivered to the lake sediments increases [*Butler*, 1999b]. Even the relatively small increases in lake water temperatures observed [Figure 2a] may enhance these effects by depleting winter oxygen concentrations further [*Hobbie et al.*, 1999] and by stimulating increases in microbial physiological activity as predicted by the Arrhenius relationship [*Vincent*, 1988]. Although a contentious topic [*Rivkin et al.*, 1996], the effect may be amplified in cold polar lakes inhabited by psychro-tolerant microbes where relatively small increases in lake water temperature produce a strong response in microbial growth rates [*Vincent*, 1988].

Ammonium-N is generated during the mineralization of organic matter and the release of large amounts from sediments has been observed during winter in some lakes on Signy Island [*Hawes*, 1983, *Ellis-Evans*, 1985, *Butler*, 1999a]. Unlike phosphate, ammonium is not influenced by the presence of iron oxyhydroxides and can readily pass across redox boundaries into the main water body [*Ellis-Evans and Lemon*, 1989]. The significant changes in water column ammonium-N concentrations in four lakes (Changing, Emerald, Tioga and Twisted Lakes) are not simple to interpret. However, evidence from Emerald Lake's sediment record suggests increasing organic content throughout the study period [*Noon et al.*, 2001], which *Jones et al.*, [2000] interpreted as indicating increased autochthonous production. The warmer lake temperatures and extended summer periods for autochthonous production and input of allochthonous nutrients will promote remineralization of organic-N to ammonium-N under winter ice cover.

A further cause of changes to some lakes and terrestrial ecosystems on Signy Island is disturbance by fur seals (*Arctocephalus gazella*). This has arisen through an explosive increase in their summer population on the island, from zero in the early 1960s to more than 20,000 in 1995 [*Heywood et al.*, 1980, *Smith*, 1990, *Hodgson et al.*, 1998], and caused a dramatic increase in nutrient input to affected lakes, which have become increasingly eutrophic with associated significant changes in water column chemistry and microbiology [*Hawes*, 1983, *Smith*, 1995, 1997; *Butler*, 1999a,b]. In particular, winter ammonium-N and orthophosphate levels have increased several-fold since the early to mid-1980s [*Ellis-Evans*, 1990, *Butler*, 1999a]. Although these nutrient increases are not directly linked with climate change, it is possible that the considerable increase in seal numbers at Signy is, at least in part, attributable to the recent reductions in the island's ice cover allowing access to suitable resting or molting areas. Therefore, the increase in seal population may itself be a secondary consequence of regional climate change, with the associated ecological consequences for seal-impacted lakes being a tertiary consequence.

CONCLUSIONS

Limnological trends in 9 lakes studied on Signy Island between 1980 and 1995 indicate that local climate change has produced amplified ecological change in these maritime Antarctic freshwater environments. Our results indicate that water temperatures have risen around three times more rapidly than local air temperatures. Importantly, significant responses by lake biota and increased nutrient levels appear to be linked to reductions in lake snow and ice cover and the deglaciation of the island brought about by increased summer air temperatures in recent years. Geographically, Signy Island is in an optimal setting for the early recognition of changes occurring in freshwater environments in response to regional climate warming. The patterns found at Signy Island have similarities with those recognized in Arctic systems and are thus likely to be seen in other polar lakes in coming years.

If the forecasts generated by recent climate models, of continued air temperature increases over the next several decades, are correct the lacustrine ecosystems on the Antarctic Peninsula may be severely affected. The trends that we have recognised provide foresight of how regional climate change may affect ecological processes in polar lakes and catchments in future. Continued deglaciation of maritime Antarctic islands is expected to occur as air temperatures continue to warm, exposing greater areas of bare ground able to absorb incident radiation and further decreasing albedo effects. The consequences of these changes will lead to further decreases in the duration of lake ice and snow cover, increases in mean and maximum water temperature and increases in the overall heat content of the lakes [Schindler, et al., 1996]. These physical changes will initially translate to a continued increase in chlorophyll-a concentration, lake primary productivity, phytoplankton biomass and the size and productivity of the bacterial plankton population caused by the enhanced light and temperature regime. This will then lead to a greater carry over of standing crop of microbial plankton between years. Photosynthetic dinoflagellates are likely to colonise lakes in which they have not previously been observed (Butler, 1999b). The timing and duration of maximum and minimum growth of phytoplankton are likely to shift as the duration of the period that lakes are ice free increases, leading to corresponding temporal shifts in

maximum numbers of heterotrophic bacteria and other organisms that make up the microbial foodweb.

Further into the future, should warming progress to such an extent that no annual ice cover forms, what may be considered to be a negative feedback may develop, whereby reduced microbial plankton numbers occur as the heat insulating properties of ice cover, and the important microbial habitat at the ice underside/lake water interface are lost. In this scenario, without lake ice cover and with year-round water mixing, oxygen would freely exchange between the atmosphere and water column and depletion of water column oxygen due to organic matter recycling will be significantly reduced. The biogeochemical cycling of organic matter and nutrients would then take place under aerobic rather than anaerobic conditions, nutrients would be unable to accumulate to significant levels and in such unenriched catchments nutrient limitation would be widespread.

Important consequences for lakes also arise from the impacts of a thickening of the seasonally thawed layer above permafrost, which will occur as a result of regional warming. A thickened 'active layer' will lead to changes in drainage through improvement of infiltration in some areas and ponding of increased volumes of surface run-off and sub-surface seepage in others. Reduction in ice-jammed ponding and peri-glacial lakes is inevitable as glaciers and permanent snow and ice fields, retreat. Run-off regime is likely to become rainfall rather than ice-melt dominated and seasonal variation will decrease. Although lake water renewal time will be shortened, the corresponding dilution effects are likely to be countered in some cases by mass movements, which will increase erosion rates and import more clays into the water column, causing increased turbidity. The benefits to primary productivity of increased turbidity such as simultaneous increases of allochthonous organic material, base cations and inorganic nutrients will be offset by decreases in penetrating radiation.

Our conclusions, like studies of lakes in other areas [Shindler et al., 1996; Fee et al., 1996] demonstrate that the net response of polar lakes to climate change involves a complexity of interactions between the different effects of temperature, hydrology and erosion in the catchment and in-lake biological activity and biogeochemical cycling. This complexity has prevented development of a clear understanding of lake catchment and in-lake processes, making robust predictions on the response of lakes to climate change scenarios impossible. Long-term monitoring of the responses of aquatic ecosystems to climate variability is expensive and frequently seen as non-cost efficient but it remains the most informative way of improving understanding of the ecological consequences of climate change.

Acknowledgments. Research support from the British Antarctic Survey (Natural Environmental Research Council) for this project over many years is gratefully acknowledged. Field sampling at Signy Island was undertaken by S. Malik, M. W. Sanders, N. L. Rose, R. Wedgwood, P. Brazier, M. G. Smithers, A. P. Caulkett, M. O. Chalmers and M. G. Edworthy, without which this research would not have been possible.

REFERENCES

Bergstrom, D. M. and S. L. Chown, Life at the front: history, ecology and change on southern ocean islands, *Trends Ecol. Evol., 14,* 472-476, 1999.

Block, W., and P. Convey, Long term variation in body water content of an Antarctic springtail - a response to climate change, *Polar Biol. 24,* 764-770, 2001.

Block, W., and P. M. Harrisson, Collembolan water relations and environmental change in the maritime Antarctic, *Global Change Biol., 1,* 347-359, 1995.

Butler, H. G., Seasonal dynamics of the planktonic microbial community in a maritime Antarctic lake undergoing eutrophication, *Journal Plankton Res., 21,* 2393-2419, 1999a.

Butler, H. G., Temporal plankton dynamics in a maritime Antarctic lake, *Arch. Hydrobiol., 146,* 311-339, 1999b.

Butler, H.G, M.G Edworthy, and J C Ellis-Evans, Temporal plankton dynamics in an oligotrophic maritime Antarctic lake. *Freshwater Biology. 43,* 215-230, 2000.

Caulkett, A. P., and J. C. Ellis-Evans, Origin and composition of settling iron aggregates in oligotrophic Sombre Lake, Signy Island, Antarctica, *Hydrobiol., 330,* 177-187, 1996.

Caulkett, A. P., and J. C. Ellis-Evans, Chemistry of streams of Signy Island, maritime Antarctic: sources of major ions, *Antarctic Science, 9,* 3-11, 1997.

Chaney, A. L. and E. P. Marbach, Modified reagents for the determination of urea and ammonia. *Clin. Chem, 8,* 130-132, 1962.

Clarke, A., and R. J. G. Leakey, The seasonal cycle of phytoplankton, macronutrients, and the microbial community in a nearshore Antarctic marine ecosystem, *Limnol. Oceanogr., 41,* 1281-1294, 1996.

Convey, P., Terrestrial ecosystem response to climate changes in the Antarctic, in *"Fingerprints" of climate change - adapted behaviour and shifting species ranges,* edited by G.-R. Walther, C. A. Burga & P. J. Edwards, pp. 17-42, Kluwer, New York, 2001.

Cullather, R. I., D. H. Bromwich, and M. L. van Woert, Interannual variations in Antarctic precipitation related to El Niño-Southern Oscillation. *J. Geophys. Res., 101,* 19109-19118. 1996.

Dillon, P. J., H. E. Evans and R. Girard, Hypolimnetic alkalinity generation in two dilute, oligotrophic lakes in Ontario, Canada, *Water Air Soil Pollution, 99,* 373-380, 1997.

Doran, P. T., J. C. Priscu, W. Berry Lyons, J.E. Walsh et al., Antarctic climate cooling and terrestrial ecosystem response, *Nature AOP,* Published online: January 2002

Ellis-Evans, J. C., Decomposition processes in Maritime Antarctic lakes, in *Antarctic Nutrient Cycles and Food Webs,* edited by W.R. Siegfried, P.R. Condy and R.M. Laws, pp 253-260, Springer-Verlag, Berlin, Heidelberg, 1985.

Ellis-Evans, J. C., Evidence for change in the chemistry of maritime Antarctic Heywood Lake, in *Antarctic Ecosystems – Ecological Change and Conservation,* edited by K. R. Kerry and G. Hempel, pp. 83-90, Springer-Verlag, Berlin, Heidelberg, 1990.

Ellis-Evans, J. C., and E. C. G. Lemon, Some aspects of iron cycling in maritime Antarctic lakes, *Hydrobiol., 172,* 149-164, 1989.

Fee, E. J., S.E. Kasian and D. Cruikshank, Effects of lake size, water clarity, and climatic variability on mixing depths in Canadian Shield lakes. *Limnol. Oceanogr.* 41, 912-920, 1996.

Fowbert, J. A. and R. I. L. Smith, Rapid population increase in native vascular plants in the Argentine Islands, Antarctic Peninsula, *Arct. Alpine Res., 26,* 290-296. 1994.

Frenot, Y., J-C Gloaguen and P. Trehen, Climate change in Kerguelen Islands and colonization of recently deglaciated areas by *Poa kerguelensis* and *P. annua,* in *Antarctic Communities: Species, Structure and Survival,* edited by B. Battaglia, J. Valencia and D. W. H. Walton, Cambridge University Press, Cambridge, 1997.

Gardiner, M. J., J. C. Ellis-Evans, M. G. Anderson and M. Tranter, Snowmelt modelling on Signy Island, South Orkney Islands. *Annals Glaciol., 26,* 161-166, 1998.

Hawes, I., Nutrients and their effects on phytoplankton populations in lakes on Signy Island, Antarctica, *Polar Biol., 2,* 115-126, 1983.

Hawes, I., Light climate and phytoplankton photosynthesis in maritime Antarctic lakes, *Hydrobiol., 123,* 69-79, 1985.

Heywood, R. B., Antarctic Ecosystems: the freshwater lakes of Signy Island and their fauna, *Phil. Trans. R. Soc. Lond., 261,* 347-362, 1967.

Heywood, R. B., H. J. G. Dartnall and J. Priddle., Characteristics and classification of the lakes of Signy Island, South Orkney Islands, Antarctica, *Freshwater Biol., 10,* 47-59, 1980.

Hobbie, J. E., B. J. Peterson, N. Bettez, L.Deegan, W. J. O'Brien, G. W. Kling, G. W. Kipphut, W. B. Bowden, and A. E. Hershey, Impact of global change on the biogeochemistry and ecology of an Arctic freshwater system, *Polar Res., 18,* 207-214, 1999.

Hodgson, D. A., N. M. Johnston, A. P. Caulkett, and V. J. Jones, Palaeolimnology of Antarctic fur seal *Arctocephalus gazella* populations and implications for Antarctic management, *Biol. Conserv, 83,* 145-154, 1998.

Jones, P. D., M. New, D. E. Parker, S. Martin, and I. G. Rigor, Surface air temperature and its changes over the last 150 years, *Rev. Geophys. 37,* 173-199, 1999.

Jones, V. J, D. A. Hodgson and A. Chepstow-Lusty. Palaeolimnological evidence for marked Holocene environmental changes on Signy Island, Antarctica, *The Holocene, 10*, 43-60, 2000.

King, J. C., Recent climate variability in the vicinity of the Antarctic Peninsula, *Int. J. Climatol., 14,* 357-369, 1994.

King, J. C., and S. A. Harangozo, Climate change in the western Antarctic Peninsula since 1945: observations and possible causes, *Ann. Glaciol., 27,* 571-575, 1998.

Laybourn-Parry, J., The microbial loop in Antarctic lakes. in *Ecosystem Processes in Antarctic ice-free Landscape,* edited by W. B. Lyons, C. Howard-Williams & I. Hawes, pp. 231-240, Balkema, Rotterdam, 1997.

Light, J. J., J. C. Ellis-Evans, and J. Priddle, Phytoplankton ecology in an Antarctic lake, *Freshwater Biol,. 11,* 11-26, 1981.

Liu, X., and B. Chen, Climatic warming in the Tibetan Plateau during recent decades. *Int. J. Climatol., 20,* 1729-1742, 2000.

Mackereth, F. J. H., Some methods of water analysis for limnologists, *Scientific Publications of the Freshwater Biological Association, 21,* 70, 1963.

Mackereth, F. J. H., J. Heron, and J. F. Talling, *Water Analysis: Some Revised Methods for Limnologists,* Freshwater Biological Association, Ambleside, 1989.

Magnuson, J. J., D. M. Robertson, B. J. Benson, R. H. Wynne, D. M. Livingstone, T. Aral, R. A Assel, R. G. Barry, V. Card, E. Kuusisto, N. G. Granin, T. D. Prowse, K. M. Stewart, and V. S. Vuglinski, Historical trends in lake and river ice cover in the northern hemisphere, *Science 289,* 1743-1746, 2000.

Marker, A. F. H., C. A. Crowther and R. J. M. Gunn, Methanol and acetone as solvents for estimating chlorophyll *a* and phaeopigments by spectrophotometry, *Arch. Hydrobiol. Beih. Ergebn. Limnol., 14,* 52-69, 1980a.

Marker, A. F. H., E. A. Nusch, H. Rai and B. Riemann, The measurement of photosynthetic pigments in freshwaters and standardization of methods: conclusions and recommendations, *Arch. Hydrobiol. Beih. Ergebn. Limnol., 14,* 91-106, 1980h.

McCarthy, J. J., O. F. Canziani, N. A. Leary, D. J. Dokken and K. S. White, *Climate Change, 2001. Impacts, adaptations and vulnerability. Contribution of Working Group II to the Third Assessment Report of the Intergovernmental Panel on Climate Change,* Chapter 16 (Cambridge Univ. Press, Cambridge), 2001.

McDonald, M. E., A. E. Hershey, and M. C. Miller, Global warming impacts on lake trout in arctic lakes, *Limnol. Oceanogr., 41,* 1102-1108, 1996.

Murphy, E. J., A. Clarke, C. Symon, and J. Priddle, Temporal variation in Antarctic sea ice: analysis of a long term fast-ice record from the South Orkney Islands, *Deep-Sea Res., 42,* 1045-1062, 1995.

Noon P. E, H. J. B. Birks, V. J. Jones and J. C. Ellis-Evans, Quantitative models for reconstructing catchment ice-extent using physical-chemical characteristics of lake sediments. *J. Palaeolimnol., 25,* 375-392, 2001.

Quayle, W. C., L. S. Peck, J. C. Ellis-Evans, H. J. Peat, and P. R. Harrigan, Extreme responses to climate change in Antarctic lakes, *Science, 295,* 645, 2002.

Ragotzkie, R.A., Heat budgets of lakes, in *Lakes Chemistry Geology Physics* edited by A. Lerman, pp. 1-19, Springer Verlag, New York, 1978.

Rivkin, R. B., M. R. Anderson, and C. Lajzerowicz, Microbial processes in cold oceans. 1. Relationship between temperature and bacterial growth rate, *Aquat. Microb. Ecol., 10,* 243-254, 1996.

Roberts, E. M., and J. Laybourn-Parry, Mixotrophic cryptophytes and their predators in the Dry Valley lakes of Antarctica, *Freshwat. Biol. 41,* 737-746, 1999.

Schindler, D. W., H. E. Welch, J. Kalff, G. J. Brunskill and N. Kritsch, Physical and chemical limnology of Char Lake, Cornwallis Island (75°N lat.), *J. Fish. Res. Bd. Can. 31,* 585-607. 1974.

Schindler, D. W., K. G. Beatty, E. J. Fee, D. R. Cruikshank, E. R. Debruyn, D. L. Findlay, G. A. Linsey, J. A. Shearer, M. P. Stainton and M. A. Turner, Effects of climatic warming on lakes of the central boreal forest, *Science, 250,* 967-970, 1990.

Schindler, D.W., S. E. Bayley, B. R. Parker, K. G. Beaty, D. R. Cruikshank, E. J. Fee, E. U. Schindler, and M. P. Stainton, The effects of climatic warming on the properties of boreal lake and streams at the Experimental Lakes Area, northwestern Ontario, *Limnol. Oceanogr., 41,* 1004-1017, 1996.

Skvarca, P., W. Rack, H. Rott and T. Ibarzábal y Donángelo, Evidence of recent climatic warming on the eastern Antarctic Peninsula, *Ann. Glaciol., 27,* 628-632, 1998.

Smith, R. I. L., Nutrient cycling in relation to biological productivity in Antarctic and sub-Antarctic terrestrial and freshwater ecosystems, in *Antarctic Nutrient Cycles and Food Webs* edited by W. R Siegfried,., P. R Condy, and R. M.. Laws, pp. 138-155, Springer-Verlag, Berlin, Heidelberg, 1985.

Smith, R. I. L., Destruction of Antarctic terrestrial ecosystems by a rapidly increasing fur seal population, *Biol. Conserv., 45,* 55-72. 1988.

Smith, R. I. L., Signy Island as a paradigm of biological and environmental change in Antarctic terrestrial ecosystems, in *Antarctic Ecosystems. Ecological Change and Conservation* edited by K. R. Kerry, and G. Hempel, pp. 32-50, Springer-Verlag, Berlin, Heidelberg, 1990.

Smith, R.I.L., Impact of an increasing fur seal population on Antarctic plant communities: resilience and recovery, in *Antarctic communities: Species, structure and Survival*, edited by B. Battaglia, J. Valencia, and D W H Walton, CUP, Cambridge, pp. 432-436, 1997.

Stott, P. A., S. F. B. Tett, G. S. Jones, M. R. Allen, J. F. B. Mitchell and G. J. Jenkins, External control of 20th century temperature by natural and anthropogenic forcings, *Science, 290,* 2133-2137, 2000.

Strickland, J. D. H., and R. T. Parsons, A practical handbook of seawater analysis, 2nd edn., *Bull. Fish. Res. Board Can., 167,* 1972.

Turner, J., J.C. King, T. A. Lachlan-Cope, and P. D. Jones, Climate change (Communication arising) Recent temperature trends in the Antarctic, *Nature, 418,* 291-292. 2002

Vincent, W. F., *Microbial Ecosystems of Antarctica,* Chapter 10, Cambridge University Press, 1988.

Watson, R. T., M. C. Zinyowera, R. H. Moss, and D. J. Dokken, Climate Change: Impacts, adaptations and mitigation of climate change: Scientific-Technical Analysis. Contribution of Working Group II to the Second Assessment Report of the Intergovernmental Panel on Climate Change, Cambridge Univ. Press, Cambridge, 1995.

Walsh, J. E., P. T. Doran, J. C. Priscu, W. B. Lyons et al., Climate change (Communication arising) Recent temperature trends in the Antarctic, *Nature, 418,* 291-292. 2002.

Helen Butler, Natural Environment Research Council Polaris House, North Star Avenue, Swindon, SN2 1EU, UK.

Peter Convey, J. Cynan Ellis-Evans, Helen J. Peat, and Lloyd S. Peck, British Antarctic Survey, Natural Environment Research Council, High Cross, Madingley Road, Cambridge, CB3 0ET, UK.

Wendy C. Quayle, CSIRO Land and Water, Research Station Road, Griffith, 2680, NSW, Australia.

LATE-HOLOCENE PENGUIN OCCUPATION AND DIET AT KING GEORGE ISLAND, ANTARCTIC PENINSULA

Steven D. Emslie

University of North Carolina, Department of Biological Sciences, Wilmington, North Carolina

Peter Ritchie and David Lambert

Allan Wilson Centre for Molecular Ecology and Evolution, Massey University, Palmerston North, New Zealand

Twelve abandoned and two active colonies of Adélie (*Pygoscelis adeliae*), Chinstrap (*P. antarctica*), and Gentoo penguin (*P. papua*) were excavated on King George Island, Antarctic Peninsula, in January/February 2002. Ancient DNA was extracted from nine samples of penguin bone from five abandoned colonies to verify species identifications. Thirty-six radiocarbon dates on penguin remains from these sites, including the first to be completed on *P. papua* from an abandoned colony in Antarctica, indicate that these three penguin species began occupying this region 500 to 600 years ago, agreeing with similar trends found at other abandoned colonies in the northern Antarctic Peninsula. This colonization event corresponds with warming trends in the Antarctic Peninsula during the Little Ice Age (A.D. 1500–1850), based on previously published isotope records from the Siple Dome ice core. Non-krill dietary remains recovered from the ornithogenic sediments at the 12 abandoned colonies indicate a high diversity of cephalopod and fish prey. Guano samples from the two active colonies further indicate that Chinstrap and Adélie penguins differ in their non-krill diet with the former species preferring lantern fish (*Electrona antarctica*) and the latter Antarctic silverfish (*Pleuragramma antarcticum*).

INTRODUCTION

Three species of pygoscelid penguins currently breed on King George Island in the Antarctica Peninsula: the Adélie (*Pygoscelis adeliae*), Chinstrap (*P. antarctica*), and Gentoo penguin (*P. papua*). Although considerable information is now known on the ecology and population biology of these species at this location [e.g., *Volkman et al.*, 1980; *Trivelpiece et al.*, 1986, 1987, 1990], little is known on their occupation history in the northern Antarctic Peninsula. *Emslie* [1995] investigated abandoned colonies of Chinstrap and Adélie penguins on King George Island and other locations. These and

subsequent investigations [*Emslie*, 2001] have indicated that no colonies older than approximately 540 years before present (B.P.) have been found in the northern Antarctic Peninsula. Although colonies as old as 5000 to 6000 years have been identified in more southern regions of the peninsula [*Emslie and McDaniel*, 2002], the absence of old colonies in the northern region remains an enigma.

In 2002, the senior author returned to King George Island to resample relict penguin colonies in the Admiralty Bay region and survey for additional abandoned colonies that might indicate older occupations by pygoscelid penguins. Active colonies also were sampled

10.1029/079ARS14

to determine their age and to recover dietary remains from fresh guano. Twelve abandoned and two active penguin colonies were sampled during this study and include the first known excavations of abandoned Gentoo penguin colonies in Antarctica. Radiocarbon dates were obtained on penguin remains recovered from all sites and ancient DNA was extracted from penguin bones to verify species identification. The results of these investigations are presented here with identification of dietary remains.

METHODS

All active and abandoned colonies sampled in this study are located on the west side of Admiralty Bay and within Site of Special Scientific Interest (SSSI) No. 8 (Figures 1–2). Excavations were conducted at five areas within this SSSI at, from north to south, Rakusa Point, Llano Point, Uchatka, Blue Dyke, and Patelnia (Figure 2A-C). The two active colonies that were sampled included one of Chinstrap penguins at Patelnia (Area D) and one of Adélie penguins at Cascade Head, Llano Point. Area D and Cascade Head refer to colony names assigned by W. Trivelpiece for specific locations within a larger colony complex. Abandoned Chinstrap penguin colonies were located at Patelnia, Blue Dyke, and Uchatka Point, while two Gentoo and seven Adélie Penguin colonies were sampled at Llano Point and Rakusa Point. Geographic Positioning System (GPS) coordinates and approximate elevation above mean sea level (± 1 m) were determined with a handheld Brunton MNS MultiNavigator for each site (Table 1).

Each site was sampled by excavating a 1x1 m test pit (TP) in 5-cm arbitrary levels, with level 1 at the top. Excavations ceased when either the bottom of the ornithogenic sediments (recognized by change in soil texture and color) or bedrock was reached. All sediments removed during excavations were quantified by level and volume, washed through three nested screens (mesh sizes, from top to bottom, of 0.64, 0.32, and 0.025 cm²), and dried and sorted following the methods of *Emslie et al.* [1998, 2002] and *Emslie and McDaniel* [2002]. Sediments recovered from the middle and lower screens were dried, weighed, and resorted in the laboratory through No. 18 (1 mm) and No. 60 (0.25 mm) sieves. These sediments were then sorted under a low-power stereomicroscope to recover organic remains (bones, otoliths, squid beaks). If no or few remains were encountered after sorting > 10% of the sample by dry weight, the remainder of the sample was left unsorted. Total dry

weights (in grams) and percent sorted are provided in Table 1.

All bones were saved from each excavated level. Penguin bones were identified to species following the methods of *Emslie* [1995]; these identifications were used to determine which species formerly occupied the abandoned colony. Dietary remains (otoliths and squid beaks) recovered from the sorted sediments were identified by W. Walker, National Marine Mammal Laboratory, Seattle, WA. For each taxon, the total number of complete and partial otoliths or squid beaks was tabulated. Minimum number of individuals (MNI) represented for each taxon was calculated following the methods of *Polito et al.* [2002] and are reported here for both partially and completely sorted samples. MNIs were corrected for amount of sediments excavated by dividing MNI values with the total liters of sediments excavated and screened from each test pit. This method was applied only to the two most abundant species of fish, *Pleuragramma antarcticum* and *Electrona antarctica*, recovered from the modern guano samples. A Chi-square analysis was used to test for significant differences in these corrected MNIs using an even distribution of MNIs between sites as the expected values.

Radiocarbon dates were obtained from penguin bone, feather, or egg membrane recovered from each site and from the 5-cm arbitrary levels. All dates were corrected and calibrated for the marine-carbon reservoir effect using the Calib 4.2 software program and the 98MARINE database [*Stuiver and Reimer*, 1993; *Stuiver et al.*, 1998] using a $\Delta R = 700 \pm 50$ B. P. [see *Emslie*, 1995, 2001]. All dates are reported in 2σ calibrated ranges in calendar years B. P. and were assigned NZA (Rafter Radiocarbon Laboratory code) numbers.

Ancient DNA was extracted from sub-fossil bones in a dedicated laboratory following the procedures of *Lambert et al.* [2002]. To construct standards for taxonomic identification, DNA was also extracted from blood samples of *P. adeliae*, *P. antarctica* and *P. papua*.

A 445 base pair (bp) portion of the mitochondrial DNA 12S rRNA gene was PCR-amplified from the three *Pygoscelis* reference samples using the primers L-12SA (5'-AAACTGGGATTAGATACCCCACTAT-3') and H-12SB (5'-GAGGGTGACGGGCGGTATGT-3'). A 193 bp sequence of the same portion was amplified from the ancient DNA samples using the PCR primers L-12SE (5'-CCCACCTAGAGGAGCCTGTTC-3') and H-12SF (5'-AAATGTAGCCCATTTCTTCC-3'). The amplified products were sequenced using the ABI PRISM® BigDye™ Terminator Cycle Sequencing kit (Applied

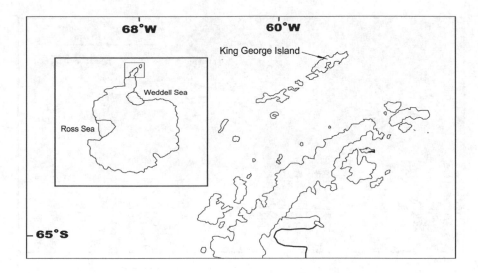

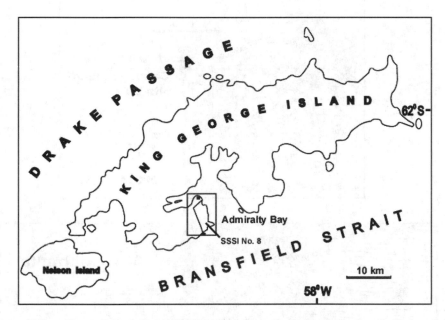

Fig. 1. Location of King George Island, South Shetland Islands, in the Antarctic Peninsula (top) and detail of King George Island (bottom) showing the location (in box) of SSSI No. 8 where abandoned and active penguin colonies were sampled.

Biosystems) and analyzed on a ABI 377A automated sequencer. The resulting DNA sequences were aligned by eye and the sub-fossil bones were identified by comparison to the reference sequences.

RESULTS

Identifiable bones of penguins, primarily juveniles, were recovered from most of the abandoned sites and indicated that three of these colonies were formerly occupied by Chinstrap, two by Gentoo, and six by Adélie penguins (Table 1). Two other abandoned sites (Sites 5 and 6 at Rakusa Point) lacked identifiable remains and are presumed to represent Adélie penguin occupations based on other sites of this species nearby. Results of the 36 radiocarbon dates are presented in Table 2. One date from Copa Site 3 Level 2 (NZA 15411) was too young to provide a valid age for calibration using the Calib 4.2

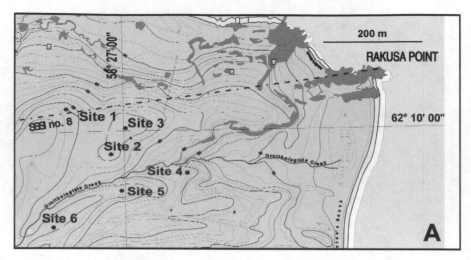

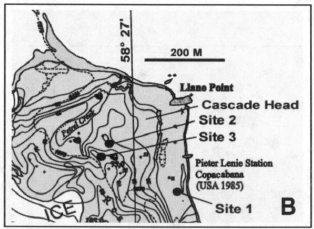

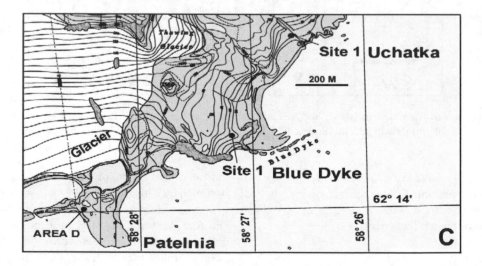

Fig. 2. A. Detail of Rakusa Point near Arctowski Station (Poland) showing the locations of Sites 1–6. B. Detail of Llano Point with Copacabana Field Station showing the locations of Sites 1–3. D. Detail of Uchatka, Blue Dyke and Patelnia showing the location of sites excavated at these areas. All maps are modified from *Padelko* [2002].

software program. All other dates except one indicate that penguin occupation in this region did not begin until approximately 500 to 600 B.P. The one exception (NZA 15300) from Rakusa Point Site 4, Level 4, is much older than any other date from all sites in this area, including four other dates from the same test pit and/or level, and is considered to be anomalous (Table 2).

The DNA sequences for the three *Pygoscelis* standards have been deposited in GenBank under the accession numbers AY234841, AY234842, and AY236364. There was 3.7–4.3% sequence difference among the three reference sequences and the DNA sequences that were retrieved from each sub-fossil bone could easily be assigned to a *Pygoscelis* species. This analysis confirmed the species identification at five of the abandoned sites (Tables 1 and 2).

Dietary remains were identified from all but two sites. Despite the apparent poor preservation of these remains at most sites, five taxa of cephalopods and 14 fish were identified in the sediments and are summarized by combining data from all levels from each test pit at each site (Table 3). At Rakusa Point, prey remains were scarce or absent and are summarized for Sites 1–4 combined; no prey remains were recovered from Sites 5–6 at this location.

Of the total prey identified, *Psychroteuthis glacialis*, *Galiteuthis glacialis*, *Pleuragramma antarcticum*, *Electrona antarctica*, and *Gymnoscopelus* cf. *G. nicholsi* were most abundant from most sites. Dietary remains recovered from guano samples from the two active colonies, Patelnia (Area D) and Cascade Head, were richer and better preserved than those from the abandoned sites. These remains allowed comparison of the modern diet in Chinstrap and Adélie penguins. Calculations of MNIs of two of the most abundant taxa represented from these sites, *Pleuragramma antarcticum* and *Electrona antarctica* (Table 3), indicate that the former species is more abundant in Adélie (MNI 0.52 per liter of sediment) compared to Chinstrap penguin (MNI 0.36 per liter) guano samples. In addition, *E. antarctica* is over three times more abundant in Chinstrap (MNI 2.8 per liter) compared to Adélie penguin (MNI 0.8 per liter) guano samples. However, these differences are not statistically significant ($\chi^2 < 1.094$, df = 1, P < 0.9).

DISCUSSION

Emslie [2001] reviewed all previous radiocarbon dates on penguin remains from the Antarctic Peninsula. A total of 63 dates from 21 abandoned colonies indicated that no sites older than approximately 540 B. P. existed in the

northern peninsula. The absence of older sites may be due to late-Holocene glacial scouring of ornithogenic soils, solifluction, or an absence of penguins in the peninsula until the past 500–600 years. However, penguin bones and guano have been recovered and dated from raised beaches and lake sediments on King George Island that indicate these birds were present in the region during the middle Holocene [*Bocheński*, 1985; *Barsch and Mäusbacher*, 1986; *Tatur et al.*, 1997, *Sun et al.*, 2000)].

Results of 36 radiocarbon dates on Adélie, Gentoo, and Chinstrap penguin remains presented here support earlier findings that no pygoscelid penguin colonies existed in the northern Antarctic Peninsula until the late Holocene. Only one date from Site 4 (NZA 15300; Table 2) at Rakusa Point provides an older age. However, four other dates from this site, including one from the same level, all converge on a calibrated age ranging from 0 to 533 B.P. In addition, all other sites excavated at Rakusa Point had similar depositional characteristics in the depth and color of the ornithogenic soils. Thus, the simplest explanation for this older date is that it is anomalous. If an earlier occupation did occur in the middle Holocene, as suggested by guano deposits in lake cores at King George Island and Hope Bay [*Zale*, 1994a, b; *Sun et al.*, 2000], then ornithogenic soils from that occupation period have not been preserved or have not yet been located in this region.

Five of the 36 dates presented here, ranging in age from 619 B. P. to the present (based on 2σ calibrated ranges), are the first known on Gentoo penguin bones from ornithogenic sediments in Antarctica. While much more needs to be done on the paleohistory of Gentoo penguins, the two abandoned sites in Admiralty Bay indicate a similar late occupation history by this species as with Chinstrap and Adélie penguins and that all three species have been contemporaneous in the Admiralty Bay region for at least the past 500 to 600 years.

The radiocarbon dates from Rakusa Point also provide a minimum age for the formation of the glacial moraines that characterize this area. Two small abandoned colonies (Rakusa Point Sites 5 and 6) are located on the top of an old lateral moraine on the north side of Ecology glacier (Figure 1) and range in age from 550–265 B. P. Although *Birkenmajer* [1997] postulated that this moraine is less than 100 years old based on lichen growth, the penguin dates indicate a much older age for their formation prior to occupation by small groups of breeding penguins.

Dietary remains from most of these sites, while sparse, demonstrate the diversity of prey consumed by

TABLE 1. Abandoned and active penguin colonies excavated on King George Island with their excavation units, GPS location, approximate elevation above mean sea level (± 1 m), total sediments excavated in liters with dry weight (in parentheses) of all sediments recovered from the 0.32 and 0.025 cm^2 mesh screens after washing, and total percent dry weight of the last sorted for dietary remains. The *Pygoscelis* species identified by bones recovered at each site, when present, also is given; identification confirmed with DNA analysis of bone are indicated by an asterisk (*).

Site	Species	Excavation units	GPS Location	Elev. (m)	Total Sediments liters (grams)	Total Sorted (%)
Uchatka	*P. antarctica	TP 1, Levels 1-5	62° 13' 15.1" S; 58° 26' 29.5"	20	143 (10,652)	69.4
		TP 2, Levels 1-4		22	114 (8452)	30.2
Blue Dyke	*P. antarctica	TP 1, Levels 1-4	62° 13' 37.6" S; 58° 26' 59.5" W	28	151 (6276)	58.9
Patelnia Area D	P. antarctica	TP 1, Levels 1-5	62° 14' 05.9" S; 58° 28' 16.9" W	7	138 (3644)	100
Patelnia Area D	P. antarctica	Modern surface	62° 14' 05.9" S; 58° 28' 16.9" W	7	25 (120)	100
Copa Site 1	P. papua	TP 1, Levels 1-4	62° 10' 45.2" S; 58° 26' 43.5" W	3	159 (13,812)	11.4
Cope Site 2	P. adeliae	TP 1, Levels 1-3	62° 10' 29.6" S; 58° 27' 03.9" W	36	82 (3951)	16.9
Copa Site 3	*P. papua	TP 1, Levels 1-3	62° 10' 30.7" S; 58° 27' 07.7" W	44	84 (3627)	19.2
Cascade Head	P. adeliae	Modern surface	62° 10' 31.6" S; 58° 26' 54.0" W	22	42 (475)	100
Rakusa Point Site 1	*P. adeliae	TP 1, Levels 1-3	62° 09' 48.6" S; 58° 28' 07.0" W	48	80 (7253)	13.7
Rakusa Point Site 2	P. adeliae	TP 1, Levels 1-5	62° 09' 51.7" S; 58° 28' 02.2" W	50	140 (15,601)	12.7
Rakusa Point Site 3	P. adeliae	TP 1, Levels 1-3	62° 09' 49.9" S; 58° 28' 00.3" W	46	84 (10,744)	14.8
Rakusa Point Site 4	*P. adeliae	TP 1, Levels 1-4	62° 09' 53.1" S; 58° 27' 51.6" W	31	108 (5970)	18.3
Rakusa Point Site 5	P. adeliae	TP 1, Levels 1-3	62° 09' 54.0" S; 58° 27' 59.7" W	48	66 (8301)	13.9
Rakusa Point Site 6	P. adeliae	TP 1, Levels 1-2	62° 09' 55.1" S; 58° 28' 05.1" W	49	44 (6168)	12.2

TABLE 2. Radiocarbon dates (in years B.P.) on penguin (*Pygoscelis* sp.) bone and eggshell membrane tissue from 12 abandoned colonies on King George Island. Stratigraphic position for each date is provided by site, test pit (TP) number, and 5-cm level. Conventional dates were corrected and calibrated for the marine-carbon reservoir effect using a $\Delta R = 700 \pm 50$ years [see *Emslie* 1995] and the MARINE98 database of *Stuiver and Reimer* (1993). The conventional date, $\delta^{13}C$ (‰) value, mean corrected date (both in radiocarbon years B. P.), and calibrated 2σ range(s) (95% confidence interval in calendar years B. P.) are provided for each sample. Multiple 2σ ranges are provided for corrected dates that intersected two or more regions of the calibration curve. All bones were identified by osteological comparison to reference skeletal material; an asterisk (*) denotes specimens further identified by DNA extraction.

Site/Provenience	Species	Material	Lab. No.	$\delta^{13}C$	Conventional Date	Calibrated Range (2σ)
Uchatka						
TP 1 Level 4	*P. antarctica	Bone	NZA 15416	−22.42	1420 ± 65	502–238
TP 1 Level 5	P. antarctica	Bone	NZA 15417	−19.13	1490 ± 65	535–273
TP 2 Level 3	*P. antarctica	Bone	NZA 15418	−18.40	1380 ± 60	478–141
TP 2 Level 4	*P. antarctica	Bone	NZA 15419	−22.92	1260 ± 55	320–0
Blue Dyke						
TP 1 Level 3	*P. antarctica	Bone	NZA 15414	−19.24	1420 ± 65	502–239
TP 1 Level 4	*P. antarctica	Bone	NZA 15415	−20.08	1300 ± 65	420–0
Patelnia						
Area D Level 5	P. antarctica	Bone	NZA 15412	−20.71	1440 ± 60	506–253
	P. antarctica	Bone	NZA 15413	−21.50	1420 ± 60	498–243
Copa						
Site 1 TP 1 Level 1	P. papua	Bone	NZA 15406	−21.20	1450 ± 55	509–264
Site 1 TP 1 Level 2	P. papua	Bone	NZA 15539	−22.36	1580 ± 55	616–376
Site 1 TP 1 Level 3	P. papua	Bone	NZA 15407	−19.77	1580 ± 55	619–391
Site 1 TP 1 Level 4	Pygoscelis sp.	Bone	NZA 15408	−19.78	1530 ± 60	555–299
Site 1 TP 1 Level 4	Pygoscelis sp.	Bone	NZA 15409	−20.74	1400 ± 70	497–142
Site 2 TP 1 Level 2	P. adeliae	Bone	NZA 15410	−22.18	1340 ± 70	461–65 or 6 − 0
Site 2 TP 1 Level 3	Pygoscelis sp.	Feather	NZA 15613	−23.74	1350 ± 55	445–115
Site 3 TP 1 Level 2	*P. papua	Bone	NZA 15411	−18.78	740 ± 55	Invalid age for calibration
Site 3 TP 1 Level 3	Pygoscelis sp.	Feather	NZA 15614	−22.09	1360 ± 55	459–133
Rakusa Point						
Site 1 TP 1 Level 2	P. adeliae	Bone	NZA 15285	−21.89	1380 ± 65	478–132
Site 1 TP 1 Level 3	*P. adeliae	Bone	NZA 15286	−21.60	1500 ± 55	511–266
	P. adeliae	Bone	NZA 15287	−22.77	1430 ± 60	504–251
Site 2 TP 1 Level 1	P. adeliae	Bone	NZA 15288	−22.13	1410 ± 55	490–243
Site 2 TP 1 Level 2	P. adeliae	Bone	NZA 15289	−20.11	1370 ± 65	473–127
Site 2 TP 1 Level 3	P. adeliae	Bone	NZA 15290	−23.77	1500 ± 65	545–281
Site 2 TP 1 Level 4	P. adeliae	Bone	NZA 15291	−16.89	1490 ± 60	536–280
Site 2 TP 1 Level 5	P. adeliae	Bone	NZA 15292	−22.54	1670 ± 60	665–458
	P. adeliae	Feather	NZA 15612	−22.50	1460 ± 55	514–269
Site 3 TP 1 Level 1	P. adeliae	Bone	NZA 15293	−21.87	1480 ± 65	534–271
Site 3 TP 1 Level 2	P. adeliae	Bone	NZA 15294	−18.80	1360 ± 55	456–131
Site 3 TP 1 Level 3	P. adeliae	Bone	NZA 15295	−23.68	1450 ± 60	510–258
Site 4 TP 1 Level 1	*P. adeliae	Bone	NZA 15296	−21.87	1290 ± 55	401–0
Site 4 TP 1 Level 2	P. adeliae	Bone	NZA 15297	−23.12	1480 ± 65	533–271
Site 4 TP 1 Level 3	*P. adeliae	Bone	NZA 15298	−23.53	1310 ± 65	433–0
Site 4 TP 1 Level 4	P. adeliae	Bone	NZA 15299	−22.78	1340 ± 55	444–113
	P. adeliae	Bone	NZA 15300	−21.19	8380 ± 60	8318–7954
Site 5 TP 1 Level 2	P. adeliae	Egg mem	NZA 15317	−21.83	1530 ± 55	550–303
Site 6 TP 1 Level 2	P. adeliae	Egg mem	NZA 15318	−21.42	1450 ± 55	510–265

TABLE 3. Taxa of cephalopods and fish identified from abandoned penguin colonies on King George Island compared to remains identified from Patelnia and Cascade Head, the active Chinstrap and Adélie Penguin colonies, respectively. The total number of identifiable specimens (including fragments) and minimum number of individuals (MNI, in parentheses) of prey (based on whole and fragmentary remains summed by level) are provided for all levels of each test pit (TP); data are combined from TP 1 and 2 at Uchatka and from Sites 1–4 at Rakusa Point.

Taxon	Uchatka TP 1-2	Blue Dyke TP 1	Patelnia Area D	Copa Site 3	Rakusa Point Sites 1-4	Patelnia (Chinstrap)	Cascade Head (Adélie)
Cephalopoda (squid and octopus)							
Psychroteuthis glacialis	7 (6)	19 (12)	15 (5)		1 (1)	1 (1)	1 (1)
Brachioteuthis sp.			2 (2)				
Galiteuthis glacialis	13 (11)	13 (8)	9 (8)		2 (2)	1 (1)	
Pholidoteuthis sp.			1 (1)				
Unident. oegopsid beak frags.	19	21	17		2	7	1
cf. Parelodone sp.			1 (1)				
Osteichthyes: Teleostei (bony fish)							
Dissostichus sp.				1 (1)			
Paranotothenia sp.			1 (1)				
Notothenia sp.			1 (1)				
Trematomus sp.	1 (1)		3 (2)				
Pleuragramma antarcticum	12 (8)	10 (5)	18 (14)		2 (2)	15 (9)	33 (22)
Harpagifer cf. H. georgianus			1 (1)				2 (2)
Pagetopsis cf. P. maculatus							
Pagetopsis sp.			2 (2)				
Chaenodraco cf. C. wilsoni		3 (3)				8 (4)	
Chaenodraco sp.			2 (2)				
Notolepis coatsi	1 (1)	1 (1)	1 (1)				
Protomyctophum bolini	1 (1)						
Electrona antarctica	254 (138)	330 (179)	512 (145)		10 (8)	140 (70)	60 (34)
Electrona carlsburgi			3 (2)			3 (3)	1 (1)
Gymnoscopelus braueri	3 (2)						
Gymnoscopelus cf. G. braueri	3 (2)	2 (1)					
Gymnoscopelus cf. G. nicholsi	27 (4)	73 (9)	59 (7)		1 (1)	4 (1)	
Unident. Notothenoidei	1		13			2	1
Unident. Channichthyidae			13				
Unident. Paralepididae	1						
Unident. Myctophidae	244	171	246		1	23	14
Unident. otolith fragment		2	172	1	1		18

pygoscelid penguins. Although krill is by far the dominant prey of choice by these species at King George Island today, fish are consumed in small to large quantities and are an important component of their diet [*Williams*, 1995]. The data from the modern samples also suggest that Chinstrap penguins prefer lantern fish (*Electrona* spp.) over Antarctic silverfish (*Pleuragramma antarcticum*) while Adélie penguins are the reverse, though these patterns from these small samples are not statistically significant. Preferences by these species for these prey previously have not been documented in stomach sampling or other dietary studies in this region [*Volkman et al.*, 1980; *Williams*, 1995]. However, addi-

tional investigation is needed to determine if a significant difference in non-krill prey selection does occur between these species at King George Island.

Last, the absence of older penguin colonies in the northern Antarctic Peninsula remains an enigma and additional research in other regions of King George Island and at other islands, including those in the Weddell Sea, are needed to determine if this occupation pattern is real or an artifact of sampling or other factors. It is possible that the northern peninsula was abandoned by penguins by the mid to late Holocene and that the record for this earlier occupation (i.e., ornithogenic sediments) has been erased by glacial scouring or other processes; only limited evidence for this occupation remains from guano deposits in lakes. If so, this abandonment was followed by reoccupation of the region beginning approximately 500 to 600 years ago, near the beginning of the Little Ice Age [LIA, A.D. 1500–1850; *Grove*, 1988]. At that time, the Antarctic Peninsula was thought to have experienced more warming than cooling periods based on the ice core record from Siple Dome [*Mosely Thompson et al.*, 1990]. However, recent evidence from marine sediments in the Palmer Deep suggest otherwise [*Warner and Domack*, 2002; see also *Domack et al.* and *Hjort et al.*, this volume].

If the LIA was characterized by colder conditions in the Antarctic Peninsula compared to today, then it is difficult to reconcile the penguin record with the climatic record. A colder period presumably would be less favorable for penguin colonization, especially if extensive sea ice and/or glacial expansion and persistent snow cover blocked access to ice-free terrain along the coast. It is possible that conditions varied throughout the peninsula during the late Holocene so that some areas, such as King George Island, were accessible to breeding penguins beginning about 600 B.P. Additional paleoenvironmental data are needed from a diversity of locations in the Antarctic Peninsula to fully address this problem.

Modern warming trends in the peninsula over the past two decades have been causing a decrease in Adélie penguins [*Smith et al., 1999*] that may lead to abandonment of this region in the future. Conversely, Gentoo penguins, which prefer warmer, sub-Antarctic regions, have been expanding farther southward in the Antarctic Peninsula over the past 20 years than they have in any period prior to this [*Smith et al.,* 1999]. Although our data indicate that Adélie, Chinstrap, and Gentoo penguins have been contemporaneous in Admiralty Bay for the past 500–600 years, their current relative population sizes and ranges are rapidly shifting with climate change. This trend could result in a new geographic distribution of pygoscelid penguins in the Antarctic Peninsula, and a loss of sympatry among these species that apparently has existed in this region for the past 600 years.

Acknowledgments. This research was funded by grants from the National Science Foundation (OPP 9909274) and the National Geographic Society (NGS Grant #7040-01). Field assistance was provided by R. Hollingshead, M. Polito, M. Romano, J. Smykla, C. Thiessen, and W. Trivelpiece. I thank S. Rakusa-Suszczewski and the Polish Academy of Sciences for their cooperation and assistance at Arctowski Station, and W. Trivelpiece for use of facilities at the Copacabana Field Station. W. Walker provided identifications of otolith and squid remains and R. Pudelko allowed use of digital maps. All radiocarbon dates were completed at the Rafter Radiocarbon Laboratory, New Zealand. The paper was improved with comments from A. Clarke, E. Domack, C. Hjort and one anonymous reviewer.

REFERENCES

Barsch, D. and R. Mäusbacher, New data on relief development of the South Shetland Island, Antarctica. *Internatl. Science Rev.*, 11, 211-218, 1986.

Birkenmajer, K., Quaternary geology at Arctowski Station, King George Island, South Shetland Islands (west Antarctica), *Studia Geologica Polonica*, 110, 91-104, 1997.

Bocheński, Z., Remains of subfossil birds from King George Island (South Shetland Islands), *Acta Zoologica Cracov*, 29, 109-116, 1985.

Domack, E. W., A. Leventer, S. Root, J. Ring, E. Williams, D. Carlson, E. Hirshorn, W. Wright, R. Gilbert, and G. Burr, Marine Sedimentary Record of Natural Environmental Variability and Recent Warming in the Antarctic Peninsula, this volume.

Emslie, S.D., Age and taphonomy of abandoned penguin colonies in the Antarctic Peninsula region, *Polar Rec.*, 31, 409-418, 1995.

Emslie, S.D., Radiocarbon dates from abandoned penguin colonies in the Antarctic Peninsula region, *Antarctic Science*,13, 289-295, 2001.

Emslie, S.D., W. Fraser, R.C. Smith, and W. Walker, Abandoned penguin colonies and environmental change in the Palmer Station area, Anvers Island, Antarctic Peninsula, *Ant. Sci.*, 10, 257-268, 1998.

Emslie, S.D. and J. McDaniel, Adélie penguin diet and climate change during the middle to late Holocene in northern Marguerite Bay, Antarctic Peninsula, *Polar Biol.*, 25, 222-229, 2002.

Grove, J.M., The Little Ice Age, 498 p., Methuen, London, 1988.

Hjort, C., Ingólfsson, Ó., Bentley, M. J., and Björck, S., Late Pleistocene and Holocene Glacial and Climate History of the Antarctic Peninsula Region: A Brief Overview of the Land and Lake Sediment Records, this volume.

Lambert, D.M., P. A. Ritchie, C. D. Millar, B. Holland, A. J. Drummond, and C. Baroni, Rates of evolution in ancient DNA from Adélie penguins. *Science*, 295, 2270-2273, 2002.

Mosley-Thompson, E., L.G. Thompson, P.M. Grootes, and N. Gunderstrup, Little Ice Age (Neoglacial) paleoenvironmental conditions at Siple Station, Antarctica. *Ann. Glaciol.*, 14, 199-204, 1990.

Pudelko R, Topographic map of the Site of Special Scientific Interest No.8 (SSSI – 8), King George Island, 1:2,500 scale. Dept. of Antarctic Biology, Polish Acad. of Sciences, Warsaw, 2002.

Polito, M., S.D. Emslie, and W. Walker, A 1000-year record of Adélie penguin diets in the southern Ross Sea, Antarctica, *Ant. Sci.*, 14, 327-332, 2002.

Smith, R.C., et al. Marine ecosystem sensitivity to climate change. *BioScience*, 49, 393-404, 1999.

Stuiver, M., and P.J. Reimer, Extended 14C data base and revised CALIB 3.0 14C age calibration program, *Radiocarbon*, 35, 215-230, 1993.

Stuiver, M., et al., INTCAL98 radiocarbon age calibration, 24,000- 0 cal BP, *Radiocarbon*, 40, 1041-1083, 1998.

Sun, L., Z. Xie, and J. Zhao, A 3,000-year record of penguin populations, *Nature*, 407, 858, 2000.

Tatur, A., A. Myrcha, J. Fabiszewski, and J. Niegodzisz, Formation of abandoned penguin colony ecosystems in maritime Antarctic, *Polar Biol.*, 17, 405-417, 1997.

Trivelpiece, W.Z., J.L. Bengston, S.G. Trivelpiece, and N.J. Volkman, Foraging behavior of Gentoo and Chinstrap Penguins as determined by new radiotelemetry techniques, *Auk* 103, 777-781, 1986.

Trivelpiece, W.Z., S.G. Trivelpiece, G.R. Geupel, J. Kjelmyr, and N.J. Volkman, Adélie and Chinstrap Penguins: their potential as monitors of the southern ocean marine ecosystem, in *Antarctic Ecosystems*, edited by Kerry, K.R., and G. Hempel, pp. 191-202, Springer-Verlag, Berlin, 1990.

Trivelpiece, W.Z., S.G. Trivelpiece, and N.J. Volkman, Ecological segregation of Adélie, Gentoo, and Chinstrap Penguins at King George Island, Antarctica, *Ecology*, 68, 351-361, 1987.

Volkman, N.J., P. Presler, and W.Z. Trivelpiece, Diets of pygoscelid penguins at King George Island, Antarctica, *Condor*, 82, 373-378, 1980.

Warner, N. R. and E. W. Domack, Millennial- to decadal-scale paleoenvironmental change during the Holocene in the Palmer Deep, Antarctica, as recorded by particle size analysis, *Paleooceanography*, 17, PAL 5-1 to 5-14, 2002.

Williams, T.D., *The Penguins*, 295 p., Oxford Univ. Press, Oxford, 1995.

Zale, R., Changes in size of the Hope Bay Adélie penguin rookery as inferred from Lake Boeckella sediment, *Ecography*, 17, 297-304, 1994.

Steven D. Emslie, University of North Carolina, Department of Biological Sciences, 601 S. College Road, Wilmington, NC 28403, USA e-mail: emslies@uncwil.edu

Peter Ritchie and David Lambert, Allan Wilson Centre for Molecular Ecology and Evolution, Institute of Molecular BioSciences, Massey University, Private Bag 11-222, Palmerston North, New Zealand e-mail: P.Ritchie@massey.ac.nz, D.M.Lambert@massey.ac.nz

Marine Geological Records

ANTARCTIC PENINSULA CLIMATE VARIABILITY
ANTARCTIC RESEARCH SERIES VOLUME 79, PAGES 183-194

RETREAT HISTORY OF THE GERLACHE-BOYD ICE STREAM, NORTHERN ANTARCTIC PENINSULA: AN ULTRA-HIGH RESOLUTION ACOUSTIC STUDY OF THE DEGLACIAL AND POST-GLACIAL SEDIMENT DRAPE

Verónica Willmott, Miquel Canals, and José L. Casamor

Department of Stratigraphy, Paleontology and Marine Geosciences, University of Barcelona, Barcelona, Spain

During the Last Glacial Maximum a large bundle structure was deposited by an ice stream flowing from Gerlache Strait to Boyd Strait across the Western Bransfield Basin. Bundles are ice-bed contact till deposits whose surface consists of sets of convex-upward, elongated, parallel to subparallel ridges. Located in the northern Antarctic Peninsula, at water depths between 500 to 1200 m, the Western Bransfield Basin bundle is up to 100 km long and 14 to 21 km wide. Here, we present ultra-high resolution topographic parametric source (TOPAS) acoustic profiles to examine the thickness and acoustic facies variations of the deglacial and postglacial drape on top of the Western Bransfield Basin bundle. Acoustic facies analysis suggests that the sediment drape is composed of hemipelagic particles from a combination of ocean surface primary productivity and subglacial sediment-laden outflows from a retreating ice stream front.

We hypothesize that the areas of thickest drape mark the earliest decoupling. Three segments are distinguished in the bundle drape: the central segment, the northern segment, and the southern segment. Drape is thickest in the central segment and thins to zero in the southern segment toward the entrance of the inner Gerlache Strait. On the basis of this observation, we infer that the ice-seabed decoupling started first in the thickest central segment, followed by the northern and southern segments, respectively. The development of an ice-free sea surface followed the same succession. Our study shows that ice stream retreat may be far from linear, and it is strongly controlled by seafloor and iceshed topography.

1. INTRODUCTION

The extent and thickness of Antarctic glaciers have changed significantly between glacial and interglacial periods [*Anderson et al.*, 2002; *Denton and Hughes*, 2002; and *Huybrechts*, 2002], but how rapidly these changes occurred remains a matter of debate. Although significant progress has been made in reconstructing the ice maxima through the identification of grounding lines, there are still many unanswered questions regarding the response of ice sheets to past climatic oscillations [*Stokes and Clark*, 2001]. One way of addressing such questions is through the study of continental margin sediment deposits. Because the Antarctic continent is 99% ice-covered, subglacial erosion and transport are the most important processes transporting terrigenous sediment to the continental margin [*Alley et al.*, 1997]. During subglacial transport, linear bedforms parallel to ice flow are produced on either the bedrock or basal till [*Clark*, 1993; and *Shaw et al.*, 2000]. Streamlining is recognized in both marine [*Anderson*, 2003; *Bartek et al.*, 1997; *Camerlenghi et al.*, 2001; *Canals et al.*, 2002; *Canals et al.*, 2000; *Eittreim et al.*, 1995; and *Rebesco et al.*, 1998] and continental settings [*Clark*, 1993; *Lonne*, 1995; and

10.1029/079ARS15

Shaw et al., 2000]. Bundles are the largest of the family of mega-scale glacial lineations. In a recent paper, *Tulaczyk et al.* [2001] hypothesize that bundles are carved by keels of ice ploughing through the subglacial sediments. Under flow acceleration and convergence, as is the case where ice streams begin, basal roughness elements in the ice undergo strain. This strain "re-shapes" the ice bottom into longitudinally aligned keels where they would generate sediment streamlining.

High latitude small ice caps are very sensitive to climatic change [*Crowley and Baum*, 1995]. This fact is extremely important for understanding the links between climatic oscillations and global sea level changes [*Barker et al.*, 1998]. All ice caps and outlet-glaciers of the north-west Antarctic Peninsula are wetbased and, in most cases, marine-based [*Anderson et al.*, 2002]. In addition, this is the region of Antarctica where precipitation is highest [*Griffith and Anderson*, 1989]. These conditions are favourable for the development of particularly mobile and unstable glacial systems which, as a consequence, are very sensitive to climatic oscillations [*Anderson and Thomas*, 1991; *Budd and Rayner*, 1993; and *Denton and Hughes*, 1986]. Because ice streams are the most dynamic components of ice caps and ice sheets, they are currently one of the most important research topics in glaciated margins, and are a key issue in understanding the consequences of climatic change [*Oppenheimer*, 1998].

Lower eustatic sea level during glacial periods increases the erosive capacity of marine-based glaciers and, in particular, that of ice streams [*Anderson*, 1999; and *Bentley*, 1994]. It is during these lowstands that erosion and subglacial transport are most effective in supplying sediment to the continental margin of the Antarctic Peninsula [*Larter and Barker*, 1991]. During deglaciation, sedimentary processes are strongly controlled by the rise of sea level, the destruction of ice shelves, and the progressive retreat of sea ice. This leads to the landward retreat of the grounding line and the opening of increasingly larger ice-free marine areas [*Anderson et al.*, 2002]. During interglacial periods, pelagic fine-grained sedimentation is dominant with an important biogenic component of planktonic origin [*Fabres et al.*, 2000; and *Harris and O'Brien*, 1998]. These sediments tend to drape and preserve previous morphologies left over from the preceding glacial maximum [*Anderson et al.*, 2002]. Seaward of the retreating grounding line, outflow deposits accumulate by turbulent transport of the suspended fine sediment in meltwater plumes [*Rebesco et al.*, 1998]. Further from the grounding line, deposition

is dominated by fine-grained suspended sediments mixed with biogenic material of planktonic origin [*Pudsey et al.*, 1994]. Subsequent glacial advance may completely erode these deposits [*Harris and O'Brien*, 1998].

The study of postglacial draping thickness and their acoustic characteristics using high to ultra-high resolution sub-bottom profiles is an insightful method for documenting ice margin retreat. Here, we present the initial results of a detailed acoustic investigation of the postglacial sediment overlying the Last Glacial Maximum bundle in the Western Bransfield Basin (WBB) [*Canals et al.*, 2000]. From these results we develop the ice retreat history of the Gerlache Strait-Boyd Strait ice stream.

2. STUDY AREA

The WBB delimits the path of the Gerlache-Boyd paleo-ice stream. With a length of 340 km, the Gerlache-Boyd paleo-ice stream transported an estimated ice volume of 8000 km^3 [*Canals et al.*, 2000]. The WBB is one of three sub-basins that constitute the north-east oriented Bransfield Basin (Plate 1). The WBB is controlled by two main structural trends, the north-east Bransfield Rift, and a north-west trend likely associated with the Hero Fracture Zone (Plate 1).

To the west, the WBB is bounded by a submarine scarp just east of Smith and Low Islands. To the east, the WBB shallows to Deception Plateau (<400 m) and Trinity Island northern shelf (<600 m), separating it from the longer Central Bransfield Basin. The maximum depth of the WBB is 1390 m (Figure 1a).

The area from Trinity Island to the Boyd Strait seafloor is occupied by a >100 km long, 15 to 25 km wide morphosedimentary bundle structure made of subparallel ridges and grooves 40 m high and 1-3 km wide in a north-west direction (Figure 1b). In cross section, the bundle shows a convex profile, with a maximum relief of 200 m off the seafloor, whereas in longitudinal section this profile is concave. The southern reach of the structure is bounded by two depressions partly filled with sediment (Figure 1a). To the north the bundle becomes progressively narrower (Figure 1b).

The WBB bundle documents the presence of a Last Glacial Maximum (LGM) paleo-ice stream that drained the Gerlache Strait and Western Bransfield Basin, including Danco Coast along the Peninsula and parts of the Palmer Archipelago, Livingston Island, and other neighboring islands [*Canals et al.*, 2000]. According to *Stokes and Clark* [2001], the WBB bundle is the most complete record of its type documented to date.

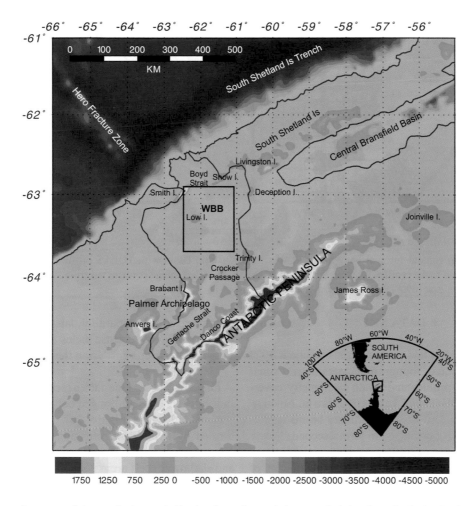

Plate 1. Location map of the north Antarctic Peninsula region and the area draining into Gerlache Strait and Western Bransfield Basin –WBB– (blue line), constructed from various digital databases. The red line shows the position of the grounding line during Last Glacial Maximum, modified from [*Banfield and Anderson,* 1995; *Bentley and Anderson,* 1998; and *Pudsey et al.,* 1994]. The box outlines Figures 1(a) and 1(b).

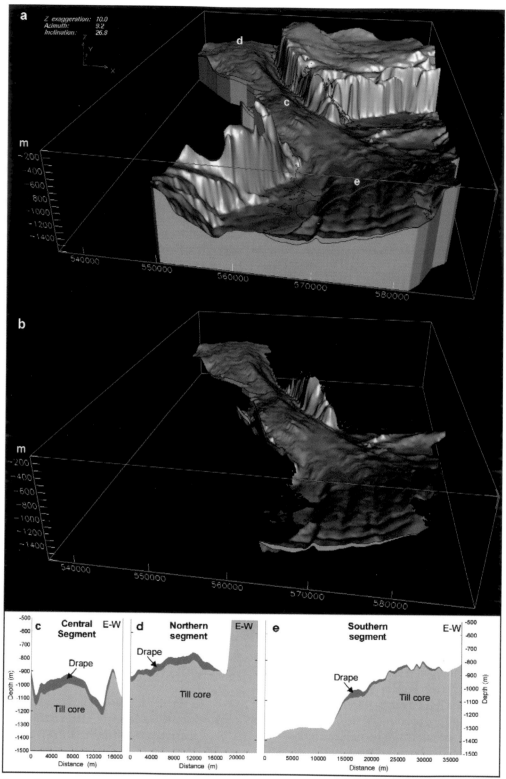

Plate 3. Three-dimensional model of the sediment drape on top of the till core of the WBB bundle with (a) sea floor topography in white—red lines show location of Plate 3 (c-e)—and (b) as a floating volume. In both (a) and (b) view is south to the north. Transverse lines are ship track artefacts. (c) Cross section in the Central segment, (d) cross section in the Northern segment and (e) Cross section in the Southern segment. In (c), (d) and (e) vertical exaggeration is magnified 25 times.

is attributed to post-glacial mass wasting deposits (debris flows or topographically controlled muddy turbidites [*Howe*, 1995]) or deposition of subglacial outflows.

5.2 Sedimentary Cover Thickness and Acoustic Basement Depth

Beginning from the Last Glacial Maximum -ca. 21 ka- (Figure 3*a*), we present an initial reconstruction of the Gerlache-Boyd ice stream and the succession of events up to the present-day.

In principle, areas that became ice-free first should exhibit a thicker postglacial sedimentary cover than zones exposed at a later stage, assuming that the postglacial sedimentation rate is uniform. This hypothesis should be testable by examining, for example, locations previously occupied by large ice streams. Studying the acoustic contrast between basal till and postglacial drape provides a means of identifying the geometry and thickness of the postglacial unit and may help to reconstruct the retreat history of the ice stream.

Supported by bathymetric and acoustic data, we propose that ice streambed decoupling, triggered by sea level rise and probably ice thinning, first occurred where the coupling zone was deepest (central segment), as expected, following the initial decoupling in the overdeepened basins (Figure 3*b*). We suggest that the lower part of the drape in this segment was formed predominately by sediment resulting from subglacial flows. No distinct acoustic facies, however, were observed over the bundle structure that can unequivocally confirm a subglacial flow origin. Subglacial outflow deposits (Facies II) are in contrast very thick in the deeper depressions bordering the bundle.

As sea level continued to rise, the input of ice from the iceshed decreased. As a result, the ice stream continued to retreat moving the grounding line landwards and opening the Boyd Strait (Figure 3*c*). Glaciers from nearby islands were rapidly decreasing as well. Progressive burial of the bundle would continue on the central segment and extend to the northern segment.

Further sea level rise and climatic improvement would then restrict the ice stream to the Gerlache Strait, perhaps ending in a floating ice shelf over the southern WBB and the deep Crocker Passage. In the protected inner Gerlache Strait, south of the Crocker Passage, a substantial ice stream portion remained grounded firmly on the seafloor (Figure 3*d*). The presence of grounded ice south of Crocker Passage is supported by the lack of sediment cover, with the exception of some drifts at the mouths of the tributary fjords [*Wright*, 2000]. Finally, this residual confined ice stream and ice tongue disappeared, resulting in the opening of the Gerlache Strait to the open ocean and its circulation. During this last deglaciation period, hemipelagic sediment (Facies II) would accumulate over the whole WBB. The proximity to the retreating ice front favours high surface productivity in summer, supporting biogenic fluxes that, together with subglacial outflow and advected suspended materials, would provide high sedimentation rates in the WBB.

Plate 2*b*, although somewhat modified by postdepositional mass wasting events, illustrates the thickness of sediment throughout the study region following deglaciation. At present, there are no dates for this hypothesized deglacial history of the WBB. We suspect, however, that the decoupling and retreat process was rapid in order to keep pace with rising sea level. Giant Jumbo cores obtained in the 2001/02 cruise season hold great promise for providing the required age model.

6. CONCLUSIONS

The thickness and acoustic facies of deglacial and postglacial drape on top of till deposits provide a record of the decoupling and retreat history of ice streams, and of the processes responsible for sediment transport and accumulation following removal of ice from sea surface.

Acoustically laminated facies (Facies II) suggest that the sediment drape consists mostly of a mixture of particles settled from the sea surface.

Deposits from mass wasting processes dominate the sediment infill of the depressions on both sides of the convex upwards bundle, as revealed by the presence of lens-shaped bodies with transparent acoustic facies (Facies III) interfingered and alternating with laminated facies (Facies II).

The thickness distribution of the drape on top of the bundle along the WBB suggests three ice stream retreat stages. The central, deepest segment was decoupled first, followed by the northern segment in the vicinity of Boyd Strait. The southern segment, off the mouth of the inner Gerlache Strait, has a thinner drape, thus indicating that it was the last to be ice-cleared.

Long sediment cores are required to determine the ages of the retreat phases of the Gerlache-Boyd ice stream.

Acknowledgments. We cordially thank the crew of R/V *Hespérides*, technicians and scientists who participated in the

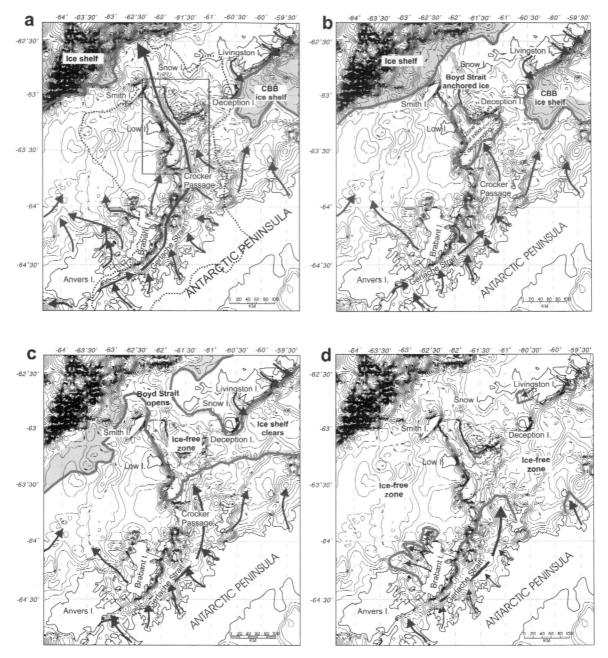

Fig. 3. Tentative reconstruction of the ice decoupling and retreat history of the Western Bransfield Basin area, including Gerlache Strait, following the Last Glacial Maximum. Bathymetry is from swath data and digital data bases. Black arrows show paleo-grounded ice flow, thick dark grey line corresponds to the grounding line, and light grey areas correspond to floating ice shelves. The boundary of the ice drainage basin feeding the Gerlache-Boyd ice stream is indicated by the dotted line. The box in (a) shows the area, of Figure 1 and Plate 2, and the box in Plate 1. Ice-flow directions and grounding line position used for the LGM are modified from *Banfield and Anderson* [1995], *Bentley and Anderson* [1998], and *Pudsey et al.* [1994], and based on shelf geomorphology. Grounding lines in the successive stages and flow directions are deduced from shelf geomorphology. (a) Ice drainage system at the LGM. (b) First deglaciation stage. The WBB central segment is the zone of initial decoupling, whereas in the Boyd Strait area the ice remains anchored. (c) Second deglaciation stage. Boyd Strait opens and central and northern segments are covered by sea ice only in winter. The grounding line retreats to a proximal area. (d) Third deglaciation stage. The ice stream is mostly confined to Gerlache Strait.

GEBRAP'96 cruise where data were acquired. The authors would like to thank Drs Angelo Camerlenghi, Robert Gilbert, Carol Pudsey, an anonymous reviewer and the editor Matthew Kirby, for their detailed and helpful comments, which greatly improved the quality of this paper.

This work has been supported by the Spanish "Programa Nacional de Investigación", projects ANT95-0889-C02-01 (GEBRAP) and REN2000-0896 / ANT (COHIMAR). We also acknowledge support from the Spanish "Programa Hispano-Norteamericano de Cooperación Científica y Tecnológica", project 99120 (GEMARANT). GRC Geociències Marines of the University of Barcelona, Spain, is supported by "Generalitat de Catalunya" autonomous government through its excellency research groups program (ref. 1999 SGR-63).V. Willmott benefited from a fellowship from the University of Barcelona.

REFERENCES

Alley, R.B., K.M. Cuffey, E.B. Evenson, J.C. Strasser, D.E. Lawson, and G.J. Larson, How glaciers entrain and transport basal sediment: physical constraints, *Quaternary Science Reviews*, *16*, 1017-1038, 1997.

Anderson, J.B., *Antarctic Marine Geology*, 289 pp., Cambridge University Press, Cambridge, 1999.

Anderson, J.B., Geomorphic and sedimentological records of grounded ice sheets and paleo-ice streams, in *Glacier-influenced sedimentation on high-latitude continental margins*, *203*, edited by J.A. Dowdeswell, and C. Ó Cofaigh, Geological Society, London, Special Publication, 2003.

Anderson, J.B., L.R. Bartek, M. Taviani, and M.A. Thomas, Seismic and sedimentologic record of glacial events on the Antarctic Peninsula shelf, in *Geological Evolution of Antarctica*, edited by M.R.A. Thomson, J.A. Crame, and J.W. Thomson, pp. 687 691, Cambridge University Press, Cambridge, 1991.

Anderson, J.B., S.S. Shipp, A.L. Lowe, J.S. Wellner, and A.B. Mosola, The Antarctic Ice Sheet during the Last Glacial Maximum and its subsequent retreat history: a review, *Quaternary Science Reviews*, *21* (1-3), 49-70, 2002.

Anderson, J.B., and M.A. Thomas, Marine ice-sheet decoupling as a mechanism for rapid, episodic sea level change: the record of such events and their influence on sedimentation, *Sedimentary Geology*, *70*, 87-104, 1991.

Banfield, L.A., and J.B. Anderson, Seismic facies investigations of the Late Quaternary glacial history of Bransfield Basin, Antarctica, in *Antarctic Research Series*, *68*, pp. 123-140, American Geophysical Union, Washington DC, 1995.

Barker, P.F., J.B. Barret, A. Camerlenghi, A.K. Cooper, F.J. Davey, E.W. Domack, C. Escutia, Y. Kristoffersen, and P.E. O'Brien, Ice sheet history from Antarctic continental margin sediments: The ANTOSTRAT approach, *Terra Antartica*, *5*, 737-760, 1998.

Bartek, L.R., J.B. Anderson, and T. Oneacre, Ice stream troughs and variety of Cenozoic seismic stratigraphic architecture

from a high southern latitude section: Ross Sea Antarctica, in *Glaciated continental margins: An Atlas of acoustic images*, edited by T.A. Davies, T. Bell, A.K. Cooper, H. Josenhans, L. Polyak, A. Solheim, M.S. Stoker, and J.A. Stravers, 1997.

Bentley, C.R., Sedimentation by the Antarctic Ice Sheet from a Glaciological Perspective, *Terra Antartica*, *1* (2), 249-250, 1994.

Bentley, C.R., and J.B. Anderson, Glacial and Marine geological evidence for the ice sheet configuration in the Weddel Sea—Antarctic Peninsula region—during the Last Glacial Maximum, *Antarctic Science*, *10*, 309-325, 1998.

Bodungen, B.V., V.S. Smetacek, M.M. Tilzer, and B. Zeitzchel, Primary production and sedimentation during spring in the Antarctic Peninsula region, *Deep Sea Research*, *33*, 177-194, 1986.

Budd, W.F., and P. Rayner, Modelling ice sheet and climatic changes throug the ice ages, in *Ice in the Climate System*, edited by W.R. Peltier, pp. 291-319, Springer, Heidelberg, 1993.

Camerlenghi, A., E.W. Domack, M. Rebesco, R. Gilbert, S.E. Ishman, A. Leventer, S.A. Brachfeld, and A. Drake, Glacial morphology and post-glacial contourites in northern Prince Gustav Channel (NW Weddell Sea, Antarctica), *Marine Geophysical Researches*, *22*, 417-443, 2001.

Canals, M., E.W. Domack, J.L. Casamor, J. Baraza, M. Farran, M. de Batist, R. Urgeles, and A.M. Calafat, A subglacial sedimentary system off the northern Antarctic Peninsula from sea-floor evidence, *Geology*, *30* (7), 603-606, 2002.

Canals, M., and GEBRAP Team, Evolución geológica del Margen Pacífico de la Antártida Occidental: Expansión del fondo Oceánico y Tectónica de Placas, Universidad de Barcelona, Barcelona, 1997.

Canals, M., R. Urgeles, and A.M. Calafat, Deep sea-floor evidence of past ice streams off the Antarctic Peninsula, *Geology*, *28* (1), 31-34, 2000.

Clark, C.D., Mega-scale glacial lineations and cross-cutting ice-flow landforms, *Earth Surface Processes and Landforms*, *18*, 1-29, 1993.

Crowley, T.J., and S.K. Baum, Is the Greenland Ice Sheet bistable?, *Paleoceanography*, *10*, 357-363, 1995.

Damuth, J.E., Use of high-frequency (3.5-12 kHz) echograms in the study of near- bottom sedimentation processes in the deep sea: A review, *Marine Geology*, *38*, 51-75, 1980.

Denton, G.H., and T.J. Hughes, Global ice-sheet system interlocked by sea level, *Quaternary Research*, *26*, 3-26, 1986.

Denton, G.H., and T.J. Hughes, Reconstructing the Antarctic Ice Sheet at the Last Glacial Maximum, *Quaternary Science Reviews*, *21*, 193-202, 2002.

Domack, E.W., E.A. Jacobson, S.S. Shipp, and J.B. Anderson, Late Pleistocene-Holocene retreat of the West Antarctic Ice-Sheet system in the Ross Sea: Part 2-Sedimentologic and stratigraphic signature, *Geological Society of America Bulletin*, *111* (10), 1517-1536, 1999.

Eittreim, S.L., A.K. Cooper, and J. Wannesson, Seismic stratigraphic evidence of ice-sheet advances on the Wilkes Land margin of Antarctica, *Sedimentary Geology*, *96*, 131-156, 1995.

Fabres, J., A.M. Calafat, M. Canals, M.A. Bárcena, and J.A. Flores, Bransfield Basin fine-grained sediments: late-Holocene sedimentary processes and Antarctic oceanographic conditions, *The Holocene*, *10* (6), 703-718, 2000.

Griffith, T.W., and J.B. Anderson, Climatic control of sedimentation in bays and fyords of the northern Antarctic Peninsula, *Marine Geology*, *85*, 181-204, 1989.

Harris, P.T., and P.E. O'Brien, Bottom currents, sedimentation and ice-sheet retreat facies successions on the Mac Robertson shelf, East Antarctica, *Marine Geology*, *151*, 47-72, 1998.

Howe, J.A., Sedimentary processes and variations in slope-current activity during the last Glacial-Interglacial episode on the Hebrides Slope, northern Rockall Trough, North Atlantic Ocean, *Sedimentary Geology*, *96* (3-4), 201-230, 1995.

Huybrechts, P., Sea-level changes at the LGM from ice-dynamic reconstructions of the Greenland and Antarctic ice sheets during the glacial cycles, *Quaternary Science Reviews*, *21*, 203-231, 2002.

Larter, R.D., and P.F. Barker, Neogene interaction of tectonic and glacial processes at the Pacific margin of the Antarctic Peninsula, *Spec. Publs. Int. Ass. Sediment.*, *12*, 165-186, 1991.

Lonne, I., Sedimentary facies and depositional architecture of ice-contact glaciomarine systems, *Sedimentary Geology*, *98*, 13-43, 1995.

Murphy, E.J., P.W. Boyd, R.J.G. Leakey, A. Atkinson, E.S. Edwards, C. Robinson, J. Priddle, S.J. Bury, D.B. Robins, P.H. Burkill, G. Savidge, N.J.P. Owens, and D. Turner, Carbon flux in ice -ocean-plankton systems of the Bellingshausen Sea during a period of ice retreat, *Journal of Marine Systems*, *17*, 207-227, 1998.

Oppenheimer, M., Global warming and the stability of the West Antarctic Ice Sheet, *Nature*, *393*, 325-331, 1998.

Pope, P.G., and J.B. Anderson, Late Quaternary glacial history of the northern Antarctic Peninsula's western continental shelf, in *Evidence from the marine record, Contributions to Antarctic Research III, Antarctic Research Series*, *57*, edited by D.H. Elliot, pp. 63-91, American Geophysical Union, Washington, D.C., 1992.

Pudsey, C.J., P.F. Barker, and R.D. Larter, Ice sheet retreat from the Antarctic Peninsula Shelf, *Continental Shelf Research*, *14*, 1647-1675, 1994.

Pudsey, C.J., and J. Evans, First survey of Antarctic sub-ice shelf sediments reveals mid-Holocene ice shelf retreat, *Geology*, *29*, 787-790, 2001.

Rebesco, M., A. Camerlenghi, and C. Zanolla, Bathymetry and Morphogenesis of the Continental Margin West of the Antarctic Peninsula, *Terra Antartica*, *5* (4), 715-725, 1998.

Shaw, J., D.M. Faragini, D.R. Kvill, and R.B. Rains, The Athabasca fluting field, Alberta, Canada: implications for the formation of large-scale fluting (erosional lineations), *Quaternary Science Reviews*, *19*, 959-980, 2000.

Stokes, C.R., and C.D. Clark, Paleo-Ice streams, *Quaternary Science Reviews*, *20*, 1437-1457, 2001.

Tulaczyk, S.M., R.P. Scherer, and C.D. Clark, A ploughing model for the origin of weak tills beneath ice streams: a qualitative treatment, *Quaternary International*, *86*, 59-70, 2001.

Wright, W.N., The Schollaert sediment drift: an ultra high resolution paleoenvironmental archive in the Gerlache Strait, Antarctica, MS thesis, Hamilton College, Clinton, NY, 2000.

Yoon, H.I., B.-K. Park, Y. Kim, and C.Y. Kang, Glaciomarine sedimentation and its paleoclimatic implications on the Antarctic Peninsula shelf over the last 15000 years, *Palaeogeography, Palaeoclimatology, Palaeoecology*, *185* (3-4), 235-254, 2002.

Miquel Canals, José L. Casamor, and Verónica Willmott, Department of Stratigraphy, Paleontology and Marine Geosciences, University of Barcelona, Barcelona, Spain. miquel@natura.geo.ub.es, jluis@natura.geo.ub.es, and veronica@natura.geo.ub.es

ANTARCTIC PENINSULA CLIMATE VARIABILITY
ANTARCTIC RESEARCH SERIES VOLUME 79, PAGES 195-204

DEGLACIAL HISTORY OF THE GREENPEACE TROUGH: ICE SHEET TO ICE SHELF TRANSITION IN THE NORTHWESTERN WEDDELL SEA

Robert Gilbert

Department of Geography, Queen's University, Kingston, Canada

Eugene W. Domack

Geology Department, Hamilton College, Clinton, New York

Angelo Camerlenghi

Instituto Nazionale di Oceanografia e di Geofisca Sperimentale, Trieste, Italy

SeaBeam bathymetry and backscatter data combined with core records from sediments from nearly 6000 km^2 of the former Larsen A Ice Shelf, eastern Antarctic Peninsula reveal the glacial and post-glacial evolution of landforms on the seafloor created by a former ice sheet and the subsequent ice shelf which disintegrated especially during the 1990s. A 500-m deep, coast-parallel trough 10 x 20 km in aerial extent was eroded by an ice stream flowing northeasterly then easterly across the continental shelf as a distributary from a larger ice stream to the south. Large-scale streamlined bed forms up to 800 m wide and 6 km long with orientation parallel to the long axis of the trough, many with surrounding horseshoe-shaped depressions attest to ice-contact processes and the action of subglacial meltwater. As the ice sheet thinned to become the partially floating and separated Larsen A and B ice shelves, its flow changed to a southwesterly direction across the shelf and large fields of smaller subglacial flutes, of mean width 44 m, height 2.2 m, and lengths of several hundred meters, were sculpted into the thin glacial sediments, crosscutting the large-scale forms. These were preserved as the ice shelf lifted from the seafloor and are covered by a thin sequence of glacimarine sediments representing both ice shelf and open sea conditions.

1. INTRODUCTION

U.S. Antarctic Program cruises NBP0003 (May 2000) and NBP0107 (December and January 2001-2002) aboard *N.B. Palmer* investigated the region of the former Larsen A Ice Shelf on the east side of the Antarctic Peninsula. The disintegration of this ice shelf, especially in January to March 1995 in response to warming climate in the Antarctic Peninsula has received considerable attention [*Rott et al.,* 1996; 1998; *Vaughn and*

Doake, 1996; *Cooper,* 1997; *Skvarca et al.,* 1999; *Scambos et al.,* 2001; and *Vaughn et al.,* 2001], while the newly revealed marine environment is providing a wealth of environmental and paleoenvironmental information [*Sloan et al.,* 1995; *Del Valle et al.,* 1998; *Domack et al.,* 2001; *Pudsey et al.,* 2001; *Camerlenghi et al.,* 2002; and *Evans and Pudsey,* 2002). The offshore section of the continental shelf to the southeast is about 300 to 500 m deep (Plate 1), and Greenpeace Trough nearer the Nordensköld Coast reaches maximum depths

10.1029/079ARS16

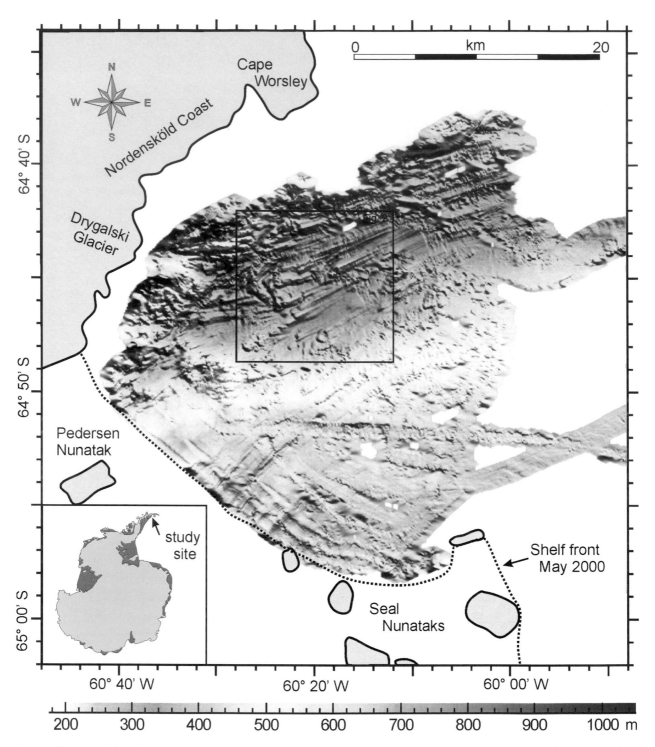

Plate 1. Shaded relief and bathymetric map of part of the region at the north end of the Larsen Ice, Shelf, Antarctic Peninsula from SeaBeam swath survey. Illumination is from 210° azimuth and 30° above the horizon. In areas where the sea floor is nearly flat, the shading amplifies small amplitude noise in the data that creates an artifact along ship's track. Red box outlines Plate 2.

of 1000 m. This report assesses former behavior of the northern portion of the Larsen Ice Sheet and Ice Shelf based on detailed mapping of seafloor features and cores recovered throughout the region.

2. METHODS

Bathymetric mapping was conducted during the winter cruise in 2000 using the *RVIB N.B. Palmer* hull-mounted SeaBeam 2112 multibeam and Bathy2000 CHIRP sonar systems [*Holik*, 1994; and *Domack et al.*, 2001] which provided bathymetry and backscatter data. Despite severe sea ice conditions which degraded the data set and prevented access to some areas, a high resolution map of most of the region beneath the former Larsen A Ice shelf was produced. Additional data collected in open water conditions during the 2001 cruise refined the map (Plate 1). The backscatter data were provided as pseudo-side-scan imagery and are used in this report to assess micro-morphological features of the seafloor.

Sediment was recovered from the seafloor with a range of instruments from grab samplers which preserved the sediment-water interface, to a kasten gravity corer which recovered sediments deposited during the period of the ice shelf as well as the final stages of the earlier ice sheet [*Domack et al.*, 2001]. The kasten core reported upon in this paper was logged and sampled aboard ship. Shear strength was measured with a hand-held vane apparatus [*Blum*, 1997]. Subsequent analyses in our laboratories and at Florida State University core archive included x-radiography, and grain size by laser diffraction techniques.

3. OBSERVATIONS AND INTERPRETATIONS

Although the first transit of the region revealed a large trough on the inner continental shelf [*Del Valle et al.*, 1998], it was first mapped with SeaBeam swath technology as part of our paleoenvironmental assessment of the ice shelf from the glacimarine record. The survey (Plate 1) shows a trough more than 10 km wide and 500 m deep extending more than 20 km along the inner continental shelf. The floor of the shelf is irregular with at least one deep sill across it and streamlined hills up to 200 m high on the valley floor. Several hanging valleys, the largest at the Drygalski Glacier, feed into the north-west side of the trough.

The north-west side of the floor of the Greenpeace Trough is marked by a small escarpment (Plate 1) probably associated with the fault forming the boundary between igneous and metamorphic crystalline rocks of the continental basement to the northwest, and the poor-

ly indurated sedimentary rocks of the Cenozoic back arc to the southeast [*BAS*, 1985]. Transverse and lateral (coast-parallel) troughs are common on glaciated shelves in Antarctica [*Vanney and Johnson*, 1979; *Anderson and Shipp*, 2001; and *Anderson et al.* 2002] and in the Northern Hemisphere [*Holtedahl*, 1970; and *Gilbert*, 1982]. They are generally ascribed to selective linear erosion [*Sugden*, 1978] during phases of maximum glaciation. In this case, significant structural control parallel to the coast may have influenced the orientation of the trough, and the soft shale that dominates till composition in the basin facilitated erosion.

3.1 Large-Scale Streamlined Forms

Crosscutting and extending throughout the floor of the trough are major streamlined forms, each 500–800 m wide and up to 6 km long. Surrounding the up-flow end of many are horseshoe-shaped depressions (Plate 2) reminiscent of small-scale rat-tails [*Sharpe and Shaw*, 1989; and *Shaw* 1994] illustrated in Figure 1. They sweep out of the hanging valleys and obliquely across the inner shelf from the Nordensköld Coast at headings of 085° to 090°. In the trough they merge with forms trending 040° to 070°, following generally the trend of the trough. The shallow continental shelf to the southeast of Greenpeace Trough is generally devoid of these features; however, several large flutes emerge from beneath the present margin of the ice shelf and trend north-east across the continental shelf (Plate 1). Gas injection air gun seismic data and grain lithologies from cores indicate the bedrock in the trough from which these features are formed is faulted, gently tilted sedimentary strata of Mesozoic age. These rocks are similar to exposures on Pedersen Island and the Sobral Peninsula except that the former are more argillaceous and less arenaceous than surface exposures [*Hathaway and Riding*, 2001]. The features are clearly recognisable in the trough because there is only a thin cover of ice-contact and post-glacial sediment [*Domack et al.*, 2001].

Identical features have recently been documented with the same SeaBeam swath mapping technology in a number of troughs around Antarctica by *Anderson and Shipp* [2001], *Anderson et al.* [2001; and 2002] *Wellner et al.* [2001] and *Lowe and Anderson* [2002] who ascribe their origin to ice-contact processes [cf. *Menzies and Rose*, 1987; and *Clark*, 1999], and refer to them as roche moutonées and drumlins. However, the horseshoe-shaped depressions (Plates 1 and 2) are characteristic of large flutes produced by turbulent, separated flow [*Allen*, 1982], in this case during major outbursts of subglacial

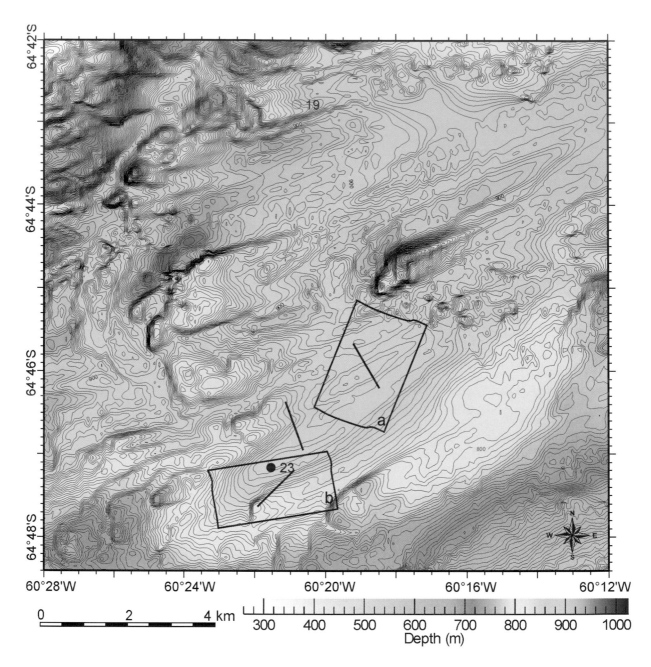

Plate 2. SeaBeam swath bathymetric map of a portion of the sea floor under the former Larsen A Ice Shelf showing detail of the large-scale streamlined forms and the location of images of fluting on the sea floor (Figure 3). The location of kasten core KC23 is shown.

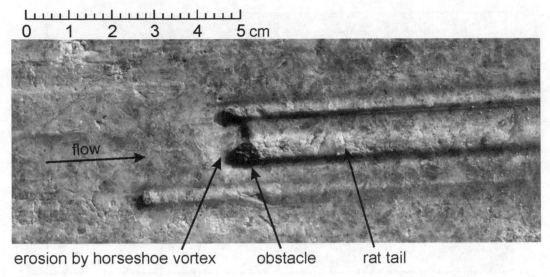

Fig. 1. Small-scale rat tail forms in bedrock created by flow under the Laurentide Ice Sheet [*Shaw*, 1994] to illustrate the similarity of form to those in the Greenpeace Trough.

meltwater [cf. *Shaw*, 1996]. The source of this water is unknown, although significant volumes of water are trapped under the Antarctic Ice Cap today [*Siegert*, 2000], and antarctic landscapes ascribed to subglacial meltwater have been described by *Sugden et al.* [1991] and *Sawagaki and Hirakawa* [1997], as have similar features created under the North American Laurentide Ice Cap [*Shaw and Gilbert*, 1990; and *Shaw et al.*, 1996]. Nye channels [*Nye*, 1973] illustrated by *Anderson et al.* [2001] in the Pine Island Trough, western Antarctica are clearly related to the presence of subglacial water, although not necessarily occurring as large floods, are absent in the Greenpeace Trough.

We hypothesise conditions during the Last Glacial Maximum (Figure 2a) following *Bentley and Anderson* [1998]. Robertson Trough carried a large ice stream from the Antarctic Peninsula across the region occupied by the present Larsen B Ice Shelf and more than 200 km eastward to the edge of the shelf. This is in accord with the inference by *Nakadaa et al.* [2000] of significantly thicker glaciers around the Weddell Sea during the Last Glacial Maximum than those around the Ross Embayment. A portion of this ice overflowed to the north, crossing a low ridge between Robertson Island/Seal Nunataks and the mainland of the Antarctic Peninsula. This ice, joined by flow from Drygalski Glacier and nearby smaller outlets continued northeasterly through Greenpeace Trough, enlarging it by selective linear erosion and sculpting the large-scale streamlined forms at least in part by subglacial fluvial processes. This distributary stream curved to the east and south-

east where it rejoined the Robertson Stream south of James Ross Island. A small ice dome or ice rise over Robertson Island and the Seal Nunataks probably facilitated the splitting of this ice drainage and contributed to its flow. Since the drainage area of the Greenpeace Trough was relatively small compared to the paleoglacial systems that produced subglacial melt features, we infer that Greenpeace Trough was plumbed by surface meltwater rather than basal melt. Further, because the paleoglacial surface across Greenpeace Trough would have been at a relatively high elevation, we also speculate that the climate was somewhat warmer than historic conditions in which the equilibrium line is only about 100 m above sea level [*Rack*, 2000].

3.2 Small-Scale Flutes

The second evidence of glacial sculpting of the seafloor is in the form of small-scale flutes recognised in pseudo-side-scan (backscatter) images produced as part of the swath survey. In the central portion of the Greenpeace Trough (Plate 2) in water depths of 860 to 920 m are a series of prominent lineations (Figure 3). They consist of straight, parallel furrows that can be traced for more than 1 km across the sediments of the seafloor. Curvature shown in Figure 3 is an artifact of changes in ship speed and direction due to the heavy pack ice present throughout this region in May 2001. The mean spacing between the 52 ridges shown in Figure 3a is 44 m with more than half (31) in the range 20 to 40 m. Their orientation varies from 150° (Figure 3a) to 225°

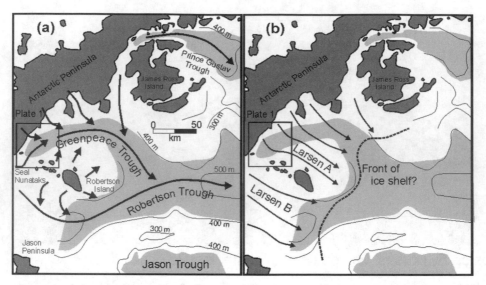

Fig. 2. Reconstruction of flow patterns of (a) the Larsen Ice Sheet during the Last Glacial Maximum when mega-flutes were sculpted in the Greenpeace trough and continental shelf [modified from *Bentley and Anderson*, 1998], and (b) the Larsen A and B Ice Shelves during the period when small-scale parallel-sided flutes were sculpted into thin glacial sediments in the Greenpeace Trough.

(Figure 3b) and 160° at an intermediate site (Plate 2). As determined from Bathy2000 profiles, these features are symmetric in cross section and have average height of 2.2 m. There is little evidence of these features in the deeper portions of the trough to the north-west, on the inner shelf, or on the extensive platform of the shallower shelf to the south-east.

We interpret these features as subglacial flutes which are common in soft sediments near both subaerial and subaqueous glacial margins elsewhere. Those described here are "parallel-sided flutes" [*Benn and Evans*, 1996] that form in till of low shear strength which deforms more readily than the overlying ice. Consequently, pressure differences in irregularities in the ice are equilibrated by flow of sediment, rather than creep of ice, and forms are maintained for significant distances down-glacier. Shear strengths in kasten core KC23 from the fluted region at Larsen Ice Shelf (Plate 2) increase from less than 10 kPa in overlying, non-ice-contact glacimarine sediments to about 20 kPa in the upper parts of the ice-contact sediments (Figure 4), indicating very limited over-consolidation due to ice pressure. *Solheim et al.* [1990] describe similar but smaller flutes on the seafloor in the Svalbard region. Here the sediments overlying bedrock are also thin (most less than 10 m), although the shear strengths are greater (up to about 80 kPa). *O'Brien et al.* [1999] also document larger flutes (1 km long, 200 m wide and 10 m high) inside a grounding line complex in Prydz Bay, Antarctica to which they ascribe a similar

origin to the features described here. We suggest that the small-scale features in Greenpeace Trough are of different origin than the "megascale glacial lineations" ascribed by *Anderson et al.* [2001; and 2002], *Canals et al.* [2000; 2002] and *Camerlenghi et al.* [2002] to the action of major ice streams elsewhere around Antarctica.

From their characteristics, the Greenpeace flutes are interpreted as the product of ice scour in the last phases of the Larsen A Ice Sheet as it thinned, floated, and subsequently lifted off the seafloor to become an ice shelf. Figure 2b shows the proposed ice conditions at this time. The ice sheets had thinned and sea level rose sufficiently that the grounding line moved back to near the present coasts along the escarpment of crystalline rock. The ridge eastward from Robertson Island was no longer over-ridden from the south and Larsen A and B became separated. Robertson Island and the Seal Nunataks no longer supported an ice dome or ice rise. The now partially floating Larsen A ice shelf spread to the south-east as it appears on modern satellite images [*Scambos et al.* 2001]. The establishment of floating ice across or between the Seal Nunataks may have taken place later, after considerably more thinning of the Larsen B Ice Shelf.

We propose that the flutes were created shortly before the ice shelf thinned sufficiently to float free from the surface, but after the ice sheet decoupled from the deepest part of the trough. Data from kasten core KC23 (Figure 4) provide evidence of their formation. Shear strength values in the shelf-contact sediments (Unit 3)

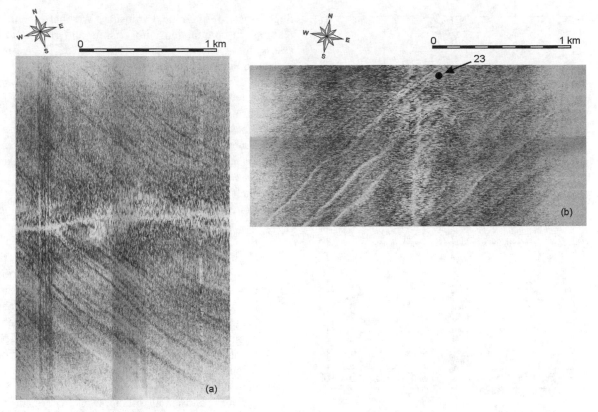

Fig. 3. Partially rectified sections of the SeaBeam swath pseudo-side-scan images of flutes. Locations are outlined on Plate 2. The images are not rectified for changes in ship's direction or speed during the transects (which varied due to heavy sea ice conditions), nor for differences in depth across the sea floor. These errors cause the lineations to appear less straight than they actually are. The location of kasten core KC23 (Figure 4) is shown.

are significantly different from those in the glacimarine units above only below about 200 cm depth (Figure 4a). This indicates that consolidation and therefore fluting occurred some time before the shelf lifted off the seafloor. The lithology of the coarse sand fraction (Figure 4b) indicates that the sediments of Unit 3 are almost completely composed of shales, most probably of local origin in the Greenpeace Trough, eroded, transported and deposited during the ice shelf phase shortly before it thinned and floated free of the trough. The sediments in the glacimarine units above are predominately quartz, feldspar, and crystalline (igneous and metamorphic) lithic fragments apparently transported from the Antarctic Peninsula and released by melting from the base of the floating ice shelf.

The transitional Unit 2 has intermediate lithic composition (Figure 4b), suggesting higher levels of debris transport but still some local influence. This implies that lift-off was at the contact of Units 3 and 2, and that it took some time for the change in composition to express itself. The abundance of grains in Unit 1 is much less

than below, suggesting a more dispersed source in the ice shelf and (after disintegration) in free-floating bergs [*Domack et al.* 2001]. However, before lift-off the change from ice sheet to ice shelf with the associated change in direction of flow did not impart new material to the sediment; it only remoulded the already deposited subglacial debris.

In the shallower water toward the edge of the continental shelf, the ice shelf remained in contact with the seafloor and erosion was sufficient that most or all of the soft sediment may have been removed. As well, iceberg scouring in open water conditions that prevailed during much of the Holocene [*Domack et al.,* 2001] and since the disintegration of the shelf in the mid 1990s probably masked the imprint of the ice shelf. Nor, apparently, are flutes present on the inner margin of the trench. Here the ice appears to have been streamed by several prominent hanging valleys, especially out from the present Drygalski Glacier, and again the scouring may have been sufficient across crystalline bedrock to remove most of the soft sediment. The flutes are also absent from the

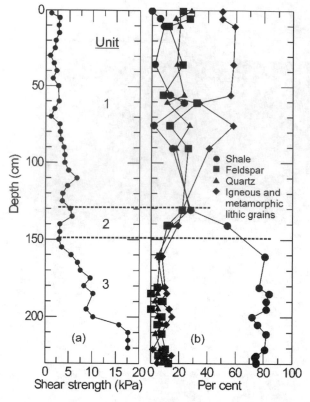

Fig. 4. (a) Shear strength and (b) mineralogy of clastic fragments larger than 1 mm in Kasten Core KC23 from Greenpeace Trough. Unit 3 was deposited when the ice shelf was in contact with the sea floor, Unit 2 during decoupling of the shelf from the sea floor, and Unit 1 beneath a floating ice shelf or in open water conditions [*Domack et al.* 2001].

deepest parts of the trough along the northwest side. It is possible that here the ice was floating in water depths greater than 1000 m and only as the depths less than about 900 m to the southwest was the ice in contact with the seafloor. A core recovered from this region exhibits stratification in Unit 3, supporting the suggestion of decoupling in the deepest part of the basin prior to lift-off in the region where the small-scale fluting is observed. The difference of 75° within the 5 km between the regions shown in Figures 2 and 3 probably relates to diversion of the shelf ice by local topography as it moved from deep to shallow water in the small valleys on the south-east side of the trough.

4. CONCLUSIONS

The advanced technology of swath bathymetric mapping has provided high-resolution images of the seafloor topography in the region beneath the former Larsen A Ice

Shelf, eastern Antarctic Peninsula. We interpret these features, in conjunction with seismic and core data, to assess the Quaternary glacial history of the eastern Antarctic Peninsula as follows.

1. A large, coast-parallel trough with depths to 1000 m on the inner continental shelf was created by glacial scour and subglacial fluvial processes beneath an ice streamflowing northeasterly across the inner shelf along the Nordensköld Coast and then easterly south of James Ross Island to the shelf edge in the Weddell Sea. Erosion of the trough was probably facilitated by juxtaposition of the weak Mesozoic sedimentary rock east of the crystalline basement rocks of the Peninsula.

2. Highly streamlined mega-flutes parallel to the trough are found in the trough, the valleys tributary to it, and, to a lesser extent, on the shelf beyond. Many are surrounded by prominent horseshoe-shaped scours. These features document a combination of direct glacial scour by the ice sheet and the action of large, probably periodic outbursts of subglacial water as is indicated beneath other parts of the Antarctic Ice Sheet and former continental glaciers elsewhere in the world. The thin cover of glacial sediment overlying bedrock throughout the region attests to the efficiency of these processes.

3. As the ice sheet thinned, probably during the late Pleistocene or early Holocene, the landmass between the Seal Nunataks and the mainland became a sufficient barrier that flow was diverted to the southwest across the shelf from the Peninsula to the Weddell Sea. In addition, the ice sheet transformed into partially floating ice shelves, the Larsen A to the north of the barrier significantly isolated from the Larsen B to the south. During this period extensive fields of smaller-scale, parallel-sided flutes were sculpted into the thin glacial sediments on the southeastern side of the trough in water depths of about 900 m. Their orientation conforms to the southeasterly flow direction of the modern Larsen A Ice Shelf in the years before it disintegrated, although valleys on the side of the trough redirected the flow of the ice shelf locally. The absence of these flutes on the shallower outer shelf indicates that the shelf was more firmly grounded in this region.

4. Seismic and core evidence [*Domack et al.*, 2001] indicates that the ice shelf continued to thin, eventually lifting off and so preserving the flutes.

Acknowledgment. The work was funded by the U.S. National Science Foundation, Office of Polar Programs (Project OPP-9814383), the Natural Sciences and Engineering Research Council of Canada, and the Programma Nazionale di Ricerche in Antartide (PNRA). We thank Captain J. Borkowski III, officers and crew of *N.B. Palmer* for their superb efforts during a logistically difficult winter cruise. Katherine Gavahan and Sheffield Corey of Raytheon Corp. provided the technical expertise to acquire the SeaBeam data, and the scientific personnel on board helped edit the data. We appreciate thoughtful reviews of our paper by T. Hughes, T. Scambos, and S. Shipp.

REFERENCES

Allen, J. R. L., *Sedimentary Structures, Their Character and Physical Basis, Volume II*, Elsevier, Amsterdam, 1982.

Anderson, J. B., J. S. Wellner, A. L. Lowe, A. B. Mosola, and S. S. Shipp, Footprint of the expanded West Antarctic Ice Sheet: Ice stream history and behavior, *Geology, 11*, 4-9, 2001.

Anderson, J. B., and S. S. Shipp, Evolution of the West Antarctic Ice Sheet, *Antarctic Res. Series, 77*, 45-57, 2001.

Anderson, J. B., S. S. Shipp, A. L. Lowe, J. S. Wellner, and A. B. Mosola, The Antarctic Ice Sheet during the Last Glacial Maximum and its subsequent retreat history: a review, *Quat. Sci. Rev., 21*, 49-70, 2002.

BAS (British Antarctic Survey), Tectonic map of the Scotia arc, 1:3,000,000 BAS Misc. 3. Cambridge, 1985,

Benn, D. I., and D. J. A. Evans, The interpretation and classification of subglacially deformed materials, *Quat. Sci. Rev., 15*, 23-52, 1996.

Bentley, M. J., and J. B. Anderson, Glacial and marine geological evidence for the ice-sheet configuration in the Weddell Sea—Antarctic Peninsula region during the Last Glacial Maximum, *Antarctic Sci., 10*, 309-325, 1998.

Blum, P., Physical Properties Handbook: A Guide to the Shipboard Measurement of Physical Properties of Deep-sea Cores, *ODP Tech. Note* 26. Available from World Wide Web: http://www-odp.tamu.edu/publications/tnotes/tn26/INDEX.HTM, 1997.

Camerlenghi, A., E. Domack, M. Rebesco, R. Gilbert, S. Ishman, A. Leventer, S. Brachfeld, and A. Drake, Glacial morphology and post-glacial contourites in northern Prince Gustav Channel (NW Weddell Sea, Antarctica), *Mar. Geophys. Res., 22*, 417-443, 2002.

Canals, M., J. L. Casamor, R. Urgeles, A. M. Calafat, E. W. Domack, J. Baraza, M. Farran, and M. DeBatist, Seafloor evidence of a subglacial sedimentary system off the northern Antarctic Peninsula, *Geology, 30*, 603-606, 2002.

Canals, M., R. Urgeles, and A. M. Calafat, Deep seafloor evidence of past ice streams off the Antarctic Peninsula, *Geology, 28*, 31-34, 2000.

Clark, C. D., Glaciodynamic context of subglacial bedform generation and preservation, *Ann. Glaciol., 28*, 23-32, 1999.

Cooper, A. P. R., Historical observations of the Prince Gustav Ice Shelf. *Polar Record, 33*, 285-294. 1997.

Del Valle, R. A., J. C. Lusky, and R. Roura, Glacial trough under Larsen Ice Shelf, Antarctic Peninsula, *Antarctic Sci., 10*, 173-174, 1998.

Domack, E., A. Leventer, R. Gilbert, S. Brachfeld, S. Ishman, A. Camerlenghi, K. Gavahan, D. Carlson, and A. Barkoukis, Cruise reveals history of Holocene Larsen Ice Shelf, *EOS (Trans Amer.Geophys. Union), 82*,13, 16-17, 2001.

Evans, J., and C. J. Pudsey, Sedimentation associated with Antarctic Peninsula ice shelves: implications for palaeoenvironmental reconstructions of glacimarine sediments, *J. Geol. Soc. (London), 159*, 233-237, 2002.

Gilbert, R., The Broughton Trough on the continental shelf of eastern Baffin Island, Northwest Territories, *Can. J. Earth Sci., 19*, 1599-1607, 1982.

Hathaway, B., and J. B. Riding, Stratigraphy and age of the Lower Cretaceous Pedersen Formation, northern Antarctic Peninsula, *Antarctic Sci., 13*, 67-74, 2001.

Holik, J. S., R/V Nathaniel B. Palmer: State of the art in marine oceanographic research for the U. S. Antarctic Program, *EOS (Trans. Amer. Geophys. Union)*, 75, 580, 1994.

Holtedahl, O., On the morphology of the west Greenland shelf with general remarks on the "marginal channel" problem, *Mar. Geol., 8*, 155-172, 1970.

Lowe, A. L., and J. B. Anderson, Reconstruction of the West Antarctic ice sheet in Pine Island Bay during the Last Glacial Maximum and its subsequent retreat history, *Quat. Sci. Rev., 21*, 1879-1897, 2002.

Menzies J., and J. Rose, Drumlins: trends and perspectives. *Episodes, 10*, 29-31, 1987.

Nakadaa, M., R. Kimuraa, J. Okunoa, K. Moriwakib, H. Miurab, and H. Maemokue, Late Pleistocene and Holocene melting history of the Antarctic ice sheet derived from sealevel variations, *Mar. Geol., 167*, 85-103, 2000.

Nye, J. F., Water at the bed of a glacier, *Int. Assoc. Sci. Hydrol. Pub., 95*, 189-194, 1973

O'Brien, P. E., L. De Sandis, P. T. Harris, E. Domack and P.G. Quilty, Ice shelf grounding zone features of western Prydz Bay, Antarctica: sedimentary processes from seismic and sidescan images, *Antarctic Sci., 11*, 78-91, 1999.

Pudsey, C. J., J. Evans, E. W. Domack, P. Morris, and R. A. Del Valle, Bathymetry and acoustic facies beneath the former Larsen-A and Prince Gustav ice shelves, north-west Weddell Sea, *Antarctic Sci., 13*, 312-322, 2001.

Rack, W., *Dynamic Behavoir and Disintegration of the Northern Larsen Ice Shelf, Antarctic Peninsula,* PhD dissertation, Leopold-Franzens Univesität, Innsbruck, Austria, 2000.

Rott, H., P. Skvarca, and T. Nagler, Rapid collapse of the Northern Larsen Ice Shelf, Antarcticia: *Science, 271*, 788-792, 1996.

Rott, H., W. Rack, T. Nagler, and P. Skvarca, Climatically induced retreat and collapse of northern Larsen Ice Shelf, Antarctic Peninsula, *Ann. Glaciol., 27*, 86-92, 1998.

Sawagaki, T., and K. Hirakawa, Erosion of bedrock by subglacial meltwater, Soya Coast, East Antarctica, *Geog. Ann., 79A*, 223-238, 1997.

Scambos, T. A., C. Hulbe, M. Fahnstrock, and J. Bohlander, The link between climate warming and break up of ice shelves in the Antarctic Peninsula, *J. Glaciol., 154*, 516-530, 2001.

Sharpe, D. R., and J. Shaw, Erosion of bedrock by subglacial meltwater, Cantley, Quebec, *Geol. Soc. Amer. Bull., 101*, 1011-1020, 1989.

Shaw, J., Hairpin erosional marks, horseshoe vortices and subglacial erosion, *Sed. Geol., 92*, 169-283, 1994.

Shaw, J., A meltwater model of Laurentide subglacial landscapes, in *Geomorphology Sans Frontieres*, edited by S. B. McCann, and D.C. Ford, John Wiley and Sons Ltd., New York, 181-236, 1996.

Shaw, J., and R. Gilbert, Evidence for large scale subglacial meltwater flood events in southern Ontario and northern New York State, *Geology, 18*, 1169-1172, 1990.

Shaw, J., B. Rains, R. Eyton, and L. Weissling, Laurentide subglacial outburst floods: landform evidence from digital elevation models, *Can. J. Earth Sci., 33*, 1154-1168, 1996.

Siegert, M. J., Antarctic subglacial lakes, *Earth Sci. Rev., 50*, 29-50, 2000.

Skvarca, P., W. Rack, H. Rott, and T.L.. Donangelo, Climatic trends and the retreat and disintegration of ice shelves on the Antarctic Peninsula: an overview, *Polar Res., 18*, 151-157, 1999.

Sloan, B. J., L. A. Lauver, and J. B. Anderson, Seismic stratigraphy of the Larsen Basen, eastern Antarctic Peninsula: America Geophysical Union, *Antarctic Res. Ser., 68*, 59-74, 1995.

Solheim, A., L. Russwurm, A. Elverhoi, and M. Nyland Berg, Glacial geomorphic features in the northern Barents Sea: direct evidence for grounded ice and implications for the pattern of deglaciation and late glacial sedimentation, in *Glacial Environments: Processes and Sediments*, edited by J. A. Dowdeswell, and J. D. Scource, *The Geological Society Special Publication No. 53*, London, 253268, 1990.

Sugden, D. E., Glacial erosion by the Laurentide ice sheet, *J. Glaciol., 20*, 367391, 1978.

Sugden, D. E., G. H. Denton, and D. R. Marchant, Subglacial meltwater channel systems and ice sheet overriding, Asgard Range, Antarctica, *Geog. Ann., 73A*, 109121, 1991.

Vanney, J. R., and G. L. Johnson, The seafloor morphology seaward of Terre Adelie (Antarctica), *Dutsche Hyrdrographica Zeistchrifter, 32*, 77-87, 1979.

Vaughan, D. G., and C. S. M. Doake, Recent atmospheric warming and retreat of ice shelves on the Antarctic Peninsula, *Nature, 379*, 328-331, 1996.

Vaughan, D. G., G. J. Marshall, U. M. Connolley, J. C. King, and R. Mulvaney, Devil is in the detail, *Science, 293*, 1777-1779, 2001.

Wellner, J. S., A.L. Lowe, S. S. Shipp, and J. B. Anderson, Distribution of glacial geomorphic features on the Antarctic continental shelf and correlation with substrate: implications for ice behavior, *J. Glaciol., 47*, 397-411, 2001.

Angelo Camerlenghi, Instituto Nazionale di Oceanografia e di Geofisca Sperimentale (OGS) Borgo Grotta Gigante, 42/c-I-34010 Sgonico (TS) Italy. acamerlenghi@ogs.trieste.it

Eugene W. Domack, Geology Department, Hamilton College, Clinton NY 13323 U.S.A. edomack@hamilton.edu

Robert Gilbert, Department of Geography, Queen's University, Kingston ON K7L 3N6 Canada gilbert@lake.geog.queensu.ca

MARINE SEDIMENTARY RECORD OF NATURAL ENVIRONMENTAL VARIABILITY AND RECENT WARMING IN THE ANTARCTIC PENINSULA

Eugene W. Domack[1], Amy Leventer[2], Stephanie Root[1], Jim Ring[3], Eric Williams[2], David Carlson[1], Emily Hirshorn[2], William Wright[1], Robert Gilbert[4], and George Burr[5]

Paleoenvironmental data from marine sediment cores collected along the western side of the Antarctic Peninsula provide a reference for Holocene climate and oceanographic changes across a 400 km north south transect. Interpretations are based upon: preserved total organic carbon, diatom abundance/assemblages, particle size, ice rafted detritus, sedimentary structures, and physical properties including water content and magnetic susceptibility. Chronology is constrained by 38 new radiocarbon dates, ^{210}Pb activity profiles, and tidal laminations. We recognize four climate/oceanographic episodes consistent with climate intervals recognized from the Northern Hemisphere including the Middle Holocene optimum (~9.0 ka to 3.2 ka) and the Neoglacial (3.2 ka to modern). These two episodes are characterized by high to low rates of sediment accumulation and high to low total organic carbon preservation, respectively. We also delineate the Medieval Warm Period (1.15 ka to 0.7 ka) and the Little Ice Age (0.7 ka to ~0.15 ka). Superimposed upon these intervals are 200-year oscillations in climate/oceanographic conditions that can be regionally correlated. High-resolution sites also record the last several decades of regional warming, the signal of which appears distinctive from the natural variability in climate/oceanographic conditions of the Neoglacial. Results support the synchronicity of climate change between the two hemispheres, a correlation that may be unique to the Antarctic Peninsula, and the impact of regional warming on marine deposystems. Despite decades of regional warming the rates and styles of glacial marine deposition are still not equivalent to characteristics of the environment recorded in the Middle Holocene climate optimum approximately 4.0 ka.

1. INTRODUCTION

Rapid environmental change in the Antarctic Peninsula has been highlighted most recently by the disintegration of the Larsen A and Prince Gustav Ice Shelves in 1995 [*Rott et al.,* 1996] and the Larsen B Ice Shelf in 2002 [*Scambos et al.,* this volume; *Morris and Vaughan,* this volume]. However, many additional observations over the last several decades demonstrate a wide variety of significant environmental changes throughout the Peninsula. For example, a total of seven ice shelves on both sides of the Peninsula have retreated [*Doake and*

[1]Geology Department, Hamilton College, Clinton, New York
[2]Geology Department, Colgate University, Hamilton, New York
[3]Physics Department , Hamilton College, Clinton, New York
[4]Department of Geography, Queen's University, Kingston Ontario, Canada
[5]NSF TAMS Facility, Department of Physics, University of Arizona, Tucson Arizona

10.1029/079ARS17

Vaughan, 1991; *Doake*, 1982; *Ward*, 1995; *Skvarca*, 1993, 1994; *Vaughan and Doake*, 1996; *Scambos et al.*, 2000]. Meteorological data document rising average surface air temperatures [*Jones*, 1990; *Sansom*, 1989; *Harangozo et al.*, 1997; *Stark*, 1994; *Smith et al.*, 1996; *Smith et al.*, 1999; *King*, 1994; *King and Harangozo*, 1998; *Jacka and Budd*, 1998] while mid-depth ocean temperatures also have warmed over the past several decades [*Gille*, 2002]. Ice core data record glaciological response to the increase in temperatures [*Thompson et al.*, 1994]. Changes in the distribution of penguin species [*Fraser et al.*, 1992] and vascular plants [*Fowbert and Lewis-Smith*, 1994; *Convey*, this volume] may be associated both directly and indirectly with the warming climate. Changes in the distribution of sea ice [*de la Mare*, 1997; *Jacobs and Comiso*, 1993] and snow cover [*Fox and Cooper*, 1998] have also been noted.

Given the overwhelming evidence for recent change, it has become increasingly important to understand the longer-term history of change in the Antarctic Peninsula region, in order to place today's events into perspective. One way to assess this longer-term history is to examine the proxy records contained in cores of marine sediment [*Smith et al.*, 1999]. Over the past thirty years, numerous cores have been retrieved from the Antarctic Peninsula continental shelf. With sediment accumulation rates that range from 0.01 mm/year to 10 cm/year, a resolution of climate variability on the scale of years to centuries is possible. Since Jumbo Piston cores now routinely reach a length of 25 m and Ocean Drilling Program Cores from the Peninsula reach 100 m in length, these high-resolution records extend back to the last glaciation. High quality marine sediment cores range from as far south as Lallemand Fjord to the northern tip of the Peninsula and eastward into the Weddell Sea (Figure 1). Collectively, these cores, with chronologies well constrained by AMS radiocarbon dating, provide the basis for a detailed reconstruction of the past 13,000 years, with a current emphasis on the mid to late Holocene. While our recent field efforts have successfully recovered suitable high-resolution cores from the northwestern Weddell Sea [*Camerlenghi et al.*, 2001; *Drake*, 2002; *Lichtenstein*, 2002; *Duran*, 2003; *Domack et al.*, 2001; *Gilbert et al.*, this volume] chronologic and proxy analyses on these records are not yet complete. Therefore our focus for this paper is restricted to the western side of the Antarctic Peninsula although we do make correlations with published observations throughout the region.

The primary goal of this paper is to present a reconstruction of climate and oceanographic change during the Holocene, based on selected proxy data from marine sediment cores (Figure 1). It represents an expanded and

updated review of earlier results [*Domack and McClennen*, 1996; *Smith et al.*, 1999] but also presents a new array of paleoenvironmental proxy data. In broad view, our new data record a timetable of major oceanographic and climatic change over this time period, with strong evidence for both the occurrence and timing of major, globally recorded, Holocene events, such as the mid-Holocene climatic optimum, the late Holocene Neoglacial, including the inferred Medieval Warm Period and the Little Ice Age. Most significantly, these events, as well as smaller scale changes, can be traced and correlated throughout the Peninsula. In addition, these data allow us to examine the finer details of climate change and the response of the Southern Ocean to these changes. For example, our proxy data help constrain changes in primary productivity and sea ice distribution over time, and the possible influence of tidal cycles on sedimentation as a function of climate. Finally, the data clearly show that the atmospheric warming trend of the past several decades has impacted sedimentation in the region and that

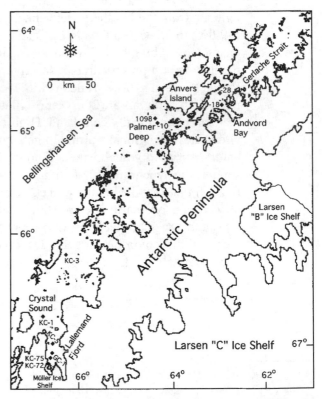

Fig. 1. Map of the northwestern side of the Antarctic Peninsula, with location of cores sites mentioned in the text. Located within the Palmer Deep but not labeled are the following cores: *Polar Duke* (*PD*)92–02 PC30 (at ODP Site–1098) and *N.B. Palmer* (*NBP*) 99–03 MC10A (at core site *NBP*99–03 JPC–10). Located within Andvord Bay but not labeled are core sites *PD*87–07 PC22 and *NBP*99–03 KC–18C (at *NBP*99–03 JPC–18).

these changes are easily recognized in the sedimentary record. However, we emphasize that not all sediment cores record the identical signal of ocean/climate episodes due to differences in resolution as controlled by sedimentation rate and bioturbation [*Smith et al., 1999*].

2. METHODS

2.1. % Total Organic Carbon

Total organic carbon (TOC) was determined on a dry weight basis using a LECO CR–412 furnace via the infrared adsorption technique. The errors associated with this method are 2–4% in accuracy of the measured standards and ± 0.05% of TOC in precision of duplicates. Samples were pretreated in 2 N HCl to remove acid soluble and inorganic carbon.

2.2. Particle Size

Particle size was determined on small samples using a Malvern laser diffraction (Master Sizer-E) system. This method has been used on Antarctic glacial marine sediments in the past with very reproducible results [*Domack et al., 1999; Warner and Domack, 2002*]. Coarse fraction (gravel) abundance was determined from x-ray radiographs following the method of *Grobe* [1987]. Counts were determined every 10 cm, for narrow u-channels of jumbo piston cores, or every 2.5 cm, for 12 cm wide kasten core archive trays. Results are expressed as the number of grains per core depth interval (i.e. 10 cm or 2.5 cm).

2.3. Sedimentary Structures

Laminations were evaluated by counting the number of laminations per 10-cm intervals on x-ray radiographs. X-ray radiographs were taken of u-channel sections or archive trays of kasten (box) cores. The negatives were scanned and then transferred to digital form using the National Institute of Health Image-Analysis Software. Gray scale intensity was detrended in order to correct for the lateral scattering of x-rays which is greater at each end of the radiographs than at the center. Visual descriptions of each core were made either on-board ship or at the time of core cutting at the Florida State University.

2.4. Physical Properties

Water content was determined by weight loss after drying in pre-weighed vials at 106°C for 48 hours. Samples were collected immediately upon opening of the cores, sealed and transported in chilled conditions to the laboratory in Hamilton College. Magnetic susceptibility was measured using a Bartington MS–2C meter and loops of varying diameter depending upon the size of the core or archive. Measurements are reported as volume susceptibility in CGS units, except for Ocean Drilling Program (ODP) site 1098 which is reported in SI units.

2.5. Diatoms

Quantitative diatom slides were prepared using a settling method [*Scherer*, 1994] that produces a random distribution of diatoms and allows calculation of the absolute abundance of diatoms. Diatoms were counted along transects under 1000X magnification using an Olympus BX60 microscope. A total of approximately 400 diatom valves were counted per slide. Only diatoms with more than half of the valve or the center of the valve present were counted. Due to the overwhelming abundance of the diatom genus *Chaetoceros* in samples from the Antarctic Peninsula, specimens from this genus were not counted. This "*Chaetoceros*-free" counting method has been used by many diatomists working with Antarctic marine sediments [*Leventer et al.*, 1996]. Absolute and relative abundances of approximately forty different species of diatoms were tracked throughout all the slides. Slides were counted every 2 cm in *NBP*99-03 MC10A, and either every 3–4 cm (0–43 cm) or every 10 cm (50–270 cm) in *NBP*99-03 KC18C.

Standard diatom taxonomic species definitions were followed, though the species *Thalassiosira antarctica* was divided into "warm" and "cold" water forms, as described by *Taylor et al.* [2001]. Water temperatures noticeably affect the development of *T. antarctica*. When this species grows in warmer water (4°C), its diameter is larger overall, the areolae are larger, and it exhibits unusual extensions of its test, termed "shoes." The colder water (–1.5°C) variety is smaller and more lightly silicified with smaller areolae and no shoes [*Villareal and Fryxell*, 1983]. Both the cold and warm water varieties indicate of open ocean conditions; however, the colder water form may also be associated with the sea ice edge [*Villareal and Fryxell*, 1983].

2.6. Radiocarbon Analyses

Presented in Table 1 are 38 unpublished radiocarbon ages determined on acid insoluble organic matter and on calcium carbonate fractions from in-situ molluscs (pelecypods and scaphopods) and foraminifera. Acid insoluble fractions (aiom) were obtained on-board ship and were processed according to established procedures [see

Domack et al., 2001; 1999]. In-situ molluscs were identified, if possible, to species level, cleaned in H_2O_2 to remove the peristracum, and ultrasonified to remove fine particulate matter. Foraminifera were hand picked from benthic assemblages and were ultrasonified in methanol to remove fine particulate matter. All carbon fractions were then sent to the University of Arizona Tandem Accelerator Laboratory for analysis. Ages are reported as uncorrected radiocarbon years in Table 1, except for adjustment due to fractionation in proportion to the values of $\delta^{13}C$ reported in Table 1. Calibration and adjustment of ages due to reworking, of organic particulates, were completed only for samples from cores KC-1, –3, –18B, and –18C, as discussed below. Calibration and reworking corrections followed methods outlined in *Stuiver et al.* [1998] and *Domack et al.* [1999; 2001].

TABLE 1: Uncorrected radiocarbon ages determined at the University of Arizona.

Laboratory Number	Cruise: *NBP*99–03 Core Depth (cm)		$\delta^{13}C$	Uncorrected Age yr BP[a]	Sedimentation Rate (cm/yr)[b]	Carbon Source[c]
AA34635	KC1	0–2	–22.90	1830 ± 40	—	aiom
AA34636	"	2	1.22	1125 ± 40	—	*Yoldia eightsii*
AA34665	"	110–112	–23.60	3195 ± 60	0.08	aiom
AA34637	"	260–262	–23.20	5040 ± 55	0.08	aiom
AA34638	KC3	0–2	–23.10	1820 ± 45	—	aiom
AA34666	"	70–72	–23.80	2660 ± 50	0.08	aiom
AA34639	"	130–132	–24.90	3580 ± 55	0.07	aiom
AA34667	"	190–192	–23.70	4180 ± 60	0.10	aiom
AA34640	JPC-10	1250	–0.71	8780 ± 70	—	*Cadulus dalli antarcticus*
AA34641	"	1320–1326	–6.42	9250 ± 65	0.16	foraminifera
AA34642	KC-18B	0–2	–23.40	1445 ± 55	—	aiom
AA34643	"	80–82	–24.10	1915 ± 40	0.17	aiom
AA34644	KC-18C	0–2	–23.40	1450 ± 40	—	aiom
AA34668	"	30–32	–23.50	1745 ± 50	0.10	aiom
AA34669	"	60–62	–24.90	1870 ± 50	0.24	aiom
AA34670	"	90–92	–24.50	2005 ± 50	0.22	aiom
AA34671	"	120–122	–23.10	2210 ± 50	0.15	aiom
AA34672	"	150–152	–22.20	2280 ± 50	0.43	aiom
AA34645	"	180–182	–23.80	2200 ± 40	—	aiom
AA34673	"	210–212	–26.40	2550 ± 50	0.22	aiom
AA34646	"	222–224	–4.94	2175 ± 40	—	mollusc
AA34674	"	240–242	–23.20	2550 ± 70	—	aiom
AA34647	"	266–268	–24.80	2605 ± 45	0.47	aiom
AA34648	JPC-18	29	1.33	1550 ± 45	—	mollusc
AA34675	"	696–697	1.15	4360 ± 55	0.24	*Yoldia eightsii*
AA34649	"	808–809	1.04	4465 ± 50	1.06	*Yoldia eightsii*
AA34676	"	960	–0.03	4755 ± 55	1.92	mollusc
AA34677	"	1262	1.20	5255 ± 60	0.60	mollusc
AA34650	"	1323	0.24	5285 ± 65	2.03	foraminifera
AA34651	"	1712	–3.76	6240 ± 55	0.78	*Thyasira dearborni*
AA34652	"	1905–1906	–2.40	6740 ± 70	0.65	mollusc
AA34655	JPC-28	97–98	0.19	1560 ± 45	—	mollusc
AA34678	"	325.5–327.5	–4.48	2040 ± 50	0.48	*Thyasira dearborni*
AA34656	"	411–412	0.91	2340 ± 45	0.28	mollusc
AA34657	"	841–843	1.02	3500 ± 50	0.37	*Yoldia eightsii*
AA34679	"	866–867	–2.88	3575 ± 60	0.32	mollusc
AA34658	"	1003–1004	1.41	3835 ± 60	0.52	mollusc
AA34659	"	1569–1571	–2.80	4430 ± 65	0.95	mollusc

[a]Ages are adjusted for $\delta^{13}C$ but uncorrected for reworking, reservoir, and calibration effects.

[b]Interval sedimentation rates (cm/yr) are calculated from uncorrected age differences except where previous intervals yield anomalous ages (see text for discussion).

[c]Aiom = acid insoluble organic matter.

2.7. ^{210}Pb Activity

Samples for ^{210}Pb activity were collected in pre-weighed vials at closely spaced intervals, sealed and transported to the laboratory in chilled condition. Water contents (porosity) were then determined and dried fractions of equal mass were placed in a gamma spectrometry well detector for ~24 hours. The detector consisted of a thallium doped sodium iodide crystal shielded by three layers of lead bricks and a thin sheet of cadmium. Counting errors were estimated via Poisson statistics on background counts that were of equal duration to, and separated by, each sample measurement. Gamma rays were counted within an energy spectral band centered at 46 keV. The resolution for ^{210}Pb activity is about 16% based upon measuring the full width of the spectral peak at half the maximum, as observed centered at 46 keV. We report the total counts per day per gram of sediment and profiles are not corrected for porosity variations. Accumulation rate estimates are therefore calculated only for wet sediment and are based on subtraction of supported ^{210}Pb activity as established at depth.

3. RESULTS AND DISCUSSION

3.1. Lallemand Fjord/Crystal Sound

In the polar setting of Lallemand Fjord (Figure 1), both terrigenous and biogenic input are relatively low, with sedimentation rates ranging from ~0.6-0.8 mm/year [Domack and McClennen, 1996; Domack et al., 1995]. In these low-resolution sediment core records, changes in primary productivity, most likely driven by changes in sea ice extent, are recorded as down core variations in the %TOC content and isotopic ratio of organic carbon (δ^{13}C) within the sediments [Shevenell et al., 1996; Taylor et al., 2001]. A ~120 km long transect of five kasten cores (KC) and one gravity core (GC), extending from northernmost Crystal Sound southward into Lallemand Fjord at the Müller Ice Shelf, was studied in order to evaluate regional changes in primary productivity over time, as recorded by %TOC (Plate 1). Over this transect, the geographical control over % TOC is obvious, with increasing values away from the ice shelf and to the north. For example, regardless of the time interval, %TOC values are lower in KC72, which is adjacent to the Müller Ice Shelf, while values on average three times higher are observed in KC3, over one hundred kilometers from the ice shelf front (Plate 1). However, clear and coherent changes in % TOC and δ^{13}C [Taylor et al., 2001] are also observed over time, demonstrating the

spatial continuity of changes in regional productivity over the past 3000 years.

The Hypsithermal, a mid-Holocene warm event that has been documented previously in Antarctica [Smith et al., 1999; Domack et al., 1991, 2001; Taylor et al., 2001] is recorded in sediments older than 2500 calendar years BP (cal yr BP) by increased % TOC at all six core sites. However, farther north in the Palmer Deep, the end of the Hypsithermal is signaled somewhat earlier, around 3.5 ka (cal yr BP), [Domack et al., 2001]. This mid-Holocene interval of heightened primary productivity certainly extended back in time earlier than most core records, as demonstrated by the 8000 yr record of core GC–1 [Shevenell et al., 1996; Taylor et al., 2001]. Hence, most of the cores only document the end of the Hypsithermal and transition to the Neoglacial at about 2500 cal yr BP. Lower %TOC characterized the Neoglacial sediments, as more persistent sea ice cover most likely limited primary production and organic carbon deposition.

By approximately 1150 cal yr BP, %TOC values display a lesser maxima, at a time corresponding to the Medieval Warm Period (MWP; ~1150–700 cal yr BP in our data, compared to 1200-800 cal yr BP [Broecker, 2001]). At this time regionally warmer conditions and decreased sea ice cover probably permitted higher levels of export production, and thus delivery of greater amounts of isotopically heavy (high δ^{13}C) organic carbon to the seafloor [Taylor et al., 2001; Gilbert et al., 2003].

Finally, in the uppermost portion of the cores, %TOC decreases. This reflects the onset of the Little Ice Age (LIA, ~700–150 cal yr BP in Plate 1; 650-140 cal yr BP [Broecker, 2001]), which is generally characterized by cooler temperatures, and greater variability of temperature and precipitation [Grove, 1988]. Ice cores from several regions of Antarctica [Mosley-Thompson and Thompson, 1982, this volume] and moraines and tills in the Antarctic Peninsula region [Clapperton and Sugden, 1989; Clapperton, 1990] provide evidence for cooling during the LIA. However, others [Mosley-Thompson et al., 1990; Mosley-Thompson, 1992] report findings from ice cores south of Lallemand Fjord that suggest warming in parts of West Antarctica and the Antarctic Peninsula during the LIA. As part of the LIA signal we also recognize evidence of the growth and advance of the Müller Ice Shelf approximately 400 cal yr BP [Domack et al., 1995]. This event corresponds to the onset of the most pronounced TOC minima (lasting ~ 300 years) observed throughout the Lallemand Fjord Crystal Sound deposystem (Plate 1). Our sediment-based chronology for these late Holocene events places them in the same temporal framework as in the northern hemisphere and hence

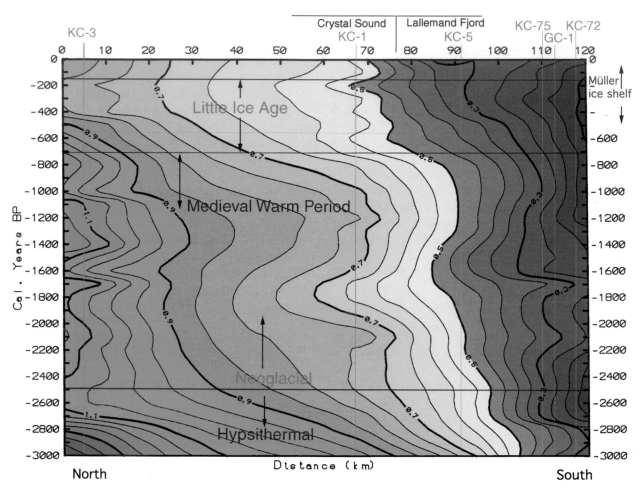

Plate 1. Spatial and temporal variability in % TOC in cores along a 120 km transect from Crystal Sound into Lallemand Fjord. Chronologic control is found in Table 1 and in references below. Contour lines are equal % of TOC with an interval of 0.05 %. TOC data for cores KC–1 and KC–3 are from *Root* [2001], and for cores GC–1, KC–5, KC–72, and KC–75 are from *Shevenell et al.* [1996]. Within the time frame represented, globally recognized climatic events, such as the Hypsithermal, Neoglacial, Medieval Warm Period, and Little Ice Age are indicated. To the right, the advance of the Müller Ice Shelf is noted [*Domack et al.,* 1995].

argues for global synchronicity in the forcing of such climate changes [*Mosley-Thompson and Thompson*, this volume]. However, the MWP is an event that has yet to be recognized in a globally synchronous time frame [*Mann et al.*, 2001, 1999; and *Hughes et al.*, 1994].

3.2. Palmer Deep

A millennial record of environmental change is preserved in an 80-cm multicore (*NBP*99-03 MC-10A) collected in a small basin perched on the northeastern slope of Palmer Deep Basin (64°53.068'S, 62°12.544'W, 937 m water depth, Figure 1 [*Domack*, 2002]). The rationale behind the collection of MC-10 A, B, and C was to recover a suite of cores (the multicore collects three short cores simultaneously) with a well-preserved sediment-water interface that could be used to complete the stratigraphic framework of a 13.4 m long Jumbo Piston Core (*NBP*99-03 JPC-10) taken at the same site. Visual inspection of the multicores at the time of recovery suggested that this was indeed the case; clear water was observed above the sediment water interface in the cores. The magnetic susceptibility records of the three cores can be correlated easily, and three distinctive siliceous laminations can be traced between the cores, supporting the integrity of the cores.

MC-10A, at 80 cm length, was the longest of the three, and was selected for detailed diatom work. Absolute dating of MC-10A has not yet been completed, so event stratigraphy was used to estimate a sediment accumulation rate for the core (Figure 2*a*). The magnetic susceptibility records of cores *NBP*99-03-MC10A, -JPC10, and ODP Leg 178, Site 1098, which has a tightly constrained chronology [*Domack et al.*, 2001], were correlated based upon magnetic susceptibility. This correlation indicates an approximate sedimentation rate of 0.1 cm/year for MC-10A; the core thus spans an 800-year history. This time span is only an approximation, as sedimentation rates likely varied over this time period. Given this chronology, MC-10A represents accumulation during a time period that includes the Medieval Warm Period, Little Ice Age and modern-day climate changes. It is important to note, however, that the chronologic definition of these events in MC-10A is less certain than in the tightly time-constrained cores from Lallemand Fjord, Crystal Sound, and the Andvord Drift (see following section).

The diatom data (Figure 2*a*) suggest warmer conditions from ~800-360 yr BP relative to the period from

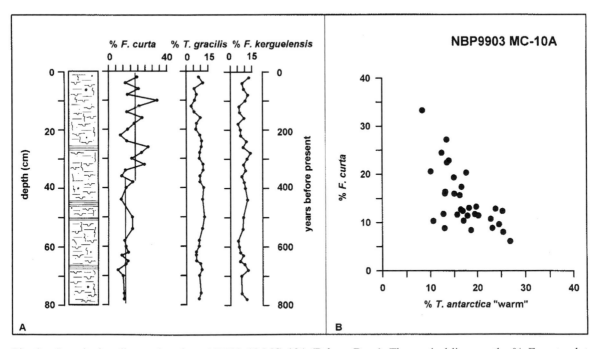

Fig. 2a. Quantitative diatom data from *NBP*99–03 MC–10A (Palmer Deep). The vertical lines on the % *F. curta* plot indicate average relative abundances of 17% and 11% above and below 36 cm, respectively. Fig. 2b. Relative abundance of the warm form of the diatom species *T. antarctica* versus *F. curta*, a sea ice indicator. Their inverse relationship suggests that icier conditions are associated with cooler surface water temperatures.

~360 yr BP to the present. This interpretation is based largely on the distinctive pattern of relative abundance exhibited by *Fragilariopsis curta*; a diatom species commonly associated with sea ice and associated ice-edge blooms [*Leventer*, 1998, and references therein]. The average abundance of this species is significantly lower below 36 cm (11%) than above, where it averages 17%. Low relative abundances of *F. curta* indicate decreased importance of primary production associated with sea ice [*Leventer and Dunbar*, 1996]. Wind, ocean currents, and temperature can influence the extent and stability of sea ice. A more thorough understanding of the changes in climatic/oceanographic conditions that drove the inferred lower sea ice extent from 800-360 yr BP can be derived from an evaluation of the distribution of other diatom species. Relative abundance of *Thalassiosira gracilis*, an indicator of windiness and its influence on water column mixing, shows no significant change in conjunction with the changes in *F. curta*, so it is unlikely that increased wind stress was a major factor leading to decreased sea ice extent. Nor does *Fragilariopsis kerguelensis*, an indicator of inputs from the Antarctic Circumpolar Current (ACC), show any significant variability through the core, suggesting that a major change in the position of the ACC was not a factor contributing to variability in sea ice cover through the period recorded in this core.

Thalassiosira antarctica, another significant contributor to sediments in the Palmer Deep, does not show a clear up-core trend in MC-10A. However, a roughly inverse relationship between the warm-water variety of *T. antarctica* and *F. curta* is observed in the core (Figure 2*b*). This suggests that increased water temperature was a factor controlling decreased sea ice extent through the period corresponding to the lower half of the core, which is roughly coincident with the end of the globally-recognized Medieval Warm Period.

Cooler and icier conditions are hypothesized to have occurred between ~360-100 yr BP in the Palmer Deep, a time period roughly coinciding with the LIA [*Grove*, 1988]. Based on the marked increase in abundance and variability of *F. curta* in the top half of core MC-10A, our data are consistent with cooling and greater environmental variability during the period of the LIA.

The top 10 cm (approximately the past 100 years) of MC-10A may document a return to a warmer climate. *F. curta* abundances decrease and are less variable, although abundances are not as low as those measured in the bottom half of the core. This may record recent regional warming but the signal is not strong, and more data are needed in order to draw definitive conclusions. This apparent lack of a strong warming signal is surprising in light meteorological records [*King et al.*, this volume; *Smith et al.*, this volume]. However, the potential moderating influence of the open Bellingshausen Sea on this locality may be responsible for dampening effects of higher regional temperatures.

3.3. Andvord Bay

Kasten Core *NBP*99-03 KC-18C was recovered from outer Andvord Bay (64°46.342' S, 62°49.746' W, 428 m water depth, Figure 1) in a sediment drift deposit [*Harris et al.* 1999]. A suite of analyses was performed on KC-18C, including measurement of water content, magnetic susceptibility, lonestone abundance, %TOC, grain size, and diatom assemblages (Figure 3). Lonestone abundance and sedimentary structures were evaluated based on x-ray radiographs of the core. Together, these data permit a detailed reconstruction of the last 1000 years, with a close look at the period corresponding with the Medieval Warm Period, Little Ice Age and modern regional warming.

An age model for this core was constructed based on a combination of 13 AMS radiocarbon dates and ^{210}Pb data (Figure 4). ^{210}Pb data were used to determine the sediment accumulation rate above 15 cm depth, and the radiocarbon data were used to determine rates below 30 cm depth. Between 15 and 30 cm, linear interpolation was used for the chronology. Our rationale for this chronology is described below.

The ^{210}Pb activity, which decreases to near constant supported levels by 15-cm depth, indicates an accumulation rate of ~5 mm/yr in the upper 15 cm of the core. Extrapolation of this rate farther down core is not reasonable, based on the multiproxy data that clearly demonstrate a significant change in the character of the sediments above 15 cm depth (Figure 4). Increased lonestone abundance, % TOC, % clay, and a warmer-water diatom flora all indicate an increase in sediment accumulation rate at the top of the core, even after correcting for porosity.

A second order polynomial (r^2 = 0.947) was fit to the calibrated radiocarbon ages below 30 cm. All the dates are based on bulk (acid insoluble) organic matter. Four dates are excluded from this chronology. Two surface dates were not used, since the ^{210}Pb chronology was utilized in this section of the core. There are also unresolved problems in interpreting ^{14}C ages for near surface sediment due to combinations of mixing and incorporation of bomb ^{14}C within the sediment profile. In addition, a mollusc date at 223 cm was not used, nor was a bulk organic matter date at 181 cm. In both cases these dates are

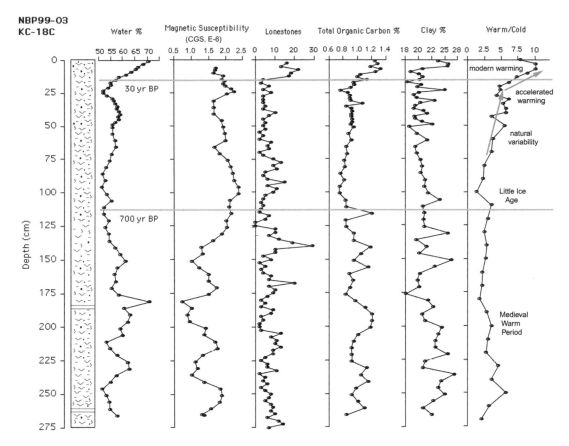

Fig. 3. Multi-proxy data set for core *NBP*99-03 KC-18C (Andvord Drift). Downcore variability in water content, volume magnetic susceptibility (CGS x 10⁻⁶), lonestone abundance, total organic carbon content, clay content, and the relative abundance of warm versus cold forms of the diatom species *Thalassiosira antarctica*. As in Plate 1, globally recognized climatic events, as the Medieval Warm Period and Little Ice Age are indicated. In addition, note the accelerated warming trend during the past several decades. See Table 1 for chronologic control.

considered outliers in comparison to the other data, and hence are not used for the longer term (century scale) chronology (Figure 4). Finally, linear interpolation of depth vs. age was used to assign ages to core depths between 15 and 30 cm. Given these constraints, the core recovered an ~1000 year record, similar to that in Palmer Deep core MC-10A. Yet, two important differences are apparent between these cores. First, at over 3 times the length of MC-10A, core KC-18C has a much higher temporal resolution with a sedimentation rate of 0.24 to 0.50 cm/yr (Figure 4). Second, the more isolated (fjord) location of KC-18C preserves a record with a more pronounced response to climate change.

In particular, the data show a dramatic change in the upper 15 cm of the core, or past 30 years of the record (Figure 3). Water content, lonestone abundance, % TOC, % clay and the relative abundance of warm water species of diatoms increase, while magnetic susceptibility

decreases in this section of the core, denoted Unit I. Increased lonestones, the product of ice rafting, suggest increased glacial ice discharge (iceberg calving) through the twentieth century. Higher organic carbon content suggests increased primary production over the past 30 years. During this time period, the relative abundance of warm versus cold water forms of the diatom *T. antarctica* also increases dramatically, suggesting increased surface water temperatures. Together, these data are strong evidence for the influence of recent warming on the Antarctic Peninsula ecosystem. While the warming has been recorded instrumentally [*Smith et al.*, this volume], and recent ice shelf decay has been attributed to regional warming [*Morris and Vaughan*, this volume; *Scambos et al.*, this volume; and, *Skvarca and DeAngelis*, this volume], these data are the first that record how changes in atmospheric temperatures are apparently influencing glacial melt and oceanic primary productivity. More specif-

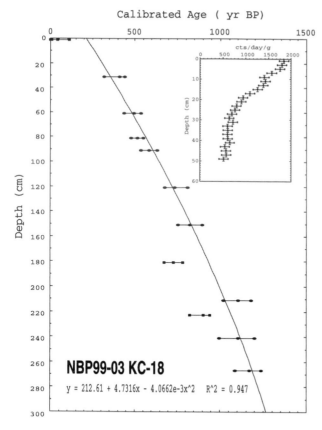

Fig. 4. Chronologic data for *NBP*99-03 KC-18, including both radiocarbon data and ²¹⁰Pb. See table I for radiocarbon data which are shown here as calibrated ages with one sigma errors [*Stuiver et al.*, 1998]. Radiocarbon ages were corrected assuming a reservoir and reworking correction of 1400 years and calibrated according to the University of Washington Quaternary Isotope Lab Radiocarbon Calibration Program Rev 4.2 [*Stuiver and Reimer*, 1993; *Stuiver et al.*, 1998]. For each calibrated date minimum and maximum range values, according to the 1 sigma error, are also indicated. A second order polynomial was fit to the calibrated ages with, y = calibrated age (years) and x = depth (cm). Inset illustrates the ²¹⁰Pb activity [*Cleary*, 2000].

ically, in this core record we note that biogenic components appear to be affected first, with changes in % TOC and the diatom flora preceding the increase in ice-rafted lonestones, which is then followed by the increase in % clay. The relative timing of these changes suggest that the biota react most quickly to warming, followed by increased influx of terrigenous material [*LoPiccolo*, 1996], more iceberg calving and then increased fine sediment in meltwater (the presumed source of clay). It is important to note, however, that these sedimentary particles remain mixed and do not separate out into distinct compositional laminae, hence indicating that the recent

rates of accumulation (~0.5 cm/yr) are still insufficient to mask bioturbation in the short term.

The remainder of the core can be divided into Unit II and Unit III, with the transition at 110 cm, or ~ 650 years before present. Generally low water content, lonestone abundance, and % TOC characterize unit II. The ratio of warm to cold *T. antarctica* is also lower in this unit. This section of the core represents the cooler and less productive period corresponding to the Little Ice Age. Decreased lonestone abundance indicates that deposition from ice rafting was less common. These data are significant in that they clearly confirm previously published marine sediment core data [*Taylor et al.*, 2001; *Shevenell et al.*, 1996; and *Shevenell and Kennett*, 2002] , which document the occurrence of Little Ice Age conditions in the Southern Ocean.

Unit III, which corresponds to the time of the Medieval Warm Period, shows a great deal of high-frequency variability. Several peaks in lonestone abundance are obvious, suggesting intervals of increased glacial melt and ice rafting. The % TOC levels are, on average, higher than unit II, with correspondingly lower magnetic susceptibility, indicating higher primary production and a greater contribution of biogenic material to the sediment column.

3.4. Regional Correlations and Holocene Climate Optimum

The questions that we now consider include:

(a) Were there periods earlier in the Holocene that record similar or even warmer conditions than those experienced in the 20th Century?

(b) If so, how are these events expressed spatially and temporally in the sedimentary record?

The answer to the first question must be directed at the conditions of the middle Holocene Climatic Optimum as inferred from both terrestrial and marine records [*Hjort et al.*, 1998 and this volume; *Shevenell et al.*, 1996; *Domack et al.*, 2001; *Taylor et al.*, 2001]. To examine this episode for marine cores in the western Antarctic Peninsula region, we present in Figure 5 a correlation diagram for long piston cores from the Palmer Deep and Gerlache Strait. The chronologic data for these correlations are listed in Table I, [also see, *Domack et al.* 1993, 1995, 2001; and *Domack and McClennen*, 1996].

We base this correlation primarily on physical correlation of distinctive magnetic susceptibility (MS) changes and key sedimentologic beds (laminated intervals) as

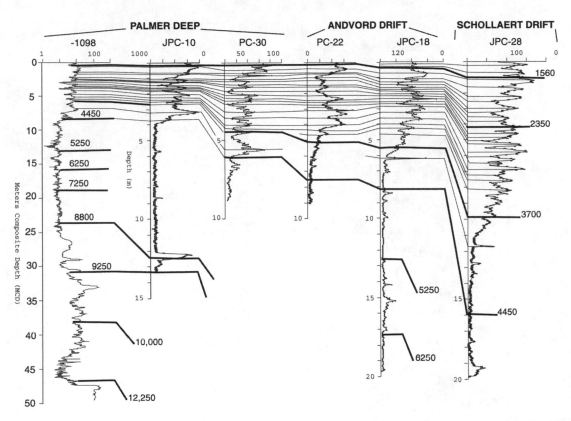

Fig. 5. Fence diagram correlating the following cores: ODP Leg 178 Site 1098, *NBP*99-03 JPC10, *PD*92-02 PC30, *PD*88-07 PC22, *NBP*99-03 JPC18, and *NBP*99-03 JPC28. Correlation is based on a combination of radiocarbon data (bold lines) and the volume magnetic susceptibility (CGS x 10^{-6}) signatures (fine lines). See Table 1 for uncorrected radiocarbon ages, and *Domack et al.*, [2001], for ODP-1098, *Leventer et al.,* [1996], for PD-92-30, and *Domack et al.*, [1993], for *PD*-88-22.

illustrated in Figures 6-8. The sediment time frame is constrained by over 100 radiocarbon dates. Magnetic susceptibility variations can be uniquely correlated from the Palmer Deep ODP Site 1098 cores to the long piston cores of the Gerlache Strait. When this is done, it can be seen that the middle reach of the Gerlache Strait preserves a higher resolution stratigraphy than most other sites, including the Palmer Deep ODP site. The site of JPC-28 (The Schollaert Drift) [*Wright*, 2000] has an overall sedimentation rate about double that in Andvord Bay and Palmer Deep.

The core chronology clearly defines most of the Middle Holocene Climatic Optimum in the Palmer Deep as occurring between 6 and 3.5 ka, compared with the two peaks of the optimum recognized by *Shevenell et al.* [1996] and *Hjort et al.*, [1998] at roughly 6 and 4 ka. Within this interval, inter-laminated diatom ooze and terrigenous silt/clay (Figures 6-8) characterize the stratigraphy. These laminated horizons alternate at a decimeter-scale with bioturbated and disrupted laminations, indica-

tive of slower rates of deposition and biological disturbance of primary physical structures.

Before this time interval, core stratigraphy is characterized by thicker bioturbated intervals (the high MS sections) and thinner laminated beds that consist of diatom ooze with minimal silt and clay. This transition in sedimentary structures is paralleled by changes in particle size distribution and ice rafted debris (IRD) concentration (Figure 7 and 8). The lower laminated intervals are more clay rich [*Wright* 2000; *Carlson*, 2001] and contain less IRD than the younger overlying section (ages < ~3200 yr. BP).

The presence of well-defined laminations of diatom ooze and terrigenous silt/clay indicates sediment from alternating sources, one biogenic (phytoplankton) and the other related to glacial meltwater [*Domack and Ishman*, 1993; *Leventer et al.*, 2002]. This style of sedimentation is not observed in the study region today (see above). Sedimentation rates based upon our radiocarbon chronology (Table 1 and Figures 7 and 8) indicate accu-

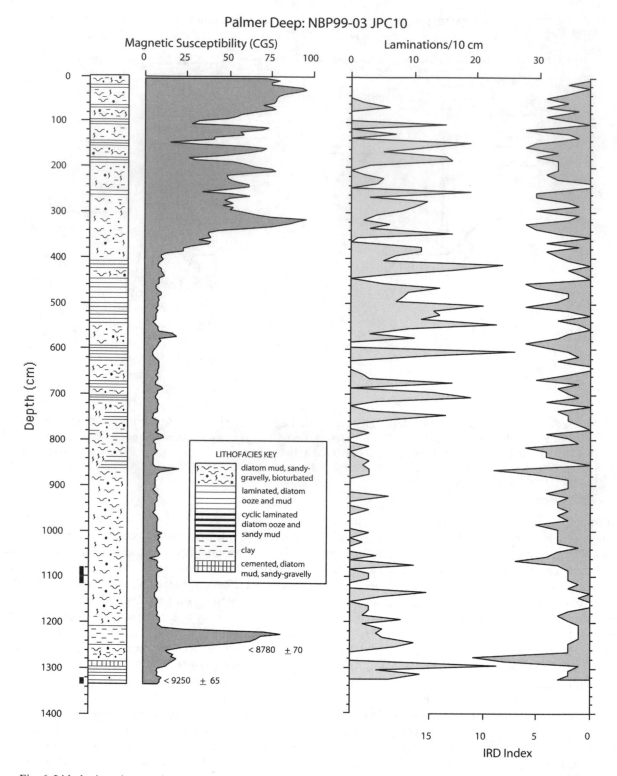

Fig. 6. Lithologic and magnetic susceptibility data for *NBP*99-03 core JPC-10 (Palmer Deep). Lithologic data are based on visual core descriptions, and laminae and ice-rafted debris counts are from x-ray radiographs. Uncorrected radiocarbon ages are indicated along side the downcore magnetic susceptibility data. Dark bars on the depth scales indicate sections for which x-ray radiographs are shown (Figure 9).

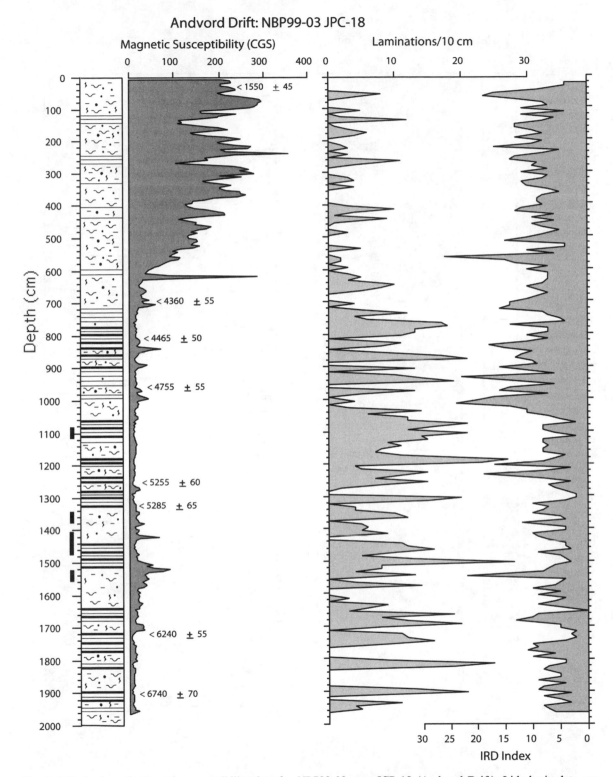

Fig. 7. Lithologic and magnetic susceptibility data for *NBP*99-03 core JCP-18 (Andvord Drift), Lithologic data are based on visual core descriptions, and laminae and ice-rafted debris counts are from x-ray radiographs. Uncorrected radiocarbon ages are indicated along side the downcore magnetic susceptibility data. Dark bars on the depth scales indicate sections for which x-ray radiographs are shown (Figure 9).

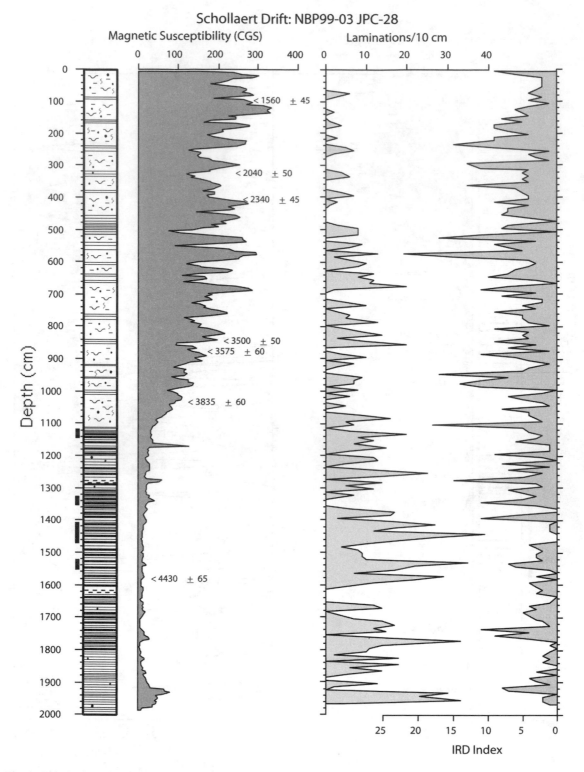

Fig. 8. Lithologic and magnetic susceptibility data for: *NBP*99-03 core JPC-28 (Schollaert Drift). Lithologic data are based on visual core descriptions, and laminae and ice-rafted debris counts are from x-ray radiographs. Uncorrected radiocarbon ages are indicated along side the downcore magnetic susceptibility data. Dark bars on the depth scales indicate sections for which x-ray radiographs are shown (Figure 9).

mulation rates of 0.95 cm/yr (JPC-28) to 2.03 cm/yr (JPC-18) during deposition of the "laminated" intervals. This compares to only 0.24 (JPC-18) to 0.28 cm/yr (JPC-28), or less, during the "bioturbated" intervals of the last 3.5 ka. But sedimentation rates were probably even greater than this during brief intervals in the middle Holocene. In support of this view, we illustrate (Figure 9) laminated and bioturbated intervals from cores JPC-10, JPC-18, and JPC-28. Some "laminated" intervals represent short-term accumulation rates that may have been as high as 2–3 cm/yr (see below), whereas in recent times they are an order of magnitude less than this.

Detailed examination of the laminated sections by x-ray radiography reveals striking centimeter-scale symmetry in the alternation of diatom ooze and silt/clay laminae. The symmetry is defined (Figure 9) by diatom ooze layers (the dark laminations) that bound thinner silt/clay lamina (the light laminations). This contrast in radiograph density (dark for the diatom ooze and light for the silt/clay as shown in negatives) can be quantified by gray scale determinations using for example, NIH Image (http://rsb.info.nih.gov/nih-image/). When this is done, and the first-order variation in x-ray intensity is removed from the scan, a striking pattern emerges (Figure 10). The gray scale variation across several laminated zones records a variety of frequencies.

A comparison of these variations to tidal signals suggests that tides may exert a strong bi-annual control on diatom ooze and silt/clay deposition. Today, the tidal amplitudes are micro to mesotidal, with amplitude ranges of between 1 and 2 m [*Laurence Padman*, personal communication, 2002]. However, the amplitude of

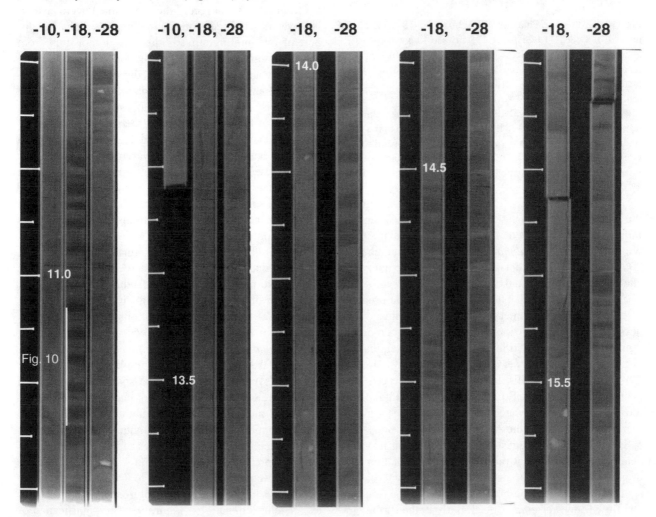

Fig. 9. Negatives of X-ray radiographs of u-channels from cores *NBP99*-03 core JPC-10, JPC-18, and JPC-28. Depths are in meters. The section for which grey scale data are presented (figure 10) is indicated. Scale bars are every 5 cm with core depth (in meters) labeled for each triplicate or pair of negatives.

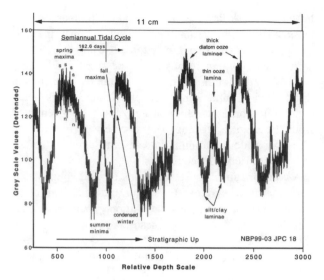

Fig. 10. Grey scale data for *NBP*99-03 JPC-18 from 11 cm interval between 11.025 to 11.135 m (see figure 8 for reference). Grey scale values generated via NIH Image from scanned x-ray radiographs. Values were detrended in a linear fashion to correct for first order density variation due to x-ray scattering. Peaks in grey scale value correspond to biogenic (diatom) rich laminae while lower grey scale values represent terrigenous (silt and clay) rich laminae as indicated to the right. Interpretation of gray scale variation according to semi-annual tidal cycle shown to left. Individual spring (s) and neap (n) tides are inferred to correlate to very thin lamina as seen in the x-ray radiograph.

the spring/neap tides is at a maximum in the spring and fall, with a pronounced minimum in the summer and winter [*Laurence Padman*, personal communication, 2002]. This type of tidal signal has been recognized in ancient rocks as well, and is referred to as the semi-annual cycle [*Kvale*, 1997]. The semi-annual cycle produces a series of laminated intervals where the laminae thickness increase in proportion to the tidal amplitude or current velocity (hence two intervals of thicker laminations per year, or one set per 182.6 modern days; [*Kvale*, 1997]). We couple this idea with the observed tidal swings in the Gerlache Strait and the annual phytoplankton bloom and meltwater cycles common to glacial marine settings along the Antarctic coast.

Specifically, we suggest that the maximum in vernal neap/spring tides causes rapid deposition of diatom ooze in repeated pulses as the tides bring coastal blooms against the frontal zone produced by estuarine (meltwater) flow [*Leventer et al.*, 2002]. Early in the spring, glacial meltwater has not yet produced a significant detrital signal due to the dominance of snowmelt over glacial melt. Later in the late spring and summer, as tidal ampli-

tudes fall, silt and clay are brought into the system from progressive glacial melting. The advance of the estuarine frontal zone (now sediment laden) replaces diatom deposition along the coast except for a brief period of higher tides in the middle summer (the thin ooze lamina recognized in our sequence; Figures 9 and 10). As autumnal tides increase in amplitude the meltwater signal decreases and the estuarine front retreats shoreward, both conditions favor resumption of diatom deposition. However, we interpret this portion of the semiannual signal to be truncated and followed by a condensed winter interval with basically little or no deposition (Figure 10). The semiannual signal resumes the following spring with sea ice retreat, increased light, and diatom blooms under high vernal tides (Figure 10). This model, with truncated and condensed fall to winter deposition, differs from the ancient tidalites that record the semiannual cycle in shallow waters with near-constant sediment supply. In support of our interpretation is the abrupt transition from silt/clay laminae to the following spring ooze laminae, suggesting an abrupt change in sediment regime. In contrast the transition from spring diatom ooze to summer silt/clay is much more gradual, suggesting gradual increases in terrigenous input as the source of sediment changes in tandem with falling tidal ranges. In our model we also recognize small-scale variations in gray scale superimposed upon the larger semi-annual cycle (Figure 10). We interpret these very fine laminae as representing individual spring and neap tides with subtle contrasts in gray scale consistent with variations in diatom versus silt/clay content.

There are other possible semiannual changes that also might produce contrasts in laminae composition. These include storm surges associated with the semi-annual oscillation of the ACL (Antarctic Circumpolar Low) or inter-annual to sub-annual variations in the shelf-ward intrusion of Circumpolar Deep Water [*van den Broeke*, 1998; *Simmonds*, this volume; *Smith et al.*, this volume]). However, these mechanisms are unlikely to produce the observed symmetry in lamination-style that is observed in the core records. We therefore sugggest that certain laminated intervals deposited within the mid-Holocene climatic optimum record a semi-annual tidal oscillation and resulted in accumulation rates of up to 2–3 cm/yr. If this interpretation is correct, the "laminated" tidalite intervals record approximately 6 to 7 years of deposition before they are replaced by bioturbated intervals with less regular laminations (Figure 9). The significance of this interpretation, in terms of paleoclimate, is that the "tidalites" imply large increases in detrital sediment supply, over and above what is found in the region today.

Although the contrast in seasonal tide amplitude is small we recognize that enhanced sediment delivery coupled with a shift in source may be a mechanism for preserving subtle changes in tidal process. This is borne out by sediment traps deployed in Lallemand Fjord that focus deposition and record sub-annual variations in sedimentation that reflect tides and/or storm surges [*Gilbert et al.*, 2003].

We suggest therefore that climatic conditions during the mid Holocene Climatic Optimum were characterized by near decadal oscillations, with surface temperatures well above freezing in the summer. We also suggest that surrounding tidewater glaciers were warm based (at pressure melting temperatures near their beds) and hence poised to respond to interannual changes in summer temperature via the production of meltwater detritus. Despite nearly 50 years of recent rising temperatures, it seems that conditions have still not reached those occurring ~4 ka BP along the western side of the Antarctic Peninsula.

4. CONCLUSIONS

Our findings support a regionally consistent pattern of paleoenvironmental change during the Holocene, with events apparently synchronous (within the limits of radiocarbon dating) with the inferred Northern Hemisphere climate changes. Specifically, we observe evidence of:

1) middle Holocene climatic optimum between about 9.0 and 3.5-2.5 ka, marked by enhanced productivity and meltwater sedimentation,

2) late Holocene (Neoglacial) interval from about 3.5-2.5 ka to 0.15 ka, marked by reduced sedimentation and productivity,

3) Medieval Warm period from about 1.15 to 0.7 ka, marked by enhanced productivity,

4) Little Ice Age from about 0.7 to 0.15 ka, marked by reduced sedimentation and productivity, and

5) modern warming interval of the last ~30 years, marked by enhanced deposition of warm water diatoms, organic carbon, and terrigenous silt-clay of meltwater origin.

The natural pace and extent of past warming episodes, in particular the middle Holocene climatic optimum, appear more extreme than the modern shift toward warmer conditions along the western Antarctic Peninsula. However, the expression of modern warming within the marine sediment record is apparently greater than natural variability of the last ~1200 years.

Acknowledgment. We thank students in the fall 2001 Paleoclimate class at Hamilton College for generating portions of the data set on KC-18. We are also indebted to Captain J. Borkowski III, his crew, and contractors from Antarctic Support Associates for their enthusiastic support during *NBP* 99-03. We also extend our thanks to T. Janecek and M. Curren for their help with core handling at the Florida State University Marine Geology Facility. We are grateful for the careful and constructive reviews and edits of D. DeMaster, S. Konfoush, R. Bindschadler, and an anonymous reviewer. Discussions with L. Padman were also most helpful. This work was supported by National Science Foundation Grants OPP-9615053 (to Hamilton College) and OPP-9714371 (to Colgate University).

REFERENCES

Broecker, W.S., Was the Medieval Warm Period global? *Science, 291*, 1497-1499, 2001.

Camerlenghi, A., E. Domack, M. Rebesco, R. Gilbert, S. Ishman, A. Leventer, S. Brachfeld, and A. Drake, Glacial morphology and post-glacial contourites in northern Prince Gustav Channel (NW Weddell Sea, Antarctica), *Marine Geophysical Researches, 22,* 417-443, 2001.

Carlson, D., High resolution study of the Andvord Drift, Western Antarctic Peninsula, Antarctica, BA thesis, 41 pp., Hamilton College, Clinton, N.Y., 2001.

Clapperton, C.M., Quaternary glaciations in the Southern Ocean and Antarctic Peninsula area, *Quaternary Science Reviews, 9*, 229-252, 1990.

Clapperton, C.M. and D.E. Sugden, Holocene glacier fluctuations in South America and Antarctica, *Quaternary Science Reviews, 7*, 185-198, 1989.

Cleary, R. Using ^{210}Pb to develop a sedimentation chronology for the Andvord Bay region, BA thesis, 48 pp., Hamilton College, Clinton, N.Y., 2000.

Convey P., Antarctic Peninsula climate change: Signals from terrestrial biology, this volume.

De la Mare, W.K. Abrupt mid-twentieth-century decline in Antarctic sea-ice extent from whaling records, *Nature, 389,* 57-60, 1997.

Doake, C.S.M., State of balance of the ice sheet in the Antarctic Peninsula, *Annals of Glaciology, 3*, 77-82, 1982.

Doake, C.S.M. and D.G. Vaughan, Rapid disintegration of the Wordie Ice Shelf in response to atmospheric warming, *Nature, 350,* 328-330, 1991.

Domack, E.W. and C.E. McClennen, Accumulation of glacial marine sediments in fjords of the Antarctic Peninsula and their use as late Holocene paleoenvironmental indicators, in *Foundations for Ecosystem Research West of the Antarctic*

Peninsula, Antarct. Res. Ser., vol. 70, edited by R. Ross, E. Hoffman, and L. Quetin, pp. 135-154, AGU, Washington, D.C., 1996.

Domack, E. W., and Ishman, S. E., Oceanographic and physiographic controls on modern sedimentation within Antarctic Fjords, *Geological Society of America Bulletin, 24,* 119-155, 1993.

Domack, E. W., Mashiotta, T. A., Burkley, L. A., and Ishman, S. E., 300 year cyclicity in organic matter preservation in Antarctic fjord sediments, in *The Antarctic Paleoenvironment: A Perspective on Global Change, Part 2,* edited by J.P. Kennett J. and D.A. Warnke, *Antarct. Res. Ser.,* vol. 60, American Geophysical Union, Washington D. C., pp. 265-272, 1993.

Domack, E.W., A.J.T. Jull, and S. Nakao, Advance of East Antarctic outlet glaciers during the Hypsithermal; implications for the volume state of the Antarctic ice sheet under global warming, *Geology, 19*(11), 1059-1062, 1991.

Domack, E.W., S.E. Ishman, A.B. Stein, C.E. McClennen, and A.J.T. Jull, Late Holocene advance of the Müller Ice Shelf, Antarctic Peninsula: Sedimentological, geochemical and palaeontological evidence, *Antarctic Science, 7*(2), 159-170, 1995.

Domack, E.W., E.A. Jacobson, S.S. Shipp, and J.B. Anderson, Late Pleistocene/Holocene retreat of the West Antarctic Ice Sheet in the Ross Sea: Part 2 – sedimentologic and stratigraphic signature, *Geological Society of America Bulletin, 111*(10), 1517-1536, 1999.

Domack, E.W., A. Leventer, R. Dunbar, F. Taylor, S. Brachfeld, C. Sjunneskog, and ODP Leg 178 Scientific Party, Chronology of the Palmer Deep site, Antarctic Peninsula: A Holocene Paleoenvironmental reference for the circum-Antarctic, *Holocene, 11,* 1-9, 2001.

Domack, E.W., A synthesis for site 1098: Palmer Deep, in *Proceedings Ocean Drilling Program, Scientific Results, 178,* edited by P.F. Barker, A. Camerlenghi, G.D. Acton, and A.T.S. Ramsay, 1-14 [CD ROM]. Available from: Texas A&M University, College Station TX 77845-9547, 2002.

Drake, A.J., Sediment analysis confirms a mid-Holocene warming event in the northwestern Weddell Sea, Antarctica, BA Thesis, Hamilton College, Clinton New York, 90 pp., 2002.

Duran, D. Sediment lithofacies from beneath the Larsen B Ice Shelf, BA thesis, Hamilton College, Clinton, New York, 2003.

Fowbert, J.A. and R.I. Lewis-Smith, Rapid population increases in aturive vascular plants in the Argentine Islands, Antarctic Peninsula, *Arctic and Alpine Research, 26*(3), 290-296, 1994.

Fox, A.J. and A.P.R. Cooper, Climate-change indicators from archival and aerial photography of the Antarctic Peninsula, *Annals of Glaciology, 27,* 636-642, 1998.

Fraser, W.R., W.Z. Trivelpiece, D.G. Ainley, and S.G. Trivelpiece, Increases in Antarctic penguin populations: Reduced competition with whales or a loss of sea ice due to environmental warming?, *Polar Biology, 11,* 525-531, 1992.

Gilbert, R., A. Chong, R.B. Dunbar, and E.W. Domack. Sediment trap records of glacimarine sedimentation at Müller Ice Shelf, Lallemand Fjord, Antarctic Peninsula. *Arctic, Antarctic, and Alpine Research,* 35, 24-33, 2003.

Gilbert, R., E.W. Domack and A. Camerlenghi, Deglacial history of the Greenpeace Trough: Ice sheet to ice shelf transition in the northwestern Weddell Sea, this volume.

Gille, S., Warming of the Southern Ocean since the 1950s, *Science, 295,* 1275-1277, 2002.

Grobe, H. A simple method for the determination of ice-rafted debris in sediment cores, *Polarforschung, 57,* 123-126, 1987.

Grove, J.M. *The Little Ice Age,* Routledge, London, 498 pp., 1988.

Harangozo, S.A., S.R. Colwell and J.C. King, An analysis of a 34-year temperature record from Fossil Bluff (71°S, 68°W), Antarctica, *Antarctic Science, 9,* 355-363, 1997.

Harris, P.T., E. Domack, P.L. Manley, R. Gilbert, and A. Leventer, Andvord Drift: A new type of inner shelf, glacial marine deposystem from the Antarctic Peninsula, *Geology, 27*(8), 683-686, 1999.

Hjort, C.A., S. Björck, O. Ingólfsson, O., and P. Moller, Holocene deglaciation and climate history of the northern Antarctic Peninsula region – a discussion of correlations between the Southern and Northern Hemispheres, *Annals of Glaciology, 27,* 110-112, 1998.

Hjort, C., Ingolfsson, O., Bentley, M. J., and Björck, S., Late Pleistocene and Holocene Glacial and Climate History of the Antarctic Peninsula Region: A Brief Overview of the Land and Lake Sediment Records, this volume.

Hughes, M.K., and H.F. Diaz, Was there a Medieval Warm Period and if so, when and where? *Climate Change, 26,* 109-142, 1994.

Jacka, T.H. and W.F. Budd, Detection of temperature and sea-ice-extent changes in the Antarctic and Southern Ocean, 1949-96, *Annals of Glaciology, 27,* 553-559, 1998.

Jacobs, S.S. and J.C. Comiso, A recent sea-ice retreat west of the Antarctic Peninsula, *Geophysical Research Letters, 20,* 1171-1174, 1993.

Jones, P.D., Antarctic temperatures over the present century – a study of the early expedition record, *Journal of Climate, 3,* 1193-1203, 1990.

King, J.C., J. Turner, G.J. Marshall, W.M. Connolley, and T.A. Lachlan-Cope, Antarctic Peninsula climate variability and its causes as revealed by analysis of instrumental records, this volume.

King, J.C., Recent climate variability in the vicinity of the Antarctica Peninsula, *Int. J. Climatol., 14*(4), 357-369, 1994.

King, J.C. and S.A. Harangozo, Climate change in the western Antarctic Peninsula since 1945: Observations and possible causes, *Annals of Glaciology, 27,* 571-575, 1998.

Kvale, E.P. and M. Mastalerz, Tidal Rhythmites: Their Implications and Applications, U.S.G.S. Short Course, 93 pp., 1997.

Leventer, A.R., Domack, E.W., Ishman, S.E., Brachfeld, S., McClennen, C.E., and Manley, P. Productivity cycles of 200-300 years in the Antarctic Peninsula region: understanding linkages among the sun, atmosphere, oceans, sea ice, and biota, *Geological Society America Bulletin, 108,* 1626-1644, 1996.

Leventer, A., The fate of sea ice diatoms and their use as paleoenvironmental indicators, in *American Geophysical Union*

Antarctic Research Series 73, Antarctic Sea Ice: Biological Processes, edited by M.P. Lizotte and K.R. Arrigo, K.R., pp. 121-137, 1998.

Leventer, A. and R.B. Dunbar, Factors influencing the distribution of diatoms and other algae in the Ross Sea, *Journal of Geophysical Research, 101*(C8), 18489-18500, 1996.

Leventer, A., E. Domack, A. Barkoukis, B. McAndrews and J. Murray, Laminations from the Palmer Deep: A diatom-based interpretation, *Paleoceanography, 17(3),* doi: 10.1029/2001PA000624, 2002.

Lichtenstein, S. J., Local circulation patterns and recent climate change in the northern Prince Gustav Channel: sedimentologic evidence from three kasten cores. BA thesis, Hamilton College, Clinton New York, 77 pp., 2002.

LoPiccolo, M. H., Productivity and meltwater cycles in Andvord Bay, Antarctica: evidence of high frequency paleoclimate fluctuations. BA thesis, Hamilton College, Clinton NY, 52 pp., 1996.

Mann, M. E., R. S. Bradley, K. R. Briffa, J. Cole, M. K. Hughes, J. M. Jones, J. T. Overpeck, H. von Storch, H. Wanner, S. L. Weber, and M. Widmann, Reconstructing the climate of the late Holocene, *Eos, 82,* 553, 2001.

Mann, M. E., R. S. Bradley, and M. K. Hughes, Northern hemisphere temperatures during the past millennium: inferences, uncertainties, and limitations. *Geophysical Research Letters, 26,* 759-762, 1999.

Morris, E. M., and D.J. Vaughan, Spatial and temporal variation of surface temperature on the Antarctic Peninsula and the limit of viability of ice shelves, this volume.

Mosley-Thompson, E., Paleoenvironmental conditions in Antarctica since A.D. 1500: Ice core evidence, in *Climate Since A.D. 1500,* edited by R.S. Bradley and P.D. Jones, pp. 572-591, Routledge, New York, 1992.

Mosley-Thompson, E. and L.G. Thompson, Nine centuries of microparticle deposition at the South Pole, *Quaternary Research, 17*(1), 1-13, 1982.

Mosley-Thompson, E., L.G. Thompson, P.M. Grootes, and N. Gunderstrup, Little Ice Age (Neoglacial) paleoenvironmental conditions at Siple station, Antarctica, *Annals of Glaciology, 14,* 199-204, 1990.

Mosley-Thompson, E. and L.G. Thompson, Ice core paleoclimate histories for the Antarctic Peninsula: Where do we go from here?, this volume.

Rebesco, M., A. Camerlenghi, and C. Zanolla, Bathymetry and morphogenesis of the continental margin west of Antarctic Peninsula, *Terra Antarctica, 5*(4), 715-728, 1998.

Root, S.A. Geochemical and sedimentological analysis of cores from Crystal Sound, Antarctic Peninsula: A paleoenvironmental analysis, BA Thesis, Hamilton College, Clinton, N.Y., 71 pp., 2001.

Rott, H., P. Skvarca, and T. Nagler, Rapid collapse of northern Larsen Ice Shelf, Antarctica, *Science, 271,* 788-792, 1996.

Rott, H., W. Rack, T. Nagler and P. Skvarca, Climatically induced retreat and collapse of the northern Larsen Ice Shelf, Antarctica Peninsula, *Annals of Glaciology, 27,* 86-92, 1998.

Sansom, J., Antarctic surface temperature time series, *Journal of Climate, 2,* 1164-1172, 1989.

Scambos, T.A., C. Hulbe, M. Fahnestock, and J. Bohlander, The link between climate warming and break-up of ice shelves in the Antarctic Peninsula, *Journal of Glaciology, 46*(154), 516-529, 2000.

Scambos, T., C. Hulbe, and M. Fahnestock, Climate-induced ice shelf disintegration in Antarctica, this volume.

Scherer, R.P., A new method for the determination of absolute abundance of diatoms and other silt-sized sedimentary particles, *Journal of Paleolimnology, 12*(2), 171-179, 1994.

Shevenell, A., E.W. Domack, and G.M. Kernan, Record of Holocene palaeoclimate change along the Antarctic Peninsula: Evidence from glacial marine sediments, Lallemand Fjord, *Pap. Proc. R. Soc. Tas., 130*(2), 55-64, 1996.

Shevenell, A.E. and J.P. Kennett, Antarctic Holocene climate change: A benthis foraminiferal stable isotope record from the Palmer Deep, *Paleoceanography,* 17(1), doi: 10.1029/2001/PA000649, 2002.

Simmonds, I., Regional and large scale influences on Antarctic peninsula climate, this volume.

Skvarca, P., Fast recession of the northern Larsen Ice Shelf monitored by space images, *Annals of Glaciology, 17,* 317-321, 1993.

Skvarca, P., Changes and surface features of the Larsen Ice Shelf, Antarctica, derived from Landsat and Kosmos mosaics, *Annals of Glaciology, 20,* 6-12, 1994.

Skvarca, P. and H. DeAngelis, Impact assessment of climatic warming on glaciers and ice shelves on northeastern Antarctic Peninsula, this volume.

Smith, R.C., W.R. Fraser, S.E. Stammerjohn and M. Vernet, Palmer long-term ecological research on the Antarctic marine ecosystem, this volume.

Smith, R. C., D. Ainley, K. Baker, E. Domack, S. Emslie, W. Fraser, J.P. Kennett, A. Leventer, E. Mosley-Thompson, S. Stammerjohn, and M. Vernet, Marine ecosystem sensitivity to climate change, *Bioscience, 49,* 393-404, 1999.

Smith, R.C., S.E. Stammerjohn and K.S. Baker, 1996, Surface air temperature variations in the western Antarctic Peninsula region, *Antarctic Research Series, 70,* 105-121, 1996.

Stark, P., Climatic warming in the central Antarctic Peninsula area, *Weather, 49*(5), 215-220, 1994.

Stuiver, M., P. Reimer, B. Bard, J.W. Beck, G.S. Burr, K.A. Hughen, B. Kromer, G. McCormack, J. Van Der Plicht, and M. Spurk, INTCAL98 radiocarbon age calibration, 24,000-0 cal. BP., *Radiocarbon, 40,* 1041-1083, 1998.

Stuiver, M. and Reimer, P. J., Extended ^{14}C data base and revised CALIB 3.0 ^{14}C calibration program, *Radiocarbon, 35,* 980-1-21, 1993.

Taylor, F., J. Whitehead, and E. Domack, Holocene paleoclimate change in the Antarctic Peninsula: Evidence from the diatom, sedimentary and geochemical record, *Marine Micropaleontology, 41,* 25-43, 2001.

Thompson, L.G., D.A. Peel, E. Mosley-Thompson, R. Mulvaney, J. Dai, P.N. Lin, M.E. Davis, and C.F. Raymond, Climate since AD 1510 on Dyer Plateau, Antarctic

Peninsula: Evidence for recent climate change, *Annals of Glaciology, 20*, 420-426, 1994.

Van den Broeke, M.R., On the interpretation of Antarctic temperature trends, *Journal of Climate, 13,* 3885-3889, 1998.

Van den Broeke, M. R., and N.P.M. Lipzig, Response of wintertime temperatures to the Antarctic oscillation: results of a regional climate model, this volume.

Vaughan, D.G. and S.M. Doake, Recent atmospheric warming and retreat of ice shelves on the Antarctic Peninsula, *Nature, 379*, 328-331, 1996.

Villareal, T.A. and G.A. Fryxell, Temperature effects on the valve structure of the bipolar diatoms *Thalassiosira antarctica* and *Porosira glacialis, Polar Biology, 2*, 163-169, 1983.

Ward, C. G., Mapping ice front changes of Müller Ice Shelf, Antarctic Peninsula, *Antarctic Science, 7,* 197-198, 1995.

Warner, N.R. and E.W. Domack, Millennial- to decadal-scale paleoenvironmental change during the Holocene in the Palmer Deep, Antarctica, as recorded by particle size analysis, *Paleoceanography, 17*(3), doi: 10.1029/2000PA000602, 2002.

Wright, W., The Schollaert Sediment Drift: An Ultra High Resolution Paleoenvironmental Archive in the Gerlache Strait, Antarctica, BA Thesis, 128 pp., Hamilton College, Clinton, N.Y., 2000.

Eugene W. Domack, Stephanie Root, David Carlson, William Wright, Geology Department, Hamilton College, 198 College Hill Road, Clinton, New York, 13323, (edomack@hamilton.edu)

Amy Leventer, Eric Williams, Emily Hirshorn, Geology Department, Colgate University, Hamilton, New York 13346, (aleventer@mail.colgate.edu)

Jim Ring, Physics Department, Hamilton College, 198 College Hill Road, Clinton, New York, 13323, (jring@hamilton.edu)

Robert Gilbert, Department of Geography, Queen's University, Kingston, ON K7L 3N6 Ontario, Canada (gilbert@lake.geog.queensu.ca)

George Burr, NSF TAMS Facility, Department of Physics, University of Arizona, Tucson Arizona 85721, (burr@u.arizona.edu)

ORIGINS AND PALEOCEANOGRAPHIC SIGNIFICANCE OF LAYERED DIATOM OOZE INTERVAL FROM THE BRANSFIELD STRAIT IN THE NORTHERN ANTARCTIC PENINSULA AROUND 2500 YRS BP

Ho ll Yoon, Byong-Kwon Park, Yeadong Kim, Cheon Yun Kang, and Sung-Ho Kang

Polar Sciences Laboratory, Korea Ocean Research and Development Institute, Ansan, Seoul 425-600, Korea

Diatom and pore water data from two piston cores from the central sub-basin and one from the western sub-basin of Bransfield Strait in the northern Antarctic Peninsula were used to elucidate the depositional mechanism of a layered diatom ooze unit observed in all three cores. The diatom ooze unit, formed by the alternation of diatom ooze layers and terrigenous layers, is enriched in organic carbon, biogenic silica, sulfide sulfur content and depleted in pore water sulfate concentration. This depletion of pore water sulfate in the diatom ooze interval is indicative of development of reducing micro-environment in which bacterially mediated sulfate reduction occurred. The negative relationship between the organic carbon and sulfate contents, however, indicates that sulfate reduction was taking place but does not control organic carbon preservation in the diatom ooze interval. Rather, the overwhelming dominance of intact *Chaetoceros* resting spores in the diatom ooze layer indicates rapid sedimentation of the diatom spores as a result of repetitive late summer blooms at the sea-ice margin on the Bransfield shelf at around 2500 yrs BP when persistent sea-ice margin might have existed on the shelf and/or shelf break. During the cold period, underflows were probably caused by cooling of the Bransfield shelf water, and these flows probably played an important role for concentrating the summer diatom blooms to produce the diatom ooze layer recorded in the sub-basins of the Bransfield Strait. Intervening terrigenous layers were mostly deposited during the winter and represent the input of reworked detrital clay by stronger bottom flows in non-bloom conditions under the sea ice.

1. INTRODUCTION

As a northern boundary of the Weddell Ice Shelf Water, the northern Antarctic Peninsula (Figure 1) is a particularly interesting region for studies of biological and modern geological processes as well as hydrological processes [*Wefer et al.*, 1988; *Dunbar*, 1985; *Dunbar et al.*, 1985]. In particular, the Bransfield Strait, which lies between the Antarctic Peninsula and the South Shetland Islands (Figure 1), is a region where oceanic (Circumpolar Deep Water) and the Antarctic continental

shelf waters (Ice Shelf Water) meet [*Whitworth III et al.*, 1994; *Amos and Lavender*, 1992; *Figueiras et al.*, 1998]. The mixture of these two water masses is weakly stratified and vertically homogeneous, with temperature and salinities lower, and oxygen concentrations higher, than those in the adjacent waters to the north and south [*Whitworth III et al.*, 1994; *Patterson and Sievers*, 1980]. During the austral summer, the Bransfield Strait not only lacks fluvial discharge systems but also receives only minor amounts of terrigenous sediments except for ice-rafting and ice marginal dumping. Modern sedimentation

10.1029/079ARS18

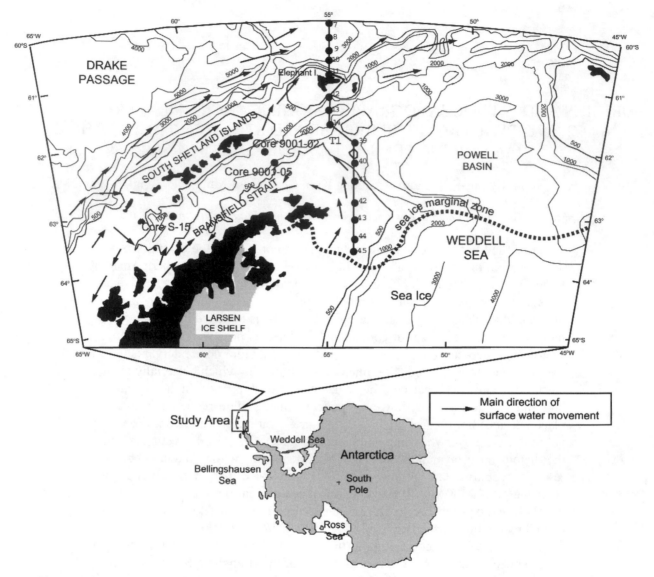

Fig. 1. Bathymetry, core locations and CTD with water sampling stations in the Bransfield Strait. Contours in meters. Station locations during the 1995 KARP cruise, which were occupied from 4 to 14 January 1995. Inferred surface water flows are shown by arrow [after *Hofmann et al.,* 1992].

under such a regime is therefore dominantly biogenic. Sediment trap and biological oceanographic studies reveal that the flux of biogenic components into the Bransfield basin is highly episodic and is tightly coupled to surface-ocean processes of primary production [*Wefer et al.,* 1988].

A recent study of three piston cores from the western and central sub-basins in the Bransfield Strait reported the occurrence of remarkably well-layered siliceous sediment intervals in the cores, that contain relatively high concentrations of organic matter (>1.5% organic carbon)

[*Yoon et al.,* 1994]. This type of siliceous sediment is generally described as laminated diatom ooze and in the Antarctic setting has been interpreted as having been deposited by vertical sedimentation and/or lateral transportation of diatom blooms from the surface ocean [*Jordan et al.,* 1991; *Leventer et al.,* 1993]. In other parts of the world's ocean, such as Santa Barbara Basin and the northeastern Arabian Sea, however, the formation of this type of diatom ooze layer has been ascribed to the inhibition of the benthic community by low concentrations of dissolved oxygen either in anoxic silled basins or beneath zones of strong upwelling, where an oxygen minimum layer intersects the shelf or slope [*Schulz et al.*, 1996; *Bull and Kemp*, 1996; *Schimmelmann and Lange*, 1996]. Therefore, it is essential that pore water chemistry analyses be completed for the diatom ooze interval in order to identify the bottom water conditions during the formation of the layered diatom ooze interval, before we accept enhanced production and flux for the origin of this interval. In this study, we examine pore water chemistry, sedimentology, and micropaleontology to provide an outline of the main mode of preserving layered diatom ooze from the Bransfield Strait, and propose a scenario for the deposition of the diatom ooze interval.

2. CLIMATE AND GLACIAL CONDITIONS

The Bransfield Strait is a region that experiences the strongest climatic gradients in Antarctica. The −3.0°C mean annual isotherm penetrates through the Bransfield Strait, separating the warmer South Shetland Islands to the north from the colder peninsula to the south [*Reynolds*, 1981]. The weather of the strait is affected predominantly by the passage of cyclonic storms that provide ample amounts of moisture for precipitation [*Deacon*, 1984]. In the northern half of the Bransfield Strait, valley and piedmont glaciers of the South Shetland Islands generally terminate on land as a result of the relatively warm climate and high levels of precipitation [*Griffith and Anderson*, 1989]. On the other hand, the Gerlache Strait region in the southern half is heavily glaciated and most glaciers are grounded below sea level [*Griffith and Anderson*, 1989].

3. SEA ICE

The annual fluctuation of sea-ice coverage affects the ice-ocean-atmosphere dynamics in the Bransfield Strait by modifying the flux of heat and salt between the ocean and atmosphere. Brine rejection associated with ice formation during oceanic cooling forms higher density surface waters, resulting in convective overturning that leads to enhanced mixing and development of deep mixed layers. In some cases, the combination of brine rejection and cooling is sufficient to form deep-water masses, such as Antarctic Bottom Water, as observed in the Weddell Sea. The modern Bransfield Strait, however, is not known as a region of deep-water formation. In the austral spring, as the sea ice melts, a meltwater lens is produced that stabilizes the upper water column and thereby restricts mixing and allows local heating to occur [*Deacon*, 1984]. This change in surface buoyancy, when coupled with warming from increased solar irradiance, causes a shallow seasonal pycnocline, resulting in diatom blooms on the surface ocean.

4. METHODS

In order to understand the origin of layered diatom ooze in the Bransfield Strait sediments, two piston cores (7 cm in diameter) were obtained from the central sub-basin and one from the western sub-basin of Bransfield Strait (Figure 1). The cores were split for X-radiography and were subsampled on board at 5 cm intervals. The subsamples were analyzed for grain size, total organic carbon, biogenic silica, sulfide sulfur, and pore water sulfate. Scanning Electron Microscope (SEM) and smear slides were used for lithologic descriptions and for identification of diatom species. For quantitative analysis of diatoms, slides were prepared using a settling technique described by *Scherer* [1995] and successfully utilized with other Antarctic marine sediments [*Leventer et al.*, 1993]. This technique helps diatom valves to be evenly distributed with minimal clumping, and produces absolute diatom concentrations per gram of sediment.

Total carbon and carbonate carbon were determined using a Carlos Erba NA-1500 Elemental Analyzer by measuring the CO_2 formed by combustion at 1100°C, and by treating with hot 10% HCl, respectively. Organic carbon was obtained by difference between total carbon and carbonate carbon. For the cores from the central sub-basin, total reduced sulfur concentration was measured on freeze-dried samples with an elemental analyzer (Carlos Erba NA 1500). Interstitial water samples for sulfate analysis were first retrieved by centrifugation, and then sulfate was determined by precipitation and weighing of $BaSO_4$ using standard seawater as a reference [*Howarth*, 1978]. Biogenic silica (bioSi) was determined by the sequential leaching method of *DeMaster* [1981]. Sulfide sulfur was measured by precipitating sulfur as barium sulfate [*Vogel*, 1975].

5. RESULTS

5.1. Chronology

Diatomaceous mud samples (approx. 1.0 g) for ^{14}C dating were taken from three depths in core S-15: the bottom of diatom ooze unit and two intervals each of distinct diatom species composition. Biogenic calcium carbonate is virtually absent from three sediments, and to analyze the 0.1-1.9% organic carbon it was necessary to use accelerator mass spectrometry. The interpretation of radiocarbon dates from the Antarctic is complex due to unusually low ^{14}C concentrations in Antarctic waters, as well as the geographical variation of water mass circulation [*Bjorck et al.,* 1991b; *Gordon and Harkness,* 1992]. Living organic material has been found to exhibit anomalously old ^{14}C dates, a phenomenon referred to as the Antarctic Reservoir Effect. For example, ^{14}C dates of recently deceased seals range from 1300 to 1770 yrs BP [*Stuiver et al.,* 1981]. Thus, measured radiocarbon dates must be corrected to account for this reservoir effect. The reservoir correction has an estimated value of 1250-1300 yrs [*Domack,* 1993; *Gordon and Harkness,* 1992] and 1000-1300 yrs [*Bjorck et al.,* 1991b] for the Antarctic Peninsula. Based on an age of 1290 yrs (*H.I. Yoon,* unpublished data) for a living gastropod recovered in a grab sample taken from Maxwell Bay, a tributary embayment in King George Island, we assume a reservoir effect of ca. 1300 yrs for the core S-15.

The date of the layered diatom ooze interval (280 cm in core depth) in core S-15 is dated to 2500 yrs BP, if we use a reservoir correction of 1300 yrs. The sedimentation rate in the present central sub-basin of the Bransfield Strait is reported to be up to 90 cm kyr^{-1} obtained by an AMS ^{14}C method [*Harden et al.,* 1992]. Based on this accumulation rate, it is inferred that the 250 cm levels of cores 9001-02 and 9001-05 from the central sub-basin is equivalent to about 2500 yrs BP, suggesting synchronous timing of deposition of the layered units from core to core at around 2500 yrs BP in the Bransfield Strait. However, we have to keep in mind that this correlation is just an estimate, since accumulation rates may vary tremendously over short distances and over time.

5.2. Lithology

Sediment cores from the Bransfield Strait consist of two lithologies; layered diatom ooze from 220-240 cm for 9001-02, 238-244 cm for 9001-05 and 290-294 cm for S-15, and homogeneous mud with interbedded turbidite layers through the rest of the cores (Figure 2).

The layered diatom ooze interval is sometimes visible in the split cores, but is much better defined by its distinctly layered appearance in X-radiographs (Figure 3a). The most diagnostic criterion is the presence of well-developed parallel (<5-10 mm) layers. Most layers are varve-like because they consist of alternating millimeter to centimeter scale of light (biogenic) and dark (terrigenous) colored layers (Figure 3a). The light layer is primarily comprised of a well-preserved monospecific diatom assemblage of *Chaetoceros* resting spores (93%)

(Figure 3c) with minor amounts of *Fragilariopsis curta* (4%), *Fragilariopsis cylindrus* (2%), and traces of *Corethron* spp, *Rhizosolenia* spp. and *Thalassiosira antarctica* (Figure 4). Lower boundaries of the biogenic layers tend to be sharp whereas the upper ones are usually gradational with increased terrigenous silt and clay upward. The dark layers have the same diatom composition as the biogenic layers, but differ in terms of the proportion of diatoms with approximately 79% *Chaetoceros* resting spores, 7% *F. curta,* 3% *F. cylindrus,* 2% *Rhizosolenia,* 3% *Corethron,* 1% *Thalassiosira antarctica,* and traces of large centric diatoms (Figure 4). Diatom valves, i.e. centric diatoms, in the dark layer are, however, partly fragmented and appear to have undergone silica dissolution (Figure 3b). Similar diatom floras were previously identified in the layered diatom ooze intervals from the central Bransfield Strait [*Yoon et al.,* 1994].

5.3. Organic Carbon, Sulfide Sulfur and Sulfate

Profiles of bioSi, organic carbon, sulfide sulfur and pore water sulfate contents from the two cores (9001-02 and 9001-05) from the central sub-basin are shown in Figures 5 and 6. BioSi, organic carbon and sulfate sulfur contents are highest in the diatom ooze interval, moderate in homogeneous mud and lowest in the turbidite.

Organic carbon content in core 9001-02 varies between 0.11% and 0.86% down to 200 cm and then increases sharply from 1.14% to 1.98% within the diatom ooze interval (Figure 5). Similarly, bioSi content varies between 9.6% and 30% down to 200 cm, below this level it increases sharply from 15% to 51% to the middle of the diatom ooze interval (Figure 5). Pore water sulfate decreases gradually from 23.2 mmoles kg^{-1} at the sediment surface to 15.8 mmoles kg^{-1} at the bottom, whereas sulfide sulfur content shows a minor fluctuation around 0.3% down to 200 cm, below which there is a conspicuously sharp increase up to 1.2% near the bottom.

Total organic carbon content for core 9001-05 is constant with depth down to 250 cm, ranging from 0.96% to 1.07%, below this level it increases slightly with a maximum value of 1.30% in the diatom ooze interval (Figure 6). BioSi content shows minor fluctuations down to 250 cm, ranging from 21% to 24%. Below this depth, bioSi content sharply increases with a maximum value of 33% in the diatom ooze interval. Sulfide sulfur content varies between 0.10% and 0.50% down to 250 cm and increases up to 0.75% at the bottom of the core, whereas pore water sulfate content decreases gradually, ranging from 22.0 to 5.00 mmoles kg^{-1} (Figure 6).

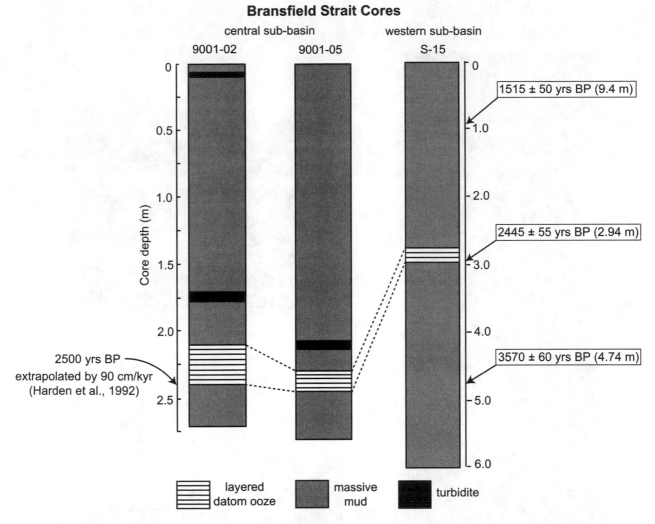

Bransfield Strait Cores

Fig. 2. Lithological logs of the cores 9001-02 and 9001-05 from the central sub-basin, and the core S-15 from the western sub-basin in the Bransfield Strait. Note the base of the layered diatom ooze unit of core S-15 has an AMS ^{14}C age of about 2500 yrs when a 1300-year reservoir correction was applied. Extrapolating the accumulation rate of 90 cm kyr[-1] obtained by *Harden et al.* [1992] from the central sub-basin in the Bransfield Strait, it is inferred that the base of the layered diatom ooze unit of the core 9001-02 is equivalent to 2500 yrs BP.

6. ORIGIN OF LAYERED DIATOM OOZE

The layered diatom ooze intervals in the sediment cores from the Bransfield Strait are characterized by high concentrations of organic carbon, biogenic silica as well as considerable preservation of sulfide sulfur, which can be accounted for by relatively small sulfate reduction (Figures 5 and 6). In contrast, the low pore water sulfate concentrations in the diatom ooze interval may reflect development of a reducing micro-environment in which bacterially mediated sulfate reduction occurred, which

would result in a decrease in the organic carbon and pore water sulfate contents and an increase in sulfide sulfur. The negative relationship between the total organic carbon and sulfate contents (Figures 5 and 6), however, seems to be in contradiction to the general result of the bacterial sulfate reduction that shows a positive trend between organic carbon and pore water sulfate contents. This indicates that sulfate reduction is partly taking place but does not control organic carbon preservation in this interval. How can this increased organic component in the diatom ooze interval where bacterial sulfate reduc-

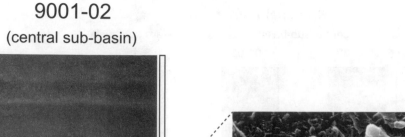

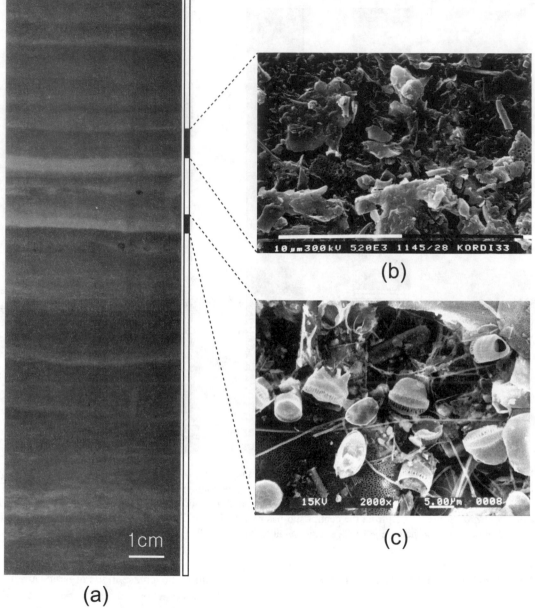

Fig. 3. (a) X-radiographs of core 9001-02 representing layered diatom ooze interval, showing well-layered diatom ooze (light layer) alternating with clayey silt (dark layer) (core 9001-02, 210-225 cm); (b) SEM micrograph of fragmented diatoms and debris in the dark layer; (c) SEM micrograph of well-preserved *Chaetoceros* resting spore in the light layer.

tion occurs be explained? Two possibilities can be considered: (1) increased flux of biogenic opal to the seafloor, and (2) lowering of oxygen minimum zone.

The first option requires increasing diatom productivity. Normally, increased diatom productivity in the overlying waters would increase both the biogenic silica and organic carbon contents in marine sediment [*Goll and Bjorklund*, 1974; *Maynard*, 1976; *Diester-Haass*, 1978], and their positive relationship would indicate a marine source of organic carbon (Figures 5 and 6). Moreover,

CORE 9001-02

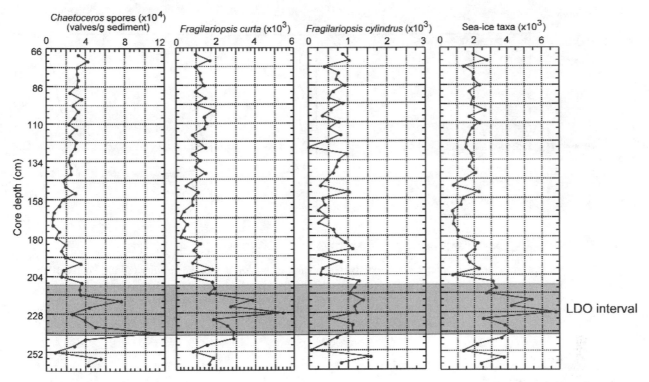

Fig. 4. Diatom analysis for core 9001-02. Note that an increase in number of sea-ice related diatom as well as diatom valves/grams of sediment in layered diatom ooze interval.

the most common diatoms in the diatom ooze (light layer), *Chaetoceros* resting spores, are generally known to dominate the termination of bloom events under the condition of nitrogen deficiency [*Davis et al.*, 1980; *French and Hargraves*, 1980]. In particular, excellent preservation of delicate surface ornamentation and setae (Figure 3c) indicates that these cells did not experience either the fragmentation or abrasion by heterotrophic grazing or dissolution by corrosive water column [*Grimm*, 1992; *Sancetta*, 1989]. Rather, the intact nature of many of diatoms attests to rapid flocculation and consequent mass sinking at the termination of bloom events which are well known processes, both from water column observation [*Alldredge and Gotschalk*, 1989] and experimental study [*Passow et al.*, 1994]. We propose, therefore, that the diatom ooze layer represents the mass sinking of diatoms produced during conditions of very high primary productivity, which in the Antarctic depends on increased stratification of upper water column and the presence of a shallow mixed layer [*Mitchell et al.*, 1991], which can occur in response to proximity to low salinity meltwater and/or protection from intense Antarctic storm activity. This situation was observed in

the northern Gerlache Strait, Antarctic Peninsula, where *Chaetoceros* resting spores dominated sediment trap and surface sediment diatom assemblages [*Leventer*, 1991]. The surface waters of the Gerlache Strait experience seasonal CO_2, nitrogen and phosphorus depletion, indicating enhanced primary production and resultant high biogenic fluxes [*Karl et al*, 1991]. However, given characteristically high primary productivity in summer stratified surface waters of the modern Gerlache Strait, why aren't these diatoms concentrated today to form diatom ooze layers in the surface layer of box core sediment from the Gerlache Strait? Rather, layered diatom ooze can be seen from the surface layer of sediment cores obtained from the open water of the eastern sub-basin of the Bransfield Strait, where *Chaetoceros* typically occur in low abundances during the months of summer stratification [*Kang and Lee*, 1995]. Given characteristically low abundance in open waters of the Bransfield Strait, how does this diatom become concentrated to form a diatom ooze layer representing mass sinking in the eastern Bransfield Strait?

An alternative mechanism for the mass sinking that does not require direct vertical flux of diatom blooms from the upper ocean may be surmised from a recent

Core 9001-02

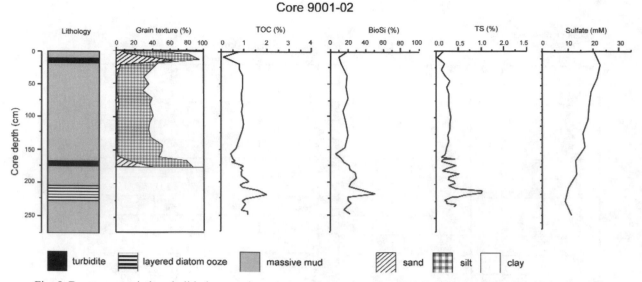

Fig. 5. Downcore variations in lithology, grain texture, total organic carbon (TOC), biogenic silica (BioSi), total sulfide sulfur (TS) and sulfate (SO$_4$) for core 9001-02. Note that all TOC, bioSi profiles vary with sediment types, showing all their values being the highest in the layered diatom ooze unit. Negative relationship of the organic carbon and porewater sulfate content is shown.

Core 9001-05

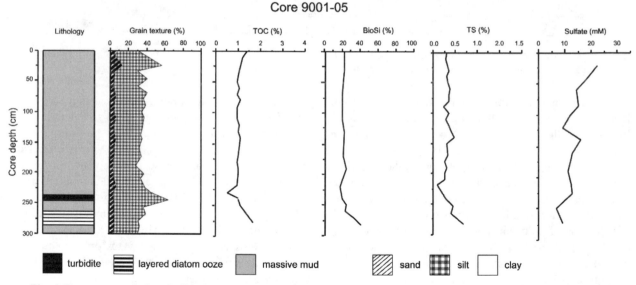

Fig. 6. Downcore variations in lithology, grain texture, total organic carbon (TOC), biogenic silica (bioSi), total sulfide sulfur (TS) and sulfate (SO$_4$) for core 9001-05. Note that all TOC, bioSi profiles vary with sediment types, showing all their values being the highest in the layered diatom ooze unit. Negative relationship of the organic carbon and sulfate content is shown.

synthesis of the hydrography and phytoplankton assemblages between the marginal ice zone (MIZ) of the northwestern Weddell Sea and open water of the eastern Bransfield Strait (Modified from Figures 1 and 7 in *Kang et al.*, 2001). The hydrography in the Bransfield area shows warm and less saline open water with temperature >0°C and salinity 33.8-34.2 psu, while that in the MIZ, where the sea ice was retreating southwards, shows cold and saline Weddell Ice Shelf Water with temperature >0°C and salinity 34.3-34.4 psu, causing a sharp thermal gradient (Figure 7a). The diatom assemblage in the MIZ also shows a marked contrast with that in the Bransfield open waters. The Weddell Sea MIZ is characterized by diatoms such as *Chaetoceros* resting spores and sea ice

Hydrographic Stations (T-1)

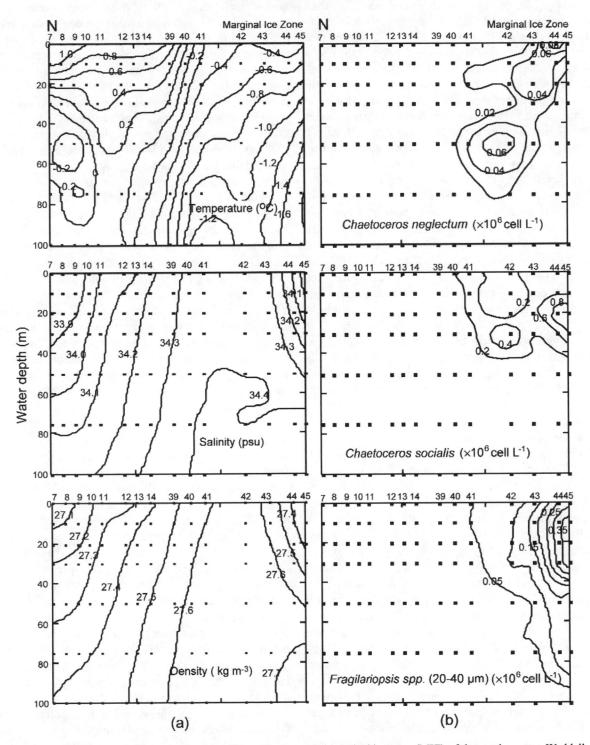

Fig. 7. Hydrography and phytoplankton assemblages between the marginal ice zone (MIZ) of the northwestern Weddell Sea and open water of the eastern Bransfield Strait (Modified from Figure 7 in *Kang et al.*, 2001).

related species of *Fragilariopsis* spp., which together account for 70% of the total phytoplankton. In contrast, the Bransfield open waters were depleted in sea ice-related diatoms (Figure 7b), indicating that *Chaetoceros* resting spores are typical of the ice-edge bloom at the northwestern Weddell Sea, and that the overwhelming dominance of the spores in the eastern sub-basin of the Bransfield Strait resulted from lateral transportation to the core site where plankton-rich, Weddell surface waters meet and mix with circumpolar deep water. The lateral transport of diatom flocs is evidenced by recent sediment trap data from the eastern Bransfield Strait, which show higher flux values for diatoms and total particles in the lower than in the upper trap [*Kim et al.*, 2000], suggesting a lateral influx of *Chaetoceros* resting spores from the Weddell Sea MIZ by downslope bottom currents. Particle increase in deep waters has also been observed by *Leventer* and *Dunbar* [1987], *Takahashi* [1987], and *Sancetta* [1992] in different regions of the world, and was explained as being a result of lateral advection, either by downslope movement or near-bottom currents. Recent hydrographic study in the eastern Bransfield Strait suggests the possible existence of intermittent downslope currents coupled with deep-water convection [*Yih et al.,* 1998]. A nepheloid layer may affect the particle increase in the lower trap. There is, however, no evidence of a thick nepheloid layer that could have affected our deeper trap, which was moored at a depth 60 m above the seafloor [*K.C. Yoo,* unpublished data].

On the basis of the modern analogue in the Weddell Sea MIZ presented above, it is not likely that the deposition of the layered diatom ooze at 2500 yrs BP in the central and western Bransfield Strait resulted from vertical sinking of diatoms by increased productivity under open water conditions, similar to the Bransfield Strait today. Since today's ice-edge bloom in Weddell Sea is dominated by *Chaetoceros* resting spores with the sea-ice taxa of *Fragilariopsis curta* and *Fragilariopsis cylindrus*, the deposition of layers of the these diatoms at 2500 yrs BP in the sub-basins of the Bransfield Strait might have required ice-edge blooms on the Bransfield shelf when the sea ice margin was further seaward on the shelf during the cold stage. During the cold phase, it is expected that the southern Bransfield shelf was covered by multiyear sea ice, and the cold and dense water was formed on the shelf as a result of freezing of seawater and a consequent increase in salinity, whereas during the warm phase, as in the modern Bransfield Strait, dense water production was likely to have diminished or ceased. Newly formed cold and dense water on the glacial Bransfield shelf sank down the continental slope drawing

diatom blooms down to the lower slope, forming a diatom-enriched underflow (i.e. bottom current). The timing of mass sedimentation of diatom blooms to the seafloor is thought to be late summer. During the austral summer in the cold stage, strong surface water stratification, the result of warmer surface temperatures, melting sea ice, and/or reduced wind stress, probably was associated with ice-edge blooms of *Chaetoceros* diatoms on the southern Bransfield shelf. In the late summer, relatively cooler temperatures and weakening of the thermocline coupled to higher wind stress and strong mixing of the water column at the sea-ice margin forced mass sedimentation of *Chaetoceros* resting spores by downward movement of dense shelf water. This hypothesis is based on sediment trap data from the eastern Bransfield basin [*Kim et al.*, 2001] that demonstrates an overwhelming dominance of *Chaetoceros* resting spores in the late summer (January-February) from the lower trap, though other studies [e.g., *Sancetta*, 1995] document the springtime flux of this species.

The intervening terrigenous layers, coupled with monospecific diatom ooze layer, contain a multispecific diatom assemblage, though still dominated by *Chaetoceros* resting spores. Diatom valves in the terrigenous layer are fragmented, indicating dissolution by corrosive water column or reworking by bottom flows. The terrigenous layers were mostly deposited during the winter. The terrigenous component may have been contributed by resuspension at the shelf and shelf break and enhanced by deep-water convection during the winter in the cold stage. In the winter, annual sea ice is formed from multiyear sea-ice margin, and most of the salt is rejected into the shelf water beneath the ice, thereby substantially increasing the water density to such an extent that the dense shelf water begins to flow away from the surface waters. Eventually, this dense shelf water is thought to leave the Bransfield shelf, causing a resuspension of terrigenous particles and diatom frustules, preserved previously on the shelf and shelf break. The lower absolute abundance of diatoms in terrigenous layers may have been due to a lower overall productivity in non-bloom conditions under the annual sea ice. The sediment trap data from the Bransfield Strait show that absolute terrigenous flux of the lower trap is much higher during the winter (June-July) than during the summer (December-January), suggesting a dominance of siliciclastic sedimentation in winter by downslope bottom flows.

In short, we propose a tentative oceanographic condition for the deposition of the intervals of alternating *Chaetoceros* ooze layer and terrigenous layer found in the lower sections of the cores from the central and western sub-basins of the Bransfield Strait at around 2500 yrs BP. During the cold phase, it is expected that the multi-year sea-ice margin existed on the southern Bransfield shelf with repeated ice-edge blooms and that cold and dense surface water was also formed on the Bransfield shelf as a result of multiyear sea-ice cover. The diatom ooze layer was, therefore, probably deposited from the bottom current driven water masses heavily laden with diatom valves in late summer during the cold phase. Relatively well preserved layers and no bioturbation suggest rapid sedimentation of the diatom ooze. In winter, annual sea-ice probably covered the Bransfield Strait, diminishing overall productivity in strait and forming abundant cold and dense shelf water (Figure 9). These conditions caused a contribution of terrigenous component to the sub-basins of the Bransfield Strait.

Alternatively, it may be suggested that the expansion of the oxygen minimum zone and a weakening bottom water circulation during deposition, as opposed to the enhanced diatom productivity, would lead to the observed high organic carbon preservation in the layered diatom ooze. However, the Bransfield Strait waters appear to have been rather well oxygenated during the deposition of the diatom ooze since the C-S (organic carbon *versus* sulfide sulfur) chart for the cores 9001-02, and 9001-05 shows that all data points fall near the normal marine sediments under oxidizing environment (Figure 8).

7. WIDER IMPLICATIONS

Paleoclimate study of lake sediments from Livingston Island revealed that the areas adjacent to the central sub-basin might have experienced more continental conditions with colder and drier climate at 2500 yrs BP than today [*Bjorck et al.*, 1991a]. This climatic decline is supported by the paleoclimate studies for fjord sediments from Antarctic Peninsula [*Shevenell et al.*, 1996; *Leventer et al.*, 1996] in which they reveal a decrease in TOC and diatom abundance around 2700 yrs BP, reflecting the formation of more extensive and seasonally persistent sea ice. The formation of multiyear sea ice around 2500 yrs BP corresponds to a climatically cold triple event ("T-Event" or minima in solar irradiance; *Stuiver and Braziunas*, 1989) around 3000-2000 yrs BP, recently documented in the GISP2 ice core [*O'Brien et al.*, 1995]. We propose that during the climatic cooling around 2500 yrs BP, multiyear sea-ice margin persisted on the southern Bransfield shelf and was subject to repeated ice-edge blooms in summer season (Figure 9).

During this cold stage, cold and dense water was also formed on the Bransfield shelf as a result of increased sea-ice coverage. This dense shelf water concentrated diatom blooms and flowed down the Bransfield slope, drawing diatoms down to the continental slope, forming an interflow, by which layered diatom ooze was deposited.

8. CONCLUSION

To elucidate the depositional mechanism of the layered diatom ooze interval observed in down core sediments from the Bransfield Strait we examined diatom and pore water data of two piston cores from the central sub-basin and one from the western sub-basin in the Bransfield Strait in the northern Antarctic Peninsula. The diatom ooze interval, formed by alternation of diatom ooze layers with terrigenous layers, is characterized by an abundance of organic carbon, biogenic silica, sulfide sulfur, and lower pore water sulfate concentration. This lack of pore water sulfate concentration in the diatom ooze

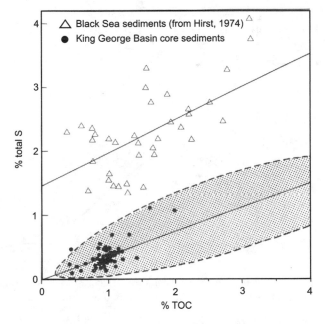

Fig. 8. Correlation between total organic carbon (TOC) and sulfide sulfur (Total S) in sediment. The data from Black Sea cores are plotted to show them to be representative for anoxic sediments having an excess amount of sulfide sulfur as evidenced by the positive intercept of the regression line onto the axis of the sulfur content. The stippled area on the diagram represents a domain for the present-day normal oxic marine sediments [after *Leventhal*, 1983]. The plot of our data inside the lobe strongly suggests that anoxic bottom water conditions had not prevailed during accumulation of the King George Basin (central sub-basin) sediments.

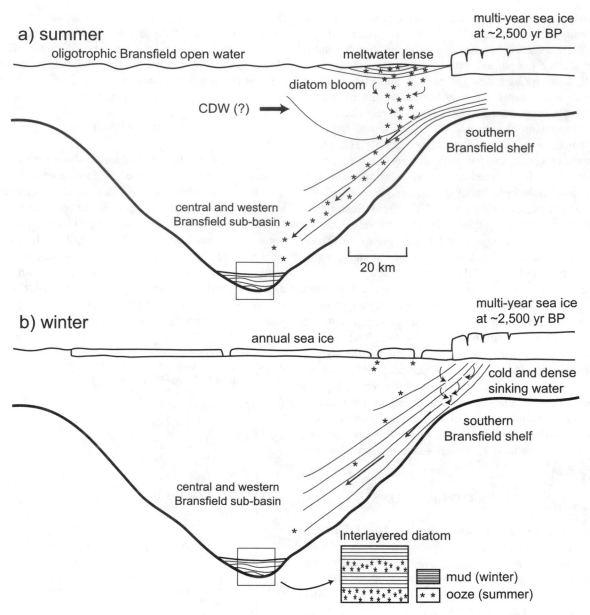

Fig. 9. Model for the depositional mechanism for the formation of alternating diatom ooze and mud layers during the cold period at around 2500 yrs BP in the central and western sub-basins in the Bransfield Strait.

interval may reflect development of reducing micro-environment in which bacterially mediated sulfate reduction occurred. The negative relationship between the total organic carbon and sulfate contents, however, indicates that sulfate reduction is partly taking place but does not control organic carbon preservation in this unit. Rather, well-preserved diatom frustules in the diatom ooze layer indicate a mass sedimentation of summer diatom bloom. Given their characteristic high abundances in late summer at the sea-ice margin in the north-western Weddell Sea, the concentration of *Chaetoceros* resting spores in the diatom ooze layer at around 2500 yrs BP from the Bransfield Strait requires rapid sinking of diatom flocs as a result of repetitive late summer blooms at the multiyear sea-ice margin in the southern Bransfield shelf during cold period and their downslope transport by bottom flows. In contrast, the intervening terrigenous layers were mostly deposited during winter, and represent the input of reworked detrital clay by stronger bottom flows in nonbloom conditions under annual sea-ice cover. Since these layers formed at around 2500 yrs BP, we suggest that cold and dense water was formed on the Bransfield shelf as a result of the formation of multiyear sea ice. Increased sea-ice formation in the Bransfield Strait around 2500 yrs BP is associated with a decrease in TOC and diatom abundance in Antarctic Peninsula fjord sediments around 2700 yrs BP [*Shevenell et al.,* 1996] and may be correlated to a cold T-event documented in the Greenland ice core at the same time [*O'Brien et al.,* 1995].

Acknowledgment. Thanks are expressed to Jae-Kyung Oh for reading an earlier manuscript for this paper. Critical comments and editorial handling by Amy Leventer and Jennifer Pike and an anonymous reviewer are also gratefully acknowledged. This research was mainly carried out by National Research Laboratory program and International Joint R&D Project (Korea-Russia and Korea-Norway), provided by Ministry of Science and Technology, and also significantly funded by KORDI (Korea Ocean Research and Development Institute) grant PP03106.

REFERENCES

Alldredge, A.L. and C.C. Gotschalk, Direct observations of the mass flocculation of diatom blooms: characteristics, settling velocities and formation of diatom aggregates, *Deep-Sea Res.,* 36, 159-171, 1989.

Amos, A.F. and M.K. Lavender, AMLR program: Dynamics of the summer hydrographic regime at Elephant Island, *Antarct. J. U.S.,* 228-230, 1992.

Bjorck, S., H. Hakansson, R. Zale, W. Karlen and B.L. Jonsson, A late Holocene lake sediment sequence from Livingstone Island, South Shetland Islands, with palaeoclimatic implications, *Antarct. Sci.,* 3, 61-72, 1991a.

Bjorck, S., C. Hjort, O. Ingolfsson and G. Skog, Radiocarbon dates from the Antarctic Peninsula-problems and potential, in *Radiocarbon dating: Recent applications and future potential,* edited by Lowe, J., *Quat. Proc. 1. Quat. Res. Assoc. Cambridge,* 55-65, 1991b.

Bull, D. and A.E.S. Kemp, Composition and origins of laminae in late Quaternary and Holocene sediments from the Santa Barbara Basin, in *Paleoclimatology and Paleoceanography from Laminated Sediments,* edited by Kemp, A.E.S., *Geol. Soc. Spec. Publ.,* 116, 143-156, 1996.

Davis, C.O., J.T. Hollibaugh, D.L.R. Seibert, W.H. Thomas and P.J. Harrison, Formation of resting spores by *Leptocylindrus danicus* (*Bacillariophyceae*) in a controlled experimental ecosystem, *J. Phyco.,* 16, 296-302, 1980.

Deacon, G., The Antarctic Circumpolar Ocean, Cambridge University Press, 1984.

DeMaster, D.J., The supply and accumulation of silica in the marine environment, *Geoch. Cosmoch. Acta.,* 45, 1715-1732, 1981.

Diester-Haass, L., Sediments as indicators of upwelling, in *Upwelling Ecosystems,* edited by Boje, R., and M. Tomczak, Springer, Berlin, 261-281, 1978.

Domack, E.W. and S. Ishman, Oceanographic and physiographic controls on modern sedimentation within Antarctic fjords, *Geol. Soc. Am. Bull.,* 106, 1175-1189, 1993.

Dunbar, R.B., Sediment trap experiments in the Antarctic continental margin, *Antarct. J. U.S.,* 19, 70-71, 1985.

Dunbar, R.B., A.J. MacPherson and G. Wefer, Water column particulate flux and seafloor deposits in the Bransfield Strait and southern Ross Sea, Antarctica, *Antarct. J. U.S.,* 20, 98-100. 1985.

Figueiras, F.G., M. Estrada, O. Lopez and B. Arbones, Photosynthetic parameters and primary production in the Bransfield Strait: relationships with mesoscale hydrographic structures, *J. Mar. Sys.,* 17, 129-141, 1998.

French, F.W. and P.E. Hargraves, Physiological characteristics of plankton diatom resting spores, *Mar. Bio. Letter,* 1, 185-195, 1980.

Goll, R.M. and K.R. Bjorklund, Radiolaria in surface sediments of the South Atlantic, *Micropaleo.,* 20, 38-75, 1974.

Gordon, J.E. and D.D. Harkness, Magnitude and geographic variation of the radiocarbon content in Antarctic marine life: implications for reservoir corrections in radiocarbon dating, *Quat. Sci. Rev.,* 11, 697-708, 1992.

Gordon, A.L. and W.D. Nowlin, The basin waters of the Bransfield Strait, *J. Phys. Oceano.,* 8, 258-264. 1978,

Griffith, T.W. and J.B. Anderson, Climatic control of sedimentation in bays and fjords of the northern Antarctic Peninsula, *Mar. Geol.,* 85, 181-204, 1989.

Grimm, K.A., C.B. Lange and A.S. Gill, Self-sedimentation of phytoplankton blooms in the geologic record, *Sed. Geol.,* 100, 151-161, 1997.

Grimm, K.A., High-resolution imaging of laminated diatomaceous sediments and their paleoceanographic significance (Quaternary, ODP Site 798, Japan Sea), in *Proceedings of the Ocean Drilling Program.* 128: College Station, Texas, Ocean Drilling Program, 547-557 pp, 1992.

Harden, S., D.J. Demaster and C.A. Nittrouer, Developing sediment geochronologies for high-latitude continental shelf deposits: A radiochemical approach, *Mar. Geol.,* 103, 69-97, 1992.

Hirst, D.M., Geochemistry of sediments from eleven Black Sea cores, in *The Black Sea, Am. Assoc. Petrol. Geol. Mem.,* 20, 430-455, 1974.

Hofmann, E.E., C.M. Lascara and J.M. Klinck, Palmer LTER: Upper-ocean circulation in the LTER region from historical sources, *Antarct. J. U.S,* 27, 239-241, 1992.

Howarth, R.W., A rapid and precise method for determining sulfate in seawater, estuarine waters and sediment pore waters, *Limnol. Oceano.,* 23, 1066-1069, 1978.

Jordan, R.W, J. Priddle, C.J. Pudsey, P.F. Barker and M.J. Whitehouse, Unusual diatom layers in Upper Pleistocene sediments from the northern Weddell Sea, *Deep-Sea Res.,* 38, 829-843, 1991.

Kang, S.H. and S.H. Lee, Antarctic phytoplankton assemblage in the western Bransfield Strait region, February 1993: composition, biomass, and mesoscale distributions, *Mar. Ecol. Prog. Series*, *129*, 253-267, 1995.

Kang, S.H., J.S. Kang, S.H. Lee, K.H. Chung, D.S. Kim and M.G. Park, Antarctic phytoplankton assemblages in the marginal ice zone of the northwestern Weddell Sea, *J. Plankton. Res.*, *23*, 333-352, 2001.

Karl, D.M., B.D. Tilbrook and G. Tien, Seasonal coupling of organic matter production and particle flux in the western Bransfield Strait, Antarctica, *Deep-Sea Res.*, *38*, 1097-1126, 1991.

Kim, D, D.-Y. Kim, J.-H. Shim and S.H. Kang, Particle flux in the eastern Bransfield Strait, Antarctica, Korea Ocean Research and Development Institute Open File Report, BSPP00001-05-1329-7, 57-70, 2000.

Leventer, A., Sediment trap diatom assemblages from the northern Antarctic Peninsula region, *Deep-Sea Res.*, *38*, 1127-1143, 1991.

Leventer, A. and R.B. Dunbar, Diatom flux in McMurdo Sound, Antarctica, *Mar. Micropaleotol.*, *12*, 49-64, 1987.

Leventer, A., R.B. Dunbar and D.J. DeMaster, Diatom evidence for late Holocene climatic events in Granite harbor, Antarctica, *Paleoceanography, 8.* 373-386, 1993.

Leventer, A., E.W. Domack, S.E. Ishman, S. Brachfeld, C.E. McClennen and P. Manly, Productivity cycles of 200-300 years in the Antarctic Peninsula region: understanding linkages among the sun, atmosphere, oceans, sea ice and biota, *Geolog. Soc. Am. Bull.*, *108*, 1626-1644, 1996.

Leventhal, J.S., An interpretation of carbon and sulfur relationships in Black Sea sediments as indicators of environments of deposition, *Geoch. Cosmoch. Acta*, *47*, 133-137, 1983.

Maynard, N., Relationship between diatoms in surface sediments of the Atlantic Ocean the biological and physical oceanography of overlying waters, *Paleobiol.*, *2*, 99-121, 1976.

Mitchell, B.G., E.A. Brady, O. Holm-Hansen, C. McClain and J. Bishop, Light limitation of phytoplankton biomass and macronutrient utilization in the Southern Ocean, 1991.

O'Brien, S.R., P.A. Mayewski, L.D. Meeker, D.A. Meese, M.S. Twickler and S.I. Whitlow, Complexity of Holocene climate as reconstructed from a Greeland ice core, *Sci.*, *270*, 1962-1964, 1995.

Passow, U., A.L. Alldredge and B.E. Logan, The role of particulate carbonhydrate exudates in the flocculation of diatom blooms, *Deep-Sea Res.*, *41*, 335-357, 1994.

Patterson, S.L. and H.A. Sievers, The Weddell-Scotia Confluence. *J. Phys. Oceano.*, *10*, 1584-1610, 1980.

Reynolds, J.M., Distribution of mean annual air temperatures in the Antarctic Peninsula, *Brit. Ant. Sur. Bull.*, *54*, 123-133, 1981.

Sancetta, C., Processes controlling the accumulation of diatoms in sediments: a model derived from British Columbian Fjords, *Paleoceanography, 4*, 235-251, 1989.

Sancetta, C., Comparison of phytoplankton in sediment trap time series and surface sediments along a productivity gradient, *Paleoceanography, 7*, 183-194, 1992.

Scherer, R.P., A new method for the determination of absolute abundance of diatoms and other silt-sized sedimentary particles, *J. Paleolimnology, 12*, 171-179, 1994.

Schimmelmann, A. and C.B. Lang, Tales of 1001 varves: a review of Santa Barbara Basin sediment studies, in *Paleoclimatology and Paleoceanography from Laminated Sediments*, edited by Kemp, A.E.S., *Geol. Soc. Spec. Publ.*, *116*, 121-141, 1996.

Schulz, K.A., U. von Rad and U. von Stackelberg, Laminated sediments from oxygen-minimum zone of the northeastern Arabian Sea, in *Paleoclimatology and Paleoceanography from Laminated Sediments*, edited by Kemp, A.E.S., *Geol. Soc. Spec. Publ.*, *116*, 185-207, 1996.

Shevenell, A.E, E.W. Domack and G.M. Kernan, Record of Holocene paleoclimate changes along the Antarctic Peninsula: evidence from glacial marine sediments, Lallemand Fjord, in *Climate succession and glacial history over the past five million years*, edited by Banks, M.R., Brown, M.J., *Roy. Soc. Tasmania, 130(2)*, 55-64, 1996.

Stuiver, M., G.H. Denton, T.J. Hughes and J.L. Fastock, History of the marine ice sheet in west Antarctica during the last glaciation: a working hypothesis, in *The Last Great Ice Sheet*, edited by Denton, G.H., Hughes T.J., *Wiley, New York*, 319-369, 1981.

Takahashi, K., Seasonal fluxes of silicoflagellates and *Actiniscus* in the subarctic Pacific during 1982-1984, *J. Mar. Res.*, *45*, 397-425, 1987.

Vogel, A.I., A Text Book of Quantitative Inorganic Analysis, Longmans, London, 1216 p, 1975.

Wefer, G., G. Fischer, D. Fuetterer and R. Gersonde, Seasonal particle flux in the Bransfield Strait, Antarctica, *Deep-Sea Res.*, *35*, 891-898, 1988.

Whitworth III, T., W.D. Nowlin Jr, A.H. Orsi, R.A., Locarnini and S.G. Smith, Weddell Sea shelf water in t he Bransfield Strait and Weddell-Scotia Confluence, *Deep-Sea Res.*, *41*, 629-641, 1994.

Yih, H., D. Kang and S. Kim, Preliminary observation along a hydrographic section in the Weddell Sea during the 1996 field season, *Kor. J. Polar Res.*, *9*, 47-54, 1998.

Yoon, H.I., M.W. Han, B.-K. Park, J.K. Oh and S.K. Chang, Depositional environment of near-surface sediments, King George Basin, Bransfield Strait, Antarctica, *Geo-Mar. Letters*, *12*, 1-19, 1994.

Yoon, H.I., B.-K. Park and Y. Kim, A possible mechanism for the formation of layered diatom ooze from Bransfield Strait, Antarctic Peninsula, International workshop on Antarctic Peninsula climate variability held in Hamilton College, April 3-5, 2002, 95 p (abstract), 2002.

Ho Il Yoon, Byong-Kwon Park, Yeadong Kim, Cheon Yun Kang and Sung-Ho Kang, Polar Sciences Laboratory, Korea Ocean Research and Development Institute, Ansan, P.O. Box 29, Seoul 425-600, Korea

ANTARCTIC PENINSULA CLIMATE VARIABILITY
ANTARCTIC RESEARCH SERIES VOLUME 79, PAGES 239-260

FORAMINIFERAL DISTRIBUTIONS IN THE FORMER LARSEN-A ICE SHELF AND PRINCE GUSTAV CHANNEL REGION, EASTERN ANTARCTIC PENINSULA MARGIN: A BASELINE FOR HOLOCENE PALEOENVIRONMENTAL CHANGE

Scott E. Ishman and Phillip Szymcek

Department of Geology, Southern Illinois University, Carbondale, Illinois

Foraminiferal analyses of surface sediment samples collected from the former Larsen-A Ice Shelf (LIS-A) and Prince Gustav Channel (PGC) region of the eastern Antarctic Peninsula margin reveal distinct biofacies/benthic foraminiferal assemblage distributions. Three biofacies are identified representing the remnant LIS-A ice edge, Greenpeace Trough, and PGC/outer LIS-A regions. Calcareous dominated assemblages include the *Angulogerina* and *Epistominella* assemblages. Arenaceous dominated assemblages include the *Miliammina* and *Portatrochammina* assemblages. High planktonic foraminiferal productivity in the absence of high organic flux to the substrate characterizes conditions associated with the ice edge biofacies and *Angulogerina* assemblage. Higher organic carbon content with high planktonic foraminiferal productivity is associated with the trough biofacies/*Epistominella* assemblage. Dissolution modifies the foraminiferal assemblages associated with the PGC/outer LIS-A biofacies. The *Miliammina* assemblage contains very few calcareous foraminifera and occurs in organic-rich siliceous sediments. The *Portatrochammina* assemblage contains higher numbers of calcareous foraminifera than the *Miliammina* assemblage.

INTRODUCTION

Historically the eastern Antarctic Peninsula has remained stable for decades. Not until recently has the major feature of this region, the Larsen Ice Shelf (LIS), shown any sign of instability. Recently the LIS, eastern Antarctic Peninsula, has undergone catastrophic collapse. Since 1996 the LIS has progressively lost significant portions of its former surface area [*Skvarca et al.,* 1999]. Today the entire northern portion, Larsen-A (LIS-A), has disappeared and a significant portion of the middle portion, Larsen-B (LIS-B), has also decayed. Some suggest this is associated with a ~2.5 °C increase in mean annual air temperatures over the past half-century [*Skvarca,* 1993; *Vaughn and Doake,* 1996; *Rott et al.,* 1996, 1998; *Doake*

et al., 1998;]. Because until quite recently this region of the Weddell Sea continental shelf was covered by ice shelf, little is know about the oceanographic and sedimentological properties of this region. The disappearance of these features has opened up a vast laboratory for the investigation of processes associated with ice shelf disintegration. Physical changes include the atmospheric exposure of surface waters that influence primary productivity and heat exchange in the Southern Ocean. Changes in atmosphere-ocean interaction impact surface and intermediate water circulation, and influence sea-ice production and the production of deep water. Changes in primary productivity associated with ice shelf decay not only affect the pelagic community but impact the benthic community by providing additional resources to the substrate. It is

10.1029/079ARS19

only recently that conditions for the former LIS-A region have been established.

Sediments recently collected from the western Weddell Sea continental shelf formerly occupied by LIS-A are being used to determine sedimentological and biological patterns associated with ice shelf collapse [*Domack et al.,* 2001; *Pudsey and Evans,* 2001; *Pudsey et al.,* 2001]. Foraminifera, single celled planktonic and benthic shelled marine organisms, have been used widely for paleoceanographic and paleoclimatic studies. Foraminiferal studies in the Antarctic are few relative to other geographic regions of the world. However, Antarctic foraminiferal studies have progressed since the early taxonomic studies allowing us to make paleoenvironmental inferences based on foraminiferal assemblage data. Modern foraminiferal distributions from the western Antarctic Peninsula [*McKnight,* 1962; *Pflum,* 1966; *Lena,* 1980; *Maiya and Inoue,* 1982; *Kellogg and Kellogg* 1987; *Ishman and Domack,* 1994;] have been more widely studied than for the eastern Antarctic Peninsula. Foraminiferal studies from the eastern Antarctic Peninsula are more limited to the eastern and deeper portions of the Weddell Sea [*Anderson,* 1975; *McKnight,* 1962] due to the lack of accessible shelf area covered by extensive ice shelves along its coast. However, the recent disappearance of LIS-A and the opening of the Prince Gustav Channel (PGC) have provided access to this region. In this paper we present results of foraminiferal analyses conducted on surface sediments from the eastern Antarctic Peninsula region, formerly covered by the LIS-A and Prince Gustav Ice Shelf, to determine associations between foraminifera and environmental conditions related to ice shelf, sedimentological and oceanographic conditions that may be useful in the interpretation of Holocene foraminiferal assemblages.

SETTING

The LIS-A and PGC are part of the Larsen Basin that occupies the eastern margin of the Antarctic Peninsula and the western Weddell Sea continental margin (Figure 1). The Larsen Basin is underlain by late Mesozoic and early Cenozoic igneous and sedimentary sequences [*Sloan et al.,* 1995]. These are overlain by middle and late Cenozoic glacial and glaciomarine sediments [*Sloan et al.,* 1995] that are capped by Holocene sediments ranging from ~1 meter to >8 meters in thickness [*Pudsey and Evans,* 2001; *Pudsey et al.,* 2001]. Sediments within the LIS-A and PGC were characterized into 9 acoustic facies

[*Pudsey et al.,* 2001] that are interpreted to represent depositional settings and bottom types ranging from open marine/glaciomarine to bedrock.

The outer margin of the continental shelf, adjacent to the LIS, is dominated by bottom waters characteristic of low salinity Ice Shelf Water with salinities not exceeding 34.6 parts per thousand (ppt) and temperatures ranging from −1.4 to < −1.8°C [*Gordon,* 1998]. Oceanographic data collected from the LIS-A and PGC regions indicate the presence of Ice Shelf Water in these inner shelf regions as well (Figure 2). To the east, this water mass is separated from warmer modified Weddell Deep Water by the Antarctic Slope Front, a pycnoclinal trough [*Gordon,* 1998]. The climate of the eastern Antarctic Peninsula region is influenced by cold easterly winds prevailing from the Antarctic Peninsula [*Schwerdtfeger,* 1975]. These cold winds are in part responsible for maintaining perennial sea ice conditions and poor thermohaline stratification that persist on the continental shelf of the western Weddell Sea. The western Weddell Sea maintains one of the highest annual average sea ice coverages in the circumantarctic [*Parkinson,* 1998] resulting in highly variable formation of small polynyas [*Comiso and Gordon,* 1998; *Markus et al.,* 1998].

MATERIAL AND METHODS

Samples

Surface sediment samples were collected in May 2000 during the *Nathaniel B. Palmer* 2000-03 (NBP00-03) oceanographic cruise to the Antarctic Peninsula. Sampling sites for this study include samples from the remnant margin of LIS-A and extend into the Greenpeace Trough onto the eastern Weddell Sea continental shelf and into the PGC (Figure 1). A Smith MacIntyre grab sampler was used to sample modern sediments. Upon recovery, each grab was observed to determine the quality of the sediment water interface. Undisturbed sediment water interfaces of the grab samples were determined from the presence of a flocculent sediment layer at the surface and the presence of undisturbed epibenthic invertebrates. One sample, SCUD24, was collected from the sledge of a remotely operated video camera. The upper 2 cms of sediment from each grab sample was collected using a 60 cc syringe that was modified by cutting off the restriction at the end creating an open barrel. The barrel end was slowly inserted into the surficial sediments while at the same time slowly retracting the plunger that functioned as a piston to

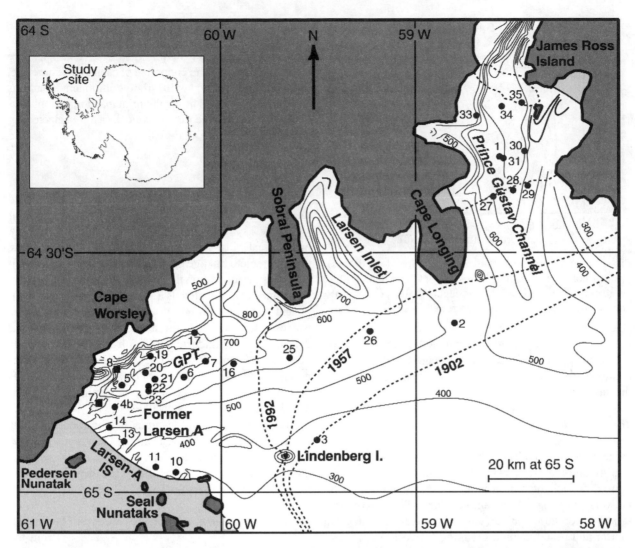

Fig. 1. Bathymetric map of the former Larsen-A Ice Shelf and Prince Gustav region. Sample locations are marked with solid circles and labeled with the sample number. Light shaded regions are ice shelves and dashed lines indicate ice shelf edge positions in 1902, 1957, and 1992. *GPT* labels the Greenpeace Trough. CTD sites 7 and 8 are indicated with solid squares and CTD 31 was collected from sample site 31 (Base map modified from Pudsey et al., 2001).

reduce compression of the sediments. This method was replicated for each sample with one of the replicates immediately treated with a 10% buffered formalin and seawater solution and stained with rose Bengal to identify those specimens containing protoplasm upon collection (living). These samples and the second replicates were refrigerated until their arrival at Southern Illinois University where they were processed for foraminiferal analyses. A total of twenty-nine sites from the LIS-A and PGC areas were used for this study (Table 1).

The modern sediments were washed through a 63 μm sieve and the ≥ 63 μm size fraction was analyzed for foraminifera. A total of 300 benthic foraminifera were collected and counted from each sample when possible. Partial specimens were counted only when identifiable. When present, planktonic foraminifers were collected and documented. Foraminifera from the modern samples with protoplasm stained by rose Bengal were tabulated as living upon collection and those not stained tabulated as dead upon collection. All of the samples collected contained stained specimens. The specimens were then identified and catalogued. Because there is a strong possibility that seasonality is a significant factor in the distribution of living and dead populations of foraminifers

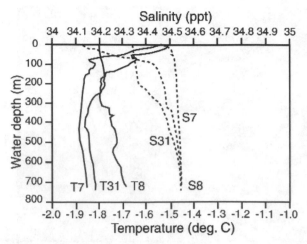

Fig. 2. Temperature (solid lines) and salinity (dashed lines) profiles collected from the LIS-A and PGC region. Profile numbers indicate CTD locality on Figure 1. Temperature values are *in-situ* and have not been corrected for post-cruise drift.

the following analyses were conducted on the total (living + dead) foraminiferal assemblages. Foraminiferal counts were converted to percent occurrence and used in the statistical analyses.

Total Organic Carbon

Samples were collected from the upper two centimeters of archive cores from the Smith-MacIntyre grabs housed at the Antarctic Research Facility, Florida State University, using 10 cc sample plugs. These sediment samples were dried pulverized and homogenized using a mortar and pestal. Samples ranging between 0.2279 and 0.2001 grams were treated to remove % $CaCO_3$ and analyzed for carbon using a LECO C-144. Weight percent organic carbon was then calculated from the total carbon values obtained.

Principal Components Analysis

Principal components analysis is an ordination method that was used to reduce the number of variables that were used in the following analyses. Here we used the correlation matrix of the benthic foraminiferal data in an eigenanalysis. The results of the eigenanalysis show two things: the eigenvalues (the variance accounted for) for each principal component axis, and secondly the eigenvectors and component loadings for each variable. The component loadings represent the relative importance of each variable along the axes, with high (> 0.50) positive and negative values indicating variables of greatest sig-

nificance along a particular axis. The positive or negative sign placed on the loadings represent the ordinal relationship of the variables with high negative and positive loadings indicating the endpoints of each axis. Therefore, the variables (taxa) with high component loadings ($\geq +0.50$ and ≤ -0.50) along the principal component axes that accounted for the trends in the data were selected for use in the remaining analyses.

Cluster Analysis

Custer analysis is a descriptive analytical method used to group variables (species or samples) based on their similarity as described using a similarity measure. A hierarchical clustering method in the Q- and R-mode using complete linkage of the Pearson correlation coefficient was performed on the benthic foraminiferal relative abundance data selected as a result of the principal components analysis using Systat©. The Q-mode method was used to determine similarities among samples, thus defining the boundaries of biofacies. The Q-mode method was applied to data sets that included both arenaceous and calcareous benthic taxa, only calcareous taxa, and only arenaceous taxa to determine if the biofacies distributions were controlled only by the abundance of arenaceous versus calcareous benthic taxa or if the biofacies were discernable from the calcareous and arenaceous taxa separately. R-mode cluster analysis grouped species and was used to identify benthic foraminiferal assemblages.

Discriminant Function Analysis

Discriminant function analysis of the benthic foraminiferal data was used to determine the correct groupings of the samples as determined by the cluster analysis, which inherently forces variables into groupings. *Post hoc* predictions are made based on classifications assigned to each sample according to its geographic location within the study region and analyzed using the discriminant function analysis of SPSS©. A discriminant function analysis is a multivariate method used to demonstrate the differences between groups and their significance. This analysis produces canonical discriminant function coefficients for each variable from which classification function coefficients are derived. The weight of the coefficient, based on its contribution to the total variance, determines the significance of the canonical variant in maximizing the distance between each group recognized with a minimum of classification error [*Johnson and Wichern*, 1982]. The results of the dis-

TABLE 1. Sample data from cruise NBP00-03. Foraminiferal species data is in relative abundance (%).

Biofacies	Ice Edge			Greenpeace Trough				
Sample	SMG10	SMG11	SMG14	SMG4b	SMG5	SMG6	SMG7	SMG13
Latitude (degrees and minutes S)	64 57.368	64 56.669	64 51.818	64 49.209	64 46.520	64 45.518	64 43.523	64 53.517
Longitude (degrees and minutes W)	60 13.392	60 19.281	60 33.438	60 32.033	60 29.684	60 10.720	60 04.771	60 28.836
Water Depth (meters)	332	350	419	668	978	733	839	323
Total Organic Carbon (weight %)	0.38	0.49	0.7	1.86	0.29	0.52	0.61	0.97
Arenaceous Benthic Species								
Adercotryma glomerata	1.97	1.32	0.98	7.77	17.31	12.50	12.01	3.28
Ammodiscus catinus	1.32	0.00	0.66	0.68	0.32	0.00	1.62	0.00
Conotrochammina alternans	0.66	1.97	1.97	0.34	1.60	0.96	0.65	0.30
Cribrostomoides arenacea	0.00	0.00	0.33	0.00	0.00	0.00	0.00	0.00
Cribrostomoides sp.	0.00	0.99	0.00	0.00	0.00	0.00	0.00	0.00
Cribrostomoides sphaeriloculus	0.33	0.00	0.00	0.34	0.00	0.00	0.00	0.00
Cyclammina pusilla	0.00	0.00	0.00	0.00	0.00	0.00	0.00	0.30
Cystammina argentea	1.97	0.66	0.00	1.35	1.28	0.32	0.65	0.30
Miliammina spp.	1.97	2.63	0.66	0.34	2.56	1.28	3.57	1.19
Portatrochammina antarctica	0.99	0.99	1.31	2.03	1.92	1.60	1.30	0.00
Portatrochammina eltaninae	2.30	1.32	0.98	2.03	0.96	2.88	1.62	1.19
Portatrochammina wiesneri	3.95	4.93	2.95	11.49	3.53	3.21	2.27	4.48
Psammosphaera fusca	0.00	0.00	0.00	0.68	0.32	0.00	0.32	0.00
Reophax ovicula	6.58	11.18	2.30	2.03	0.96	0.96	1.30	0.30
Reophax subdentaliniformis	0.66	6.25	0.33	0.68	0.32	0.00	0.00	0.30
Rhabdammina abyssorum	8.22	8.88	1.97	1.35	2.24	0.32	0.00	0.00
Saccammina tabulata	0.33	0.00	0.00	0.00	0.32	0.00	0.00	0.00
Textularia antarctica	1.32	3.29	0.33	5.41	3.21	0.64	2.92	2.39
Textularia wiesneri	1.32	2.30	2.62	4.05	4.17	0.96	0.97	1.49
Trochammina bullata	0.33	0.00	0.00	0.00	0.00	0.00	0.00	0.00
Trochammina conica	0.00	0.33	0.00	0.34	0.00	0.00	0.00	0.00
Trochammina glabra	2.63	6.58	3.93	3.38	0.96	0.64	2.92	2.39
Trochammina intermedia	0.00	0.00	0.00	0.00	0.32	0.00	0.32	0.00
Trochammina pygmeae	1.32	0.00	0.00	0.00	0.00	0.00	0.00	0.00
Trochammina sp1	0.00	0.00	0.66	0.00	0.00	0.00	0.00	0.00
Trochammina sp2	0.00	0.00	0.00	0.00	0.00	0.00	0.00	0.00
Verneulina minuta	0.00	0.00	0.00	0.00	0.00	0.32	0.65	0.30
Calcareous Benthic Species								
Astrononion echolsi	0.33	0.00	0.98	0.00	0.00	0.32	0.32	0.00

TABLE 1 (continued).

Biofacies	Ice Edge			Greenpeace Trough				
Sample	SMG10	SMG11	SMG14	SMG4b	SMG5	SMG6	SMG7	SMG13
Latitude (degrees and minutes S)	64 57.368	64 56.669	64 51.818	64 49.209	64 46.520	64 45.518	64 43.523	64 53.517
Longitude (degrees and minutes W)	60 13.392	60 19.281	60 33.438	60 32.033	60 29.684	60 10.720	60 04.771	60 28.836
Water Depth (meters)	332	350	419	668	978	733	839	323
Total Organic Carbon (weight %)	0.38	0.49	0.7	1.86	0.29	0.52	0.61	0.97
Astrononion stelligerum	0.99	0.00	0.00	0.00	0.00	0.00	0.00	0.00
Bulimina chapmani	0.00	0.00	0.00	0.34	0.00	0.00	0.00	0.00
Cassidulina spp.	0.33	0.00	0.33	1.01	1.60	1.28	1.62	1.19
Cibicides lobatulus	2.30	0.33	2.95	0.34	0.32	0.00	0.65	2.09
Cibicides refulgens	1.32	0.33	3.93	1.69	0.00	0.00	0.00	0.00
Ehrenbergina glabra	1.64	0.99	1.31	1.01	0.00	0.96	0.65	0.60
Epistominella exigua	4.28	0.00	4.59	10.14	10.26	15.71	10.39	11.64
Eponides tumidulus	0.99	0.33	1.31	1.35	0.32	0.64	3.25	2.39
Globocassidulina biora	15.13	10.86	29.18	18.92	12.82	16.67	12.66	16.12
Melonis affinis	0.99	0.00	0.98	0.68	1.60	1.60	1.62	1.79
Miliolids	0.33	0.33	1.97	0.00	0.32	0.00	0.00	0.30
Nodosarids	3.62	2.96	2.62	0.68	0.96	1.28	2.27	5.97
Nonionella iridea	3.95	1.64	8.52	16.55	25.64	28.53	23.05	24.18
Pyrgo depressa	0.33	0.00	0.33	0.34	0.00	0.00	0.00	0.00
Rosalina glabra	0.00	0.00	0.00	0.00	0.00	0.00	0.00	0.00
Spiroloculina spp.	0.00	0.66	0.00	0.00	0.00	1.60	3.25	3.58
Stainforthia concava	0.00	0.00	0.98	1.35	3.53	3.21	5.84	1.79
Angulogerina earlandi	18.75	18.09	7.87	0.00	0.00	0.00	0.32	4.78
Angulogerina pauperata	5.59	9.54	8.20	0.00	0.00	0.00	0.00	4.48
Triloculina spp.	0.00	0.00	1.31	1.35	0.32	1.60	0.97	0.90
Uvigerina sp.	0.99	0.33	0.66	0.00	0.00	0.00	0.00	0.00
Planktonic Species								
Globigerina sp.	28.95	11.11	22.54	21.43	28.57	20.00	33.33	42.11
Neogloboquadrina pachyderma	71.05	88.89	77.46	78.57	71.43	80.00	66.67	57.89
Percent Calcareous Benthic	61.84	46.38	78.03	55.74	57.69	73.40	66.88	81.79
Percent Arenaceous	38.16	53.62	21.97	44.26	42.31	26.60	33.12	18.21
Total Benthic	304	304	305	296	312	312	308	335
Total Planktonic	76	9	71	14	7	5	3	19
Total Counts	380	313	376	310	319	317	311	354
Species Richness	35	27	33	31	28	24	29	26

TABLE 1 (continued).

Biofacies				Greenpeace Trough			
Sample	SMG17	SMG19	SMG20	SMG21	SMG22	SMG23	SCUD24
Latitude (degrees and minutes S)	64 39.793	64 42.778	64 44.984	64 45.827	64 46.632	64 47.144	64 47.959
Longitude (degrees and minutes W)	60 07.662	60 20.846	60 22.345	60 19.450	60 21.557	60 21.566	60 19.895
Water Depth (meters)	719	879	899	912	868	901	912
Total Organic Carbon (weight %)	0.64	0.48	0.56	0.67	0.66	NA	NA
Arenaceous Benthic Species							
Adercotryma glomerata	13.33	15.61	11.29	9.63	12.66	10.16	7.42
Ammodiscus catinus	2.22	0.00	0.65	0.00	0.32	0.00	0.00
Conotrochammina alternans	0.00	1.33	2.26	2.48	2.27	3.28	0.97
Cribrostomoides arenacea	0.00	0.00	0.32	0.00	0.65	0.00	0.00
Cribrostomoides sp.	0.00	0.00	0.00	0.00	0.00	0.00	0.00
Cribrostomoides sphaeriloculus	0.00	0.00	0.00	0.00	0.00	0.00	0.32
Cyclammina pusilla	0.00	0.00	0.00	0.00	0.00	0.00	0.00
Cystammina argentea	0.00	0.66	2.26	3.11	1.95	1.97	0.97
Miliammina spp.	0.63	1.00	2.90	3.73	3.57	2.62	0.00
Portatrochammina antarctica	4.13	0.66	0.65	0.62	3.25	1.97	2.26
Portatrochammina eltaninae	1.90	0.66	1.29	0.93	1.95	3.28	0.65
Portatrochammina wiesneri	2.86	0.33	9.35	1.55	2.27	1.64	4.16
Psammosphaera fusca	0.00	0.33	2.26	0.00	0.00	0.00	0.00
Reophax ovicula	1.90	0.66	0.65	1.24	0.97	0.66	1.29
Reophax subdentaliniformis	0.00	0.00	0.32	0.00	0.00	0.00	0.97
Rhabdammina abyssorum	1.27	0.66	0.32	0.00	3.25	2.30	0.00
Saccammina tabulata	0.63	0.00	0.00	0.00	0.65	0.00	0.32
Textularia antarctica	0.63	4.65	3.23	2.17	1.30	2.95	4.52
Textularia wiesneri	3.17	0.66	13.55	0.00	0.00	0.33	1.61
Trochammina bullata	0.32	0.00	0.32	0.31	0.00	0.00	0.00
Trochammina conica	0.00	0.00	0.00	0.00	0.00	0.00	0.00
Trochammina glabra	0.63	0.33	0.65	1.24	1.95	0.00	3.87
Trochammina intermedia	0.00	0.66	0.00	1.55	0.65	0.33	0.00
Trochammina pygmeae	0.00	0.00	0.00	0.00	0.00	0.00	0.00
Trochammina sp1	0.32	0.00	0.00	0.00	0.00	0.00	0.00
Trochammina sp2	0.63	0.00	0.00	0.00	0.00	0.00	0.00
Verneulina minuta	0.32	0.00	0.32	0.31	0.00	0.00	0.00
Calcareous Benthic Species							
Astrononion echolsi	0.32	0.00	0.00	0.62	0.65	0.33	0.32

TABLE 1 (continued).

Biofacies			Greenpeace Trough				
Sample	SMG17	SMG19	SMG20	SMG21	SMG22	SMG23	SCUD24
Latitude (degrees and minutes S)	64 39.793	64 42.778	64 44.984	64 45.827	64 46.632	64 47.144	64 47.959
Longitude (degrees and minutes W)	60 07.662	60 20.846	60 22.345	60 19.450	60 21.557	60 21.566	60 19.895
Water Depth (meters)	719	879	899	912	868	901	912
Total Organic Carbon (weight %)	0.64	0.48	0.56	0.67	0.66	NA	NA
Astrononion stelligerum	0.00	0.00	0.00	0.00	0.00	0.00	0.00
Bulimina chapmani	0.00	0.00	0.00	0.00	0.00	0.00	0.00
Cassidulina spp.	1.59	1.33	0.65	1.24	0.97	2.30	1.29
Cibicides lobatulus	1.27	0.00	0.00	0.31	0.32	0.33	0.97
Cibicides refulgens	1.27	0.00	0.00	0.31	0.32	0.33	0.65
Ehrenbergina glabra	0.00	0.00	0.00	0.31	0.00	0.33	0.00
Epistominella exigua	12.38	8.97	9.03	7.76	11.04	3.93	14.52
Eponides tumidulus	2.54	0.00	1.94	0.93	1.95	1.97	1.29
Globocassidulina biora	11.43	18.60	18.39	16.77	11.69	23.28	16.77
Melonis affinis	2.86	2.66	1.94	3.73	2.92	1.97	3.87
Miliolids	0.00	0.00	0.00	0.00	0.00	0.00	0.00
Nodosarids	2.54	1.66	0.00	3.73	3.25	2.62	3.87
Nonionella iridea	21.90	32.23	11.61	24.22	20.78	24.59	17.74
Pyrgo depressa	0.00	0.00	0.00	0.31	0.00	0.00	0.00
Rosalina glabra	0.00	0.00	0.00	1.24	0.00	0.00	0.00
Spiroloculina spp.	3.81	1.99	0.00	3.42	2.60	1.31	1.94
Stainforthia concava	3.17	2.66	3.55	5.28	4.87	2.95	4.52
Angulogerina earlandi	0.00	0.00	0.00	0.00	0.00	0.00	0.00
Angulogerina pauperata	0.00	0.00	0.00	0.00	0.00	0.00	0.00
Triloculina spp.	0.00	1.66	0.32	0.93	0.97	2.30	0.65
Uvigerina sp.	0.00	0.00	0.00	0.00	0.00	0.00	0.00
Planktonic Species							
Globigerina sp.	55.56	60.00	0.00	0.00	0.00	0.00	40.00
Neogloboquadrina pachyderma	44.44	40.00	100.00	100.00	100.00	100.00	60.00
Percent Calcareous Benthic	65.08	71.76	47.42	71.12	62.34	68.52	68.40
Percent Arenaceous	34.92	28.24	52.58	28.88	37.66	31.48	29.33
Total Benthic	315	301	310	322	308	305	310
Total Planktonic	9	5	3	5	3	5	15
Total Counts	324	306	313	327	311	310	325
Species Richness	28	23	26	29	28	26	26

TABLE 1 (continued).

Biofacies	PGC/Outer LIS-A							
Sample	SMG1	SMG2	SMG3	SMG16	SMG25	SMG26	SMG27	SMG28
Latitude (degrees and minutes S)	64 17.625	64 38.387	64 53.533	64 43.897	64 43.314	64 39.564	64 22.934	64 22.018
Longitude (degrees and minutes W)	58 34.678	58 37.911	59 30.694	59 55.745	59 38.459	59 13.226	58 36.976	58 30.942
Water Depth (meters)	768	504	385	713	628	564	684	794
Total Organic Carbon (weight %)	0.96	0.71	0.085	0.67	0.51	0.56	0.82	NA
Arenaceous Benthic Species								
Adercotryma glomerata	8.41	1.83	18.63	12.66	19.33	25.68	6.67	6.82
Ammodiscus catinus	0.93	0.00	2.29	0.97	0.67	1.71	1.27	0.32
Conotrochammina alternans	1.40	2.74	1.96	6.17	1.33	1.03	2.86	3.25
Cribrostomoides arenacea	0.00	0.00	0.00	0.00	0.33	0.00	0.63	0.00
Cribrostomoides sp.	3.74	5.48	2.29	0.00	0.00	0.00	0.00	0.00
Cribrostomoides sphaeriloculus	0.00	0.00	0.00	0.00	1.00	0.68	0.32	0.32
Cyclammina pusilla	0.00	0.91	0.00	0.00	0.00	0.00	0.00	0.00
Cystammina argentea	5.14	4.11	3.27	3.90	3.67	1.03	10.16	3.90
Miliammina spp.	5.14	1.83	6.21	4.87	2.67	4.45	9.84	12.34
Portatrochammina antarctica	9.35	7.76	0.98	6.17	4.00	3.42	11.43	11.36
Portatrochammina eltaninae	0.00	2.28	2.94	2.60	2.67	1.37	1.59	0.65
Portatrochammina wiesneri	15.42	5.94	13.40	12.99	12.00	25.68	12.38	8.77
Psammosphaera fusca	0.00	10.50	0.00	0.00	0.33	0.68	0.00	0.00
Reophax ovicula	5.14	8.68	5.88	2.27	2.00	1.03	4.76	4.55
Reophax subdentaliniformis	16.82	5.94	7.52	0.32	1.33	0.34	4.76	6.17
Rhabdammina abyssorum	8.41	26.48	13.40	6.49	5.33	5.48	10.48	2.60
Saccammina tabulata	0.47	6.39	0.00	0.00	0.33	0.00	0.63	0.00
Textularia antarctica	0.00	3.20	7.52	8.12	7.67	2.40	3.49	5.19
Textularia wiesneri	0.00	0.91	3.92	2.27	1.33	0.34	3.81	26.30
Trochammina bullata	3.74	0.00	0.00	0.00	0.00	0.00	0.00	0.00
Trochammina conica	1.40	0.91	1.63	0.00	1.67	0.68	2.54	0.00
Trochammina glabra	3.27	3.65	7.52	5.84	3.33	5.48	9.52	2.60
Trochammina intermedia	0.00	0.46	0.00	0.00	0.00	0.00	2.54	2.27
Trochammina pygmeae	0.00	0.00	0.00	0.00	0.00	0.00	0.00	0.00
Trochammina sp1	0.00	0.00	0.00	0.00	0.00	0.68	0.00	0.00
Trochammina sp2	0.00	0.00	0.00	0.00	0.00	0.00	0.00	0.00
Verneulina minuta	0.00	0.00	0.65	0.65	1.00	0.68	0.32	0.00
Calcareous Benthic Species								
Astrononion echolsi	0.00	0.00	0.00	0.97	2.67	2.74	0.00	2.27

ISHMAN AND SZYMCEK: FORAMINIFERAL DISTRIBUTIONS 247

TABLE 1 (continued).

Biofacies					PGC/Outer LIS-A			
Sample	SMG1	SMG2	SMG3	SMG16	SMG25	SMG26	SMG27	SMG28
Latitude (degrees and minutes S)	64 17.625	64 38.387	64 53.533	64 43.897	64 43.314	64 39.564	64 22.934	64 22.018
Longitude (degrees and minutes W)	58 34.678	58 37.911	59 30.694	59 55.745	59 38.459	59 13.226	58 36.976	58 30.942
Water Depth (meters)	768	504	385	713	628	564	684	794
Total Organic Carbon (weight %)	0.96	0.71	0.085	0.67	0.51	0.56	0.82	NA
Astrononion stelligerum	0.00	0.00	0.00	0.00	0.00	0.00	0.00	0.00
Bulimina chapmani	0.00	0.00	0.00	0.00	0.00	0.00	0.00	0.00
Cassidulina spp.	0.00	0.00	0.00	0.00	0.00	0.00	0.00	0.00
Cibicides lobatulus	0.00	0.00	0.00	0.65	0.00	1.03	0.00	0.00
Cibicides refulgens	0.00	0.00	0.00	1.62	1.00	1.71	0.00	0.00
Ehrenbergina glabra	0.00	0.00	0.00	0.00	0.00	0.00	0.00	0.00
Epistominella exigua	0.00	0.00	0.00	6.17	6.33	4.79	0.00	0.00
Eponides tumidulus	0.00	0.00	0.00	2.27	4.33	0.00	0.00	0.00
Globocassidulina biora	0.00	0.00	0.00	1.30	2.00	2.74	0.00	0.00
Melonis affinis	0.93	0.00	0.00	2.92	5.67	1.03	0.00	0.00
Miliolids	0.93	0.00	0.00	0.00	0.00	0.34	0.00	0.00
Nodosarids	0.00	0.00	0.00	2.60	1.67	1.37	0.00	0.32
Nonionella iridea	0.00	0.00	0.00	1.30	1.67	0.34	0.00	0.00
Pyrgo depressa	0.00	0.00	0.00	0.00	0.00	0.00	0.00	0.00
Rosalina glabra	0.00	0.00	0.00	0.00	0.00	0.00	0.00	0.00
Spiroloculina spp.	0.00	0.00	0.00	0.00	2.33	0.00	0.00	0.00
Stainforthia concava	9.35	0.00	0.00	2.60	0.00	0.68	0.00	0.00
Angulogerina earlandi	0.00	0.00	0.00	0.97	0.33	0.34	0.00	0.00
Angulogerina pauperata	0.00	0.00	0.00	0.32	0.00	0.00	0.00	0.00
Triloculina spp.	0.00	0.00	0.00	0.00	0.00	0.00	0.00	0.00
Uvigerina sp.	0.00	0.00	0.00	0.00	0.00	0.00	0.00	0.00
Planktonic Species								
Globigerina sp.	0.00	0.00	0.00	5.88	0.00	0.00	0.00	0.00
Neogloboquadrina pachyderma	0.00	0.00	0.00	94.12	100.00	100.00	0.00	100.00
Percent Calcareous Benthic	11.21	0.00	0.00	23.70	28.00	17.12	0.00	2.60
Percent Arenaceous	88.79	100.00	100.00	76.30	72.00	82.88	100.00	97.40
Total Benthic	214	219	306	308	300	292	315	308
Total Planktonic	0	0	0	17	10	4	0	1
Total Counts	214	219	306	325	310	296	315	309
Species Richness	18	19	17	27	30	30	20	18

TABLE 1 (continued).

Biofacies			PGC/Outer LIS-A			
Sample	SMG29	SMG30	SMG31	SMG33	SMG34	SMG35
Latitude (degrees and minutes S)	64 21.361	64 16.875	64 17.764	64 11.959	64 10.995	64 10.471
Longitude (degrees and minutes W)	58 26.637	58 26.985	58 34.561	58 41.857	58 34.140	58 28.505
Water Depth (meters)	690	843	779	587	865	651
Total Organic Carbon (weight %)	0.78	0.86	NA	0.68	0.79	0.52
Arenaceous Benthic Species						
Adercotryma glomerata	3.63	6.27	10.21	8.20	5.84	2.82
Ammodiscus catinus	0.33	0.00	0.00	0.00	0.00	0.00
Conotrochammina alternans	2.64	1.32	2.82	0.39	0.32	0.00
Cribrostomoides arenacea	0.00	0.66	0.00	0.00	0.00	0.00
Cribrostomoides sp.	0.00	0.00	3.52	5.47	2.60	1.57
Cribrostomoides sphaeriloculus	0.00	0.00	0.00	0.00	0.00	0.00
Cyclammina pusilla	0.00	0.00	0.00	0.00	0.00	0.00
Cystammina argentea	0.99	0.00	7.39	1.17	0.65	0.00
Miliammina spp.	29.04	24.75	1.06	3.91	8.44	38.56
Portatrochammina antarctica	5.94	3.96	10.92	6.25	3.25	5.33
Portatrochammina eltaninae	0.33	1.65	4.23	1.56	0.97	0.94
Portatrochammina wiesneri	7.59	10.89	28.17	12.11	10.06	12.85
Psammosphaera fusca	0.00	0.00	0.35	0.00	0.00	0.00
Reophax ovicula	3.63	3.63	3.87	1.17	1.95	10.03
Reophax subdentaliniformis	6.93	11.88	3.52	1.95	7.47	9.72
Rhabdammina abyssorum	0.99	1.98	4.23	6.25	0.97	0.63
Saccammina tabulata	0.00	0.00	0.35	2.73	0.00	0.00
Textularia antarctica	3.96	4.62	3.17	5.08	4.87	1.57
Textularia wiesneri	26.07	23.43	7.39	1.17	45.78	6.58
Trochammina bullata	0.00	0.00	0.00	0.00	0.00	0.00
Trochammina conica	1.65	0.33	3.52	0.39	0.65	1.25
Trochammina glabra	1.98	3.30	1.76	1.56	2.92	4.70
Trochammina intermedia	2.31	0.99	2.11	0.00	2.27	2.82
Trochammina pygmeae	0.00	0.00	0.00	0.00	0.00	0.00
Trochammina sp1	0.00	0.00	0.00	0.00	0.00	0.00
Trochammina sp2	0.00	0.00	0.00	0.00	0.00	0.00
Verneulina minuta	0.00	0.33	1.41	0.00	0.00	0.00
Calcareous Benthic Species						
Astrononion echolsi	0.33	0.00	0.00	2.34	0.00	0.31

TABLE 1 (continued).

Biofacies			PGC/Outer LIS-A			
Sample	SMG29	SMG30	SMG31	SMG33	SMG34	SMG35
Latitude (degrees and minutes S)	64 21.361	64 16.875	64 17.764	64 11.959	64 10.995	64 10.471
Longitude (degrees and minutes W)	58 26.637	58 26.985	58 34.561	58 41.857	58 34.140	58 28.505
Water Depth (meters)	690	843	779	587	865	651
Total Organic Carbon (weight %)	0.78	0.86	NA	0.68	0.79	0.52
Astrononion stelligerum	0.00	0.00	0.00	1.56	0.00	0.00
Bulimina chapmani	0.00	0.00	0.00	0.00	0.00	0.00
Cassidulina spp.	0.00	0.00	0.00	0.00	0.00	0.00
Cibicides lobatulus	0.00	0.00	0.00	0.78	0.00	0.00
Cibicides refulgens	0.00	0.00	0.00	0.00	0.00	0.00
Ehrenbergina glabra	0.00	0.00	0.00	0.00	0.00	0.00
Epistominella exigua	0.00	0.00	0.00	16.41	0.00	0.00
Eponides tumidulus	0.00	0.00	0.00	0.78	0.00	0.00
Globocassidulina biora	0.66	0.00	0.00	0.00	0.00	0.00
Melonis affinis	0.33	0.00	0.00	3.91	0.65	0.00
Miliolids	0.00	0.00	0.00	0.00	0.00	0.00
Nodosarids	0.33	0.00	0.00	3.52	0.00	0.31
Nonionella iridea	0.00	0.00	0.00	0.00	0.00	0.00
Pyrgo depressa	0.00	0.00	0.00	0.00	0.00	0.00
Rosalina glabra	0.00	0.00	0.00	0.00	0.00	0.00
Spiroloculina spp.	0.00	0.00	0.00	0.39	0.32	0.00
Stainforthia concava	0.33	0.00	0.00	1.56	0.00	0.00
Angulogerina earlandi	0.00	0.00	0.00	5.47	0.00	0.00
Angulogerina pauperata	0.00	0.00	0.00	3.13	0.00	0.00
Triloculina spp.	0.00	0.00	0.00	0.78	0.00	0.00
Uvigerina sp.	0.00	0.00	0.00	0.00	0.00	0.00
Planktonic Species						
Globigerina sp.	0.00	0.00	0.00	48.72	0.00	0.00
Neogloboquadrina pachyderma	0.00	0.00	0.00	51.28	0.00	0.00
Percent Calcareous Benthic	1.98	0.00	0.00	40.63	0.97	0.63
Percent Arenaceous	98.02	100.00	100.00	59.38	99.03	99.37
Total Benthic	303	303	284	256	308	319
Total Planktonic	0	0	0	39	0	0
Total Counts	303	303	284	295	308	319
Species Richness	21	16	19	28	18	16

criminant analysis simply reveal whether or not the *post hoc* predictions were acceptable within the level of significance defined for the analysis. It does not however infer that the groupings defined are unique. The benefit of this analysis is its statistical significance using Wilk's Lambda test statistic where values of lambda that are near zero denote high discrimination between groups.

For classification purposes, three *post hoc* classifications were used to represent geographic position of the sample site within the overall study region; 1 for the PGC/outer LIS-A, 2 for the Greenpeace Trough, and 3 for the LIS-A ice edge. This approach was used to recognize distinct geographic distributions of biofacies.

RESULTS

Modern Foraminiferal Distributions

A total of 51 foraminiferal taxa (some of the groupings include several species and are indicated by spp.) were identified including 27 arenaceous, 22 calcareous benthic and 2 planktonic taxa with species diversity, as measured by species richness (number of taxa), ranging from 16 to 35 taxa (Table 1). Calcareous benthic foraminifera were the dominant component of the assemblages adjacent to the LIS-A ice edge and in the Greenpeace Trough regions of the study area (Table 1). Arenaceous foraminifera dominated the assemblages found in the PGC and outer LIS-A areas (Table 1). Planktonic foraminifera, dominated by *Neogloboquadrina pachyderma*, were abundant in the assemblages found adjacent the LIS-A ice edge, decreasing in numbers in the Greenpeace Trough and greatly reduced in the PGC and outer LIS-A (Table 1).

Total Organic Carbon

Results of the total organic carbon analysis of the surface sediment samples give a range from 0.29 to 1.86 weight % organic carbon (Table 1) with a median value of 0.66 and mean of 0.66. These values are consistent with previously published values determined for the same region [*Domack et al.,* 2001; *Pudsey and Evans,* 2001; *Pudsey et al.,* 2001], which ranged between 0.40 to 1.00 weight % organic carbon.

Principal Components Analysis

Results of the R-mode (species) principal components analysis (PCA) indicate 4 principal component axes,

PCA1-PCA4, that explain 53.32 % of the variance in the data (Table 2). Using the results of this analysis, 38 of the original 51 benthic foraminiferal taxa having component loadings $\geq +0.50$ or ≤ -0.50 (Table 2) were selected for use in the subsequent analyses.

Cluster Analysis

The Q- and R-mode cluster analyses were successful in describing the foraminiferal distributions by resolving distinct biofacies and assemblages, respectively. Initial inspection of the Q-mode cluster dendrograms (Figure 3) clearly illustrates two distinct groupings that represent samples dominated by either calcareous benthic or arenaceous assemblages. Closer examination of the cluster dendrograms generated from the total and calcareous benthic assemblage data (Figures 3a, b) indicates that the clusters composed of samples dominated by calcareous benthic assemblages can be further resolved into samples from the LIS-A ice edge; SMG10, SMG11, and SMG14, and Greenpeace Trough; SMG4-SMG7, SMG13, SMG17, SMG19-SMG23, and SCUD24 (Figures 3a, b and 4), representing the ice edge and trough biofacies. One inconsistency in the results was sample SMG13 that clustered with the trough biofacies when geographically it is positioned at the ice edge of LIS-A; this will be discussed in a following section. Results from the arenaceous taxa cluster analysis also separated the samples of the ice-edge from the trough samples (Figure 3c).

The third biofacies identified is represented by samples dominated by arenaceous foraminiferal assemblages from the PGC/outer LIS-A region; samples SMG1-SMG3, SMG16, SMG25-SMG31, SMG33 and SMG34 (Figures 3, 4). This PGC/outer LIS-A biofacies is clearly distinguished from the total and calcareous benthic assemblage Q-mode cluster analyses. When using only the arenaceous foraminiferal assemblage data this biofacies was less well resolved with several of the PGC/outer LIS-A samples, SMG3, SMG16, SMG25, SMG26, SMG31, and SMG33, grouped with samples from the trough biofacies, and with the ice shelf biofacies samples, SMG10, SMG11, and SMG14, grouping with the remainder of the PGC/outer LIS-A samples. However, this was useful in that it resolved a group of samples from the PGC, SMG28-30, SMG34, and SMG35 (Figure 3), which contain a distinct foraminiferal assemblage, the *Miliammina* assemblage, identified in the R-mode cluster analysis below.

The R-mode cluster analysis resulted in defining two major assemblages (Figure 5), one dominated by arena-

TABLE 2. Results of the R-mode principal components analysis showing component loadings for the species used in the cluster and discriminant function analyses.

	PCA1	PCA2	PCA3	PCA4
Nonionella iridea	−0.9107	0.0928	−0.1648	0.173
Cassidulina spp.	−0.8839	0.1023	−0.1383	0.1564
Epistominella exigua	−0.8506	0.0022	0.1663	0.0459
Globocassidulina biora	−0.7696	−0.3727	−0.1523	0.2276
Spiroloculina spp.	−0.7637	0.1114	0.087	0.0668
Triloculina spp.	−0.7252	−0.0346	−0.1783	0.1582
Melonis affinis	−0.6695	0.134	0.3949	0.0459
Stainforthia concava	−0.6596	0.2034	−0.1599	0.0216
Trochammina conica	0.6592	0.3238	0.2478	−0.064
Nodosarids	−0.6537	−0.3966	0.0892	0.0347
Portatrochammina wiesneri	0.6512	0.2363	0.479	0.0429
Portatrochammina antarctica	0.6341	0.3945	0.1176	−0.1052
Reophax subdentaliniformis	0.6098	0.18	−0.4054	−0.2027
Trochammina glabra	0.6036	−0.0912	0.2955	0.0302
Eponides tumidulus	−0.5893	−0.031	0.4544	0.0641
Reophax ovicula	0.5883	−0.113	−0.298	−0.5614
Trochammina intermedia	0.5406	0.4058	−0.3269	0.3085
Miliammina spp.	0.5351	0.2552	−0.3686	0.328
Uvigerina sp.	0.1248	−0.9105	−0.0614	0.0322
Angulogerina pauperata	0.0594	−0.8292	−0.1232	−0.0453
Angulogerina earlandi	0.112	−0.8214	−0.1033	−0.0759
Ehrenbergina glabra	−0.2137	−0.7876	−0.1322	0.1073
Cibicides lobatulus	−0.2941	−0.7847	0.1737	0.0514
Miliolids	0.0774	−0.6787	−0.0492	0.0463
Cibicides refulgens	−0.0753	−0.6399	0.4543	
Trochammina pygmeae	0.0811	−0.5938	−0.0205	−0.028
Pyrgo depressa	−0.1338	−0.5915	−0.0541	0.1716
Verneulina minuta	0.1424	0.2811	0.7082	0.0383
Cribrostomoides sphaeriloculus	0.1446	−0.0556	0.6751	0.1132
Adercotryma glomerata	−0.2492	0.4077	0.659	0.1709
Astrononion echolsi	0.0719	−0.0373	0.6141	0.0807
Ammodiscus catinus	0.069	−0.0939	0.5637	0.0421
Portatrochammina eltaninae	−0.0268	0.1523	0.5003	−0.109
Saccammina tabulata	0.0416	0.0622	−0.0353	−0.9115
Rhabdammina abyssorum	0.3435	−0.0012	0.0241	−0.8783
Cyclammina pusilla	−0.0264	0.0032	−0.1564	−0.8236
Psammosphaera fusca	0.0367	0.0961	−0.0731	−0.8189
Cribrostomoides sp.	0.376	0.1455	−0.1167	−0.6297
Percent of total variance explained:	20.0872	14.2711	9.7112	9.2458

ceous taxa and the second dominated by calcareous benthic taxa. These major groups can be further resolved into assemblages that are associated with the biofacies defined in the Q-mode cluster analyses. The major group dominated by calcareous benthic taxa can be subdivided into two assemblages (Figure 5): one composed of *Epistominella exigua, Eponides tumidulus, Melonis affinis, Spiroloculina* sp., Nodosarids, *Triloculina* spp., *Globocassidulina biora, Nonionella iridea, Cassidulina* spp., and *Stainforthia concava*; and the second composed of Miliolids, *Cibicides refulgens, C. lobatulus,*

Angulogerina pauperata, A. earlandi, Ehrenbergina glabra, Pyrgo depressa, and *Uvigerina* sp. The first of these assemblages is characterized by abundant *Epistominella exigua, Nonionella iridea* and *Globocassidulina biora* and will be referred to herein as the *Epistominella* assemblage. The *Epistominella* assemblage occurs in samples of the trough biofacies. The second assemblage, the *Angulogerina* assemblage, is characterized by high numbers of *Angulogerina earlandi* and *A. pauperata* with *Cibicides* spp. as subsidiary species, and occurs in samples representing the ice edge biofacies.

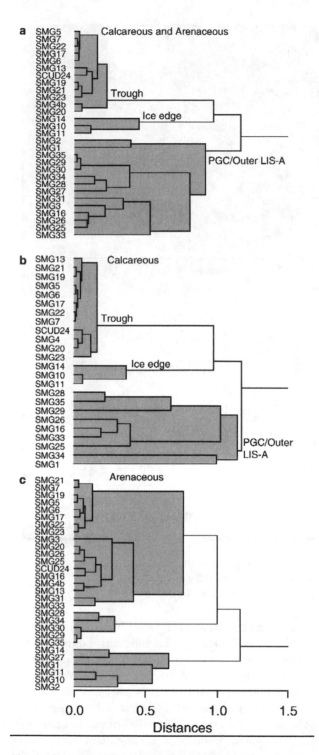

Fig. 3. Cluster dendrograms from the results of the Q-mode cluster analyses using (a) the total benthic foraminiferal data, (b) only the calcareous benthic foraminiferal data, and (c) only the arenaceous foraminiferal data. The three biofacies, ice edge, trough, and PGC/outer LIS-A, are shaded and labeled on dendrograms a and b.

The arenaceous foraminiferal assemblage can be further resolved into two assemblages (Figure 5) with less well-defined associations with the biofacies than those of the calcareous foraminiferal assemblages. The first of these, the *Miliammina* assemblage, is dominated by *Miliammina* spp. This assemblage also includes *Trochammina intermedia*, *T. conica*, *T. glabra*, *Portatrochammina antarctica*, and *P. weisneri*. This assemblage is associated with PGC/outer LIS-A samples SMG28, SMG29, SMG30, SMG34, and SMG35 that were grouped in the Q-mode cluster analyses using arenaceous and total foraminiferal data (Figures 3a, c). The second arenaceous assemblage, the *Portatrochammina* assemblage is characterized by abundant *Portatrochammina* spp., and *Rhabdammina abyssorum* with *Psammosphaera fusca*, *Saccammina tabulata*, *Cyclammina pusilla*, *Cribrostomoides* sp., *Reophax ovicula* and *Reophax subdentaliniformis* also present. This assemblage also contains calcareous benthic taxa that include *E. exigua* and *Astrononion echolsi*, and is most closely associated with the PGC/outer LIS-A biofacies.

Discriminant Function Analysis

Results of the discriminant function analysis demonstrated that the three *post hoc* classifications assigned to the samples were correct (Figure 6). The three sample groups described from the Q-mode cluster analysis, LIS-A ice edge, Greenpeace Trough, and PGC/outer LIS-A are clearly differentiated along Functions 1 and 2 (Figure 6). The eigenanalysis shows that 100% of the variance in the data is explained by 2 Functions (Table 3). The Wilks' Lambda test statistic indicates high discrimination between the groups (lambda values of 0.000 and 0.060) at significance levels of 0.0001 and 0.046 (Table 3).

DISCUSSION

Foraminiferal distribution studies from Weddell Sea surface sediments have been restricted to its eastern margin and the continental shelf slope (depths > 600 m) of the western Weddell Sea [*Anderson*, 1975]. *Anderson* [1975] recognized six faunal facies, based on core-top sediments, in the Weddell Sea that were strongly controlled by water mass distributions and their properties with respect to $CaCO_3$ dissolution. These results reflected the glacial conditions present during the 1960's and 1970's when the samples were collected; with the extent of the LIS nearer the continental shelf break (Figure 1). The western Weddell Sea biofacies were dominated by

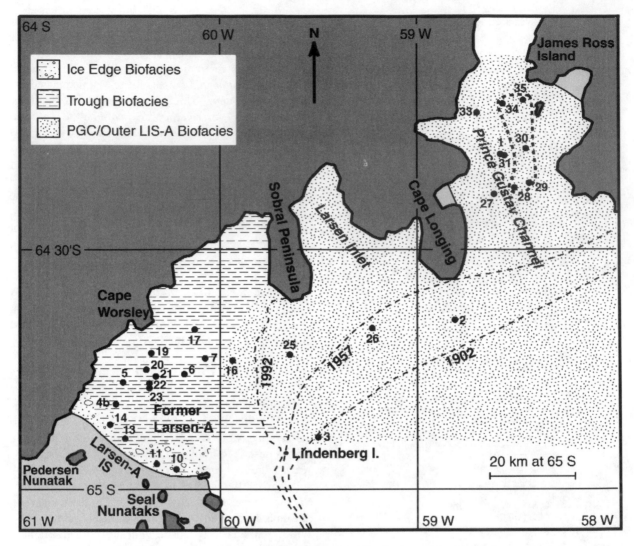

Fig. 4. Map of the study region showing the sample sites and the distribution of the biofacies, ice edge, trough, and PGC/outer LIS-A, defined from the Q-mode cluster analyses. The area delineated by the heavy dashed line in the Prince Gustav Channel indicates the region containing the *Miliammina* assemblage described in Figure 5.

arenaceous assemblages while the eastern Weddell Sea and deep-water biofacies (>3700 m) were dominated by calcareous benthic taxa. *Anderson* [1975] also noted the multibathic nature of the calcium carbonate compensation depth (CCD) in the Weddell Sea was related to water mass distributions, where non-corrosive Fresh Shelf Water included biofacies containing abundant calcareous benthic foraminifera that were restricted to the eastern Ronne-Filchner Ice Shelf edge, and the presence of Warm Deep Water was associated with biofacies lacking a calcareous benthic component. He concluded that the severity of glacial conditions influences oceanographic conditions in the Weddell Sea and thus is reflected in its foraminiferal biofacies distributions.

Foraminiferal studies from other circumantarctic localities also indicate multibathic distributions of calcareous and arenaceous benthic foraminiferal assemblages. On the George V-Adelie continental shelf arenaceous foraminferal assemblage occurrence was associated with the presence of corrosive saline shelf water [*Milam and Anderson*, 1985]. Distributions of distinct calcareous and arenaceous foraminiferal assemblages found in the Ross Sea were attributed to a shallow CCD limiting the calcareous assemblages to depths shallower than between 550 and 620 meters [*Kennett*, 1966, 1968; *Ward et al.*, 1987].

As discussed in the results, the foraminiferal biofacies in the LIS-A and PGC can be clearly delineated based on

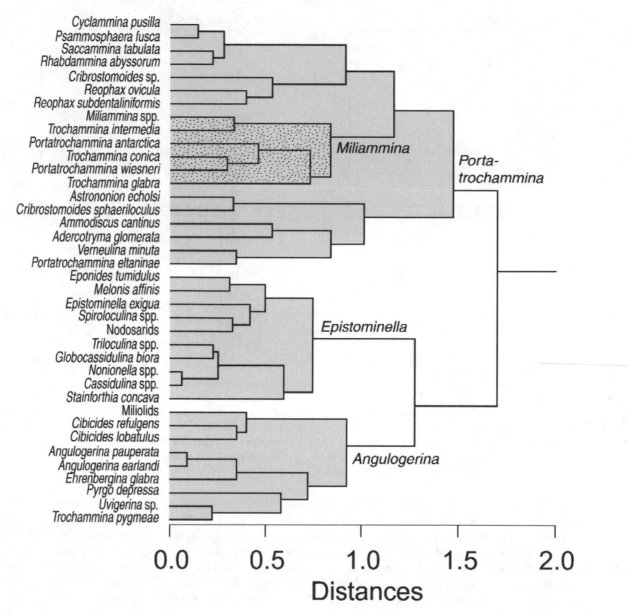

Fig. 5. Cluster dendrogram from the results of the R-mode cluster analysis. The dendrogram shows the benthic foraminiferal assemblages (shaded regions), *Portatrochammina*, *Epistominella*, and *Angulogerina* and their species compositions. The *Miliammina* assemblage is identified by the stippled region.

their calcareous, arenaceous and total assemblage compositions. Their distribution are independent of depth (Figure 7), thus supporting the multibathic CCD conditions in the Weddell Sea and circumantarctic regions. Conductivity, temperature, and depth (CTD) data from the LIS-A and PGC (Figure 2) indicate bottom waters in both regions having the temperature and salinity characteristics of Ice Shelf Water [*Carmack and Foster*, 1975; *Foldvick et al.*, 1985; *Gammelsrød et al.*, 1994;], similar to the Fresh Shelf Water of the eastern Weddell Sea con-

tinental shelf [*Anderson*, 1975]. However, the occurrence of both calcareous and arenaceous foraminiferal assemblages in the study region indicates that factors other than bottom water mass are influencing the distribution of the foraminiferal biofacies discussed below.

Ice Edge Biofacies

The ice edge biofacies (Figures 3, 4) is characterized by the *Angulogerina* assemblage that contains taxa with

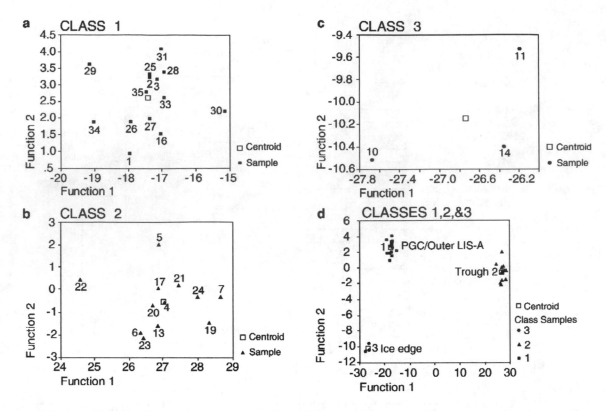

Fig. 6. Results of the discriminant analysis showing three distinct sample groups defined as classes based on *post hoc* classifications using the results of the Q-mode cluster analyses. (a) Class 1 are samples (solid squares) from the PGC/outer LIS-A biofacies; (b) Class 2 are samples (solid triangles) from the trough biofacies; and (c) Class 3 are samples (solid circles) from the LIS-A ice edge biofacies. The classes are clearly differentiated along Functions 1 and 2 (d). The open squares are the centroids for each class.

morphotypes indicative of infaunal habitat preference [*Corliss*, 1985; *Corliss and Chen*, 1988; *Gooday*, 1993]. Taxa within this assemblage, including *Angulogerina earlandi*, *A. pauperata* and *Ehrenbergina glabra* have been associated with ice-edge facies in the Ross Sea [*Osterman and Kellogg*, 1979], with Fresh Shelf Water Facies on the eastern continental shelf of the Weddell Sea [*Anderson*, 1975], and *E. glabra* with stable, organic rich microhabitats in Explorer's Cove, McMurdo Sound [*Bernhard*, 1987]. Included in this assemblage is *Uvigerina* sp., a taxon with an infaunal morphology that is typically associated with upwelling systems and organic rich sediments throughout the world's oceans

TABLE 3. Test statistics from the discriminant function analysis.

Function	Canonical Correlation	Wilk's Lamda	Significance
1	.999	.000	.000
2	.970	.060	.046

[*Phleger and Soutar*, 1973; *Lutze and Coulbourn*, 1984; *Morigi et al.*, 2001]. Samples from the ice edge biofacies have relatively low total organic carbon (TOC) ranging from 0.38 to 0.70 weight % TOC (Table 1) with an average of 0.52%, and low diatom abundance ranging from 5.4 to 9.6 millions of valves/gram of sediment with an average of 7.5 millions of valves/gm sediment [*Leventer and Rubin*, unpublished data]. Low TOC and diatom abundance indicate low organic flux to the seafloor in the LIS-A ice edge region. Sediment size data [*Domack*, unpublished data] indicate higher weight percent sand and coarser sediments, averaging 72.9% less than sand size, for samples from the ice edge biofacies. We suggest that high current activity at the LIS-A ice edge is effective in transporting fine organics and diatoms away from the ice front. Benthic foraminiferal evidence supports this contention in the presence of *Cibicides lobatulus* and *C. refulgens* in the ice edge biofacies. These taxa are epibenthic with tests having coarsely perforated evolute sides indicating an attached mode of life, inhabiting areas of high bottom current activity [*Hald and*

Steinsund, 1992; *Wollenberg and Mackensen,* 1998; *Polyak et al.,* 2002]. In addition, high numbers of infaunal foraminifera occurring in the *Angulogerina* assemblage/ice edge biofacies may reflect the ability of infaunal taxa to occupy horizons with greater TOC availability in its absence at the surface.

Abundant planktonic foraminifers, dominated by *Neogloboquadrina pachyderma,* are present in samples containing the *Angulogerina* assemblage. Their abundance possibly indicates high ice front productivity associated with open marine conditions that result from the formation of seasonal leads and narrow polynyas at the ice edge front [*Markus et al.,* 1998]. *Barbieri et al.* [1999] found similar foraminiferal associations between benthic foraminiferal assemblages abundant in *Angulogerina* spp. and occurrences of *N. pachyderma* in late Pleistocene-Holocene organic-rich, diatomaceous sediments from the Ross Sea. They suggested that high planktonic foraminiferal productivity enhanced carbonate preservation in the sediments. However, we suggest from our samples that the abundance of calcareous benthic and planktonic foraminifer tests in the absence of typical dissolution features of calcareous tests, such as broken or missing final chambers, etched and opaque test walls, and abundant test fragments, indicates little alteration through dissolution of the *Angulogerina* assemblage.

Trough Biofacies

The trough biofacies includes all of the samples within the Greenpeace Trough and one sample from the LIS-

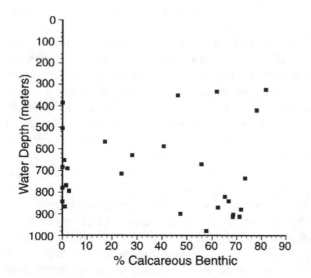

Fig. 7. Plot showing percent calcareous benthic foraminifera of total assemblage versus water depth.

A ice edge, SMG13 (Figure 4). This biofacies is characterized by the *Epistominella* assemblage that is dominated by *Epistominella exigua,* a taxa that occurs in deep Atlantic and Southern Ocean sediments [*Mackensen et al.,* 1993, 1990]. In the deep sea *E. exigua* is a significant component of benthic foraminiferal assemblages that exploit phytodetritus. A significant component of the *Epistominella* assemblage, *Globocassidulina biora,* is similar to another taxon associated with phytodetritus, *G. sugblobosa* [*Gooday,* 1988, 1993]. Phytodetritus exploiting foraminifera are considered opportunists (r-strategists) [*Gooday and Lambshead,* 1989]. The abundance of opportunistic foraminifera in the *Epistominella* assemblage indicates that, like other opportunists, it represents an assemblage adapted to a food-limited environment with seasonal nutrient input [*Grassel and Morse-Porteous,* 1987; *Gooday,* 1993]. It is difficult to determine whether the samples from the trough biofacies containing the *Epistominella* assemblage were rich in phytodetritus at the time of collection. Because phytodetritus typically exists as a relatively thin layer of light "fluff" at the sediment-water interface, it is easily lost by being dispersed by the energy of the sampling device. TOC for the trough biofacies samples range from 0.29 to 1.86% TOC with an average of 0.73%. This is slightly greater than the average TOC values in the ice edge biofacies (0.52%). It should be noted that sample SMG13 had a value of 0.92% TOC, considerably higher than the other ice edge samples. Diatom abundance in the trough biofacies averaged 6.0 millions of valves/grams of sediment, which is not significantly different from their abundance in the ice edge biofacies [*Leventer,* personal communication]. Sediment size distribution within the trough biofacies is finer grained than the ice edge biofacies with an average sediment size 96.78 weight % less than sand size [Domack, unpublished data]. The higher average TOC values and finer grained sediment distribution may indicate that seasonal phytodetritus flux in this particular region is controlling the distribution of the *Epistominella* assemblage/trough biofacies. However, because of the strong seasonality, timing and method of sampling any associations between phytodetritus and the distribution of the trough biofacies remains unclear.

The abundance of calcareous benthic taxa in the trough biofacies indicates little carbonate dissolution in this region as well. The dominant taxa, *Epistominella exigua,* occurs in deep sea sediments and is a taxa resistant to dissolution. *E. exigua* and the other calcareous benthic foraminifera found in the trough biofacies samples are well preserved with a majority of the tests translucent, indicating little dissolution. The planktonic foraminifer *Neogloboquadrina pachyderma* occurs in the trough biofacies in lower abundances than in the ice edge biofacies. A number of the *N. pachyderma* specimens lack their final chambers and have surface features suggesting dissolution. The lower numbers and dissolution features of the planktonic foraminifers may suggest preferential preservation of the calcareous benthic at the expense of the more dissolution susceptible planktonic tests.

PGC/Outer LIS-A Biofacies

The PGC/outer LIS-A biofacies contains two distinct benthic foraminiferal assemblages that are dominated by arenaceous taxa, the *Miliammina* assemblage and the *Portatrochammina* assemblage. *Miliammina* spp. dominate the *Miliammina* assemblage with calcareous taxa composing less than 3 percent. This assemblage is restricted to a small set of samples located in the PGC, SMG27, SMG28, SMG29, SMG30, SMG34, and SMG35. These samples are fine grained with an average of 93 weight percent sediment less than sand size. These samples also contain higher TOC content with a minimum of 0.52% and the remainder ranging from 0.78 to 0.96 weight percent TOC, and an order of magnitude greater diatom abundance than the LIS-A region, with values ranging from 17.4 to 37.7 millions of valves/grams of sediment. *Miliammina* rich foraminiferal assemblages are known from several circumantarctic localities. *Milam and Anderson* [1985] identified three arenaceous foraminiferal assemblages with high abundances of *Miliammina* from the George V-Adelie continental shelf, the shelf arenaceous, shallow basin and deep basin assemblages. The occurrence of these arenaceous assemblages was associated with the presence of Saline Shelf Water and the distribution of organic-rich siliceous muds and oozes. High abundances of *Miliammina* in the southwestern Weddell Sea and the western Antarctic Peninsula [*Anderson*, 1975; *Ishman and Domack*, 1994; *Ishman and Sperling*, 2002] were associated with corrosive Saline Shelf Water. However, CTD data from the PGC (Figure 2) indicates the presence of lower salinity Ice Shelf Water, thus the distribution of the *Miliammina* assemblage in the PGC is more likely associated with the increased abundance of fine grained, diatom-rich sediments with elevated organic content.

The *Portatrochammina* assemblage occurs in the remainder of the samples of the PGC/outer LIS-A biofacies, and is dominated by arenaceous foraminifera with calcareous foraminifera a minor component. Arenaceous taxa include *Portatrochammina* spp., *Textularia* spp. and *Cystammina argentea*, which are also common components of the shallow water arenaceous facies from the southwestern Weddell Sea continental shelf [*Anderson*, 1975]. Calcareous taxa include minor abundances of planktonic and the calcareous benthic foraminifera *Epistominella exigua* and *Cibicides refulgens*. *E. exigua* is commonly found at abyssal depths and is resistant to corrosive bottom water conditions. In the Weddell Sea *E. exigua* was part of the deep water calcareous-arenaceous facies and *C. refulgens* part of the lysoclinal facies [*Anderson*, 1975], both indicating resistance to dissolution. One sample containing an increased abundance of calcareous foraminifera, SMG-33, was recovered from the PGC region. The assemblage from this sample contained the typical arenaceous taxa of the *Portatrochammina* assemblage but also included *E. exigua*, *A. earlandi*, and abundant planktonic foraminifera. Sample SMG-33 is near the outlet of the Sjögren Glacier and we suggest that this site is influenced by seasonal freshwater input increasing surface productivity and providing a haven for the dissolution resistant calcareous benthic taxa. The *Portatrochammina* assemblage represents an assemblage whose composition is modified by dissolution. In regions lacking surface carbonate productivity, i.e. the absence of planktonic foraminifera, the occurrences of calcareous benthic taxa are greatly reduced and arenaceous taxa are dominant.

CONCLUSIONS

Spatial distributions of foraminifera in the former LIS-A and PGC regions of the eastern Antarctic Peninsula demonstrate clear biofacies/assemblage changes associated with environmental gradients established with the collapse of the LIS-A. The presence of calcareous benthic dominated assemblages characterizes foraminiferal biofacies associated with planktonic foraminiferal productivity, and sediments with low diatom abundance. Benthic foraminiferal assemblages associated with the ice edge and trough biofacies are differentiated based on the abundance of infaunal taxa and attached epiphytes, abundant in the ice edge biofacies, and opportunists, abundant in the trough biofacies. These differences are possibly related to current energy and the spatial and temporal availability of organics in the sediments from these regions. Biofacies of the PGC and outer LIS-A regions of the study area are dominated by arenaceous taxa. The composition of the assemblages from these regions is influenced by dissolution with mixed calcareous and arenaceous assemblages containing dissolution resistant calcareous benthic and planktonic foraminifers. This indicates the importance of planktonic foraminiferal production to the preservation of calcareous benthic assemblages on the Antarctic continental shelf. Arenaceous assemblages with abundant *Miliammina* are associated with fine-grained, diatom-rich sediments with elevated total organic carbon. These assemblages represent regions associated with high siliceous primary productivity and deposition.

Acknowledgments. This research was funded by the National Science Foundation, Grant # OPP-0003633 to Scott Ishman. The authors would like to thank Eugene Domack for his invitation to participate on NBP00-03 and the science that is evolving from it. A special thanks to Captain Joe Borkowski, the crew of the *Nathanial B. Palmer*, the scientific and Raytheon staff on NBP00-03 for their support. The authors would like to thank Anthony Rathbun and three anonymous reviewers for their constructive comments that greatly improved this manuscript. Finally, we acknowledge Amy Leventer and Anna Rubin for providing diatom data, and Eugene Domack for providing sediment size and CTD data referred to in the manuscript.

REFERENCES

Anderson, J.B., Ecology and distribution of foraminifera in the Weddell Sea of Antarctica. *Micropaleontology*, *21*, 69-96, 1975.

Barbieri, R., S. D'Onofrio, R. Melis, and F. Westall, r-Selected foraminifera with associated bacterial colonies in Upper Pleistocene sediments of the Ross Sea (Antarctica): implications for calcium carbonate preservation. *Palaeogeography, Palaeoclimatology, Palaeoecology*, *149*, 41-57, 1999.

Bernhard, J.M., 1987, Foraminiferal biotopes in Explorers Cove, McMurdo Sound, Antarctica. *Journal of Foraminiferal Research*, *17*, 286-297, 1987.

Carmack, E.C., and T.D. Foster, Circulation and distribution of oceanographic properties near the Filchner Ice Shelf. *Deep Sea Research*, *22*, 77-90, 1975.

Comiso, J.C., and A.L. Gordon, Interannual Variability in Summer Sea Ice Minimum, Coastal Polynyas, and Bottom Water Formation in the Weddell Sea, in *Antarctic Sea Ice Physical Processes, Interactions, and Variability*, *Antartcic Research Series, 74*, edited by M. Jeffries, pp. 293-316, AGU, Washington, D.C., 1998.

Corliss, B., Microhabitat of benthic foraminifera within deepsea sediments, *Nature, 314*, 435-438, 1985.

Corliss, B.H. and C. Chen, Morphotype patterns of Norwegian Sea deep-sea benthic foraminifera and ecological implications, *Geology, 16*, 716-719, 1988.

Doake, C.S.M., J.F.J. Corr, H. Rott, P. Skvarca, and G. M. Young, Breakup and conditions for stability of the northern Larsen Ice Shelf, Antarctica, *Nature*, *391*, 778-780, 1998.

Domack, E., A. Leventer, R. Gilbert, S. Brachfeld, S. Ishman, A. Camerlenghi, K. Gavahan, D. Carlson and A. Barkoukis, Cruise reveals history of Holocene Larsen Ice Shelf. *EOS, AGU Transactions*, *28*, 13, 16-17, 2001.

Foldvik, A., T. Gammelsrød, and T. Tøorresen, Circulation and water masses on the southern Weddell Sea shelf, in *Oceanology of the Antarctic Continental Shelf, Antarctic Research Series*, *43*, edited by S.S Jacobs, pp. 5-20, AGU, Washington, D.C., 1985.

Gammelsrød, T., A. Foldvik, O.A. Nøst, Ø. Skagseth, L.G. Anderson, E. Fogelqvist, K. Olsson, T. Tanhua, E.P. Jones, and S. Østerus, Distribution of water masses on the continental shelf in the southern Weddell Sea, in *The Polar Oceans and Their Role in Shaping the Global Environment*, *Geophysical Monograph*, *85*, edited by O.M. Johannessen, R.D. Muench, and J.E. Overlands, pp. 159-176, AGU, Washington, D.C., 1994.

Gooday, A.J., A response by benthic Foraminifera to the deposition of phytodetritus in the deep sea, *Nature, 332*, 70-73, 1988.

Gooday, A.J., Deep-sea benthic foraminiferal species which exploit phytodetritus: characteristic features and controls on distribution, *Marine Micropaleontology*, *22*, 187-205, 1993..

Gooday, A.J. and P.J.D. Lambshead, The influence of seasonally deposited phytodetritus on benthic foraminiferal populations in the bathyal northeast Atlantic, *Marine Ecology Progress Series*, *58*, 53-67, 1989.

Gordon, A.L., Western Weddell Sea thermohaline stratification, in *Ocean, Ice, and Atmosphere: Interactions at the Antarctic Continental Margin, Antarctic Research Series, 75*, edited by S.S. Jacobs and R.F. Weiss, pp. 215-240, AGU, Washington, D.C., 1998.

Grassel, J.F. and L.S. Morse-Porteous, Macrofaunal colonization of disturbed deep-sea environments and the structure of deep-sea benthic communities. *Deep-Sea Research*, *34*, 1911-1950, 1987.

Hald, M. and P.I. Steinsund, Distribution of surface sediment benthic foraminifera in the southwestern Barents Sea. *Journal of Foraminiferal Research, 22*, 347-362, 1992.

Ishman, S.E., and E.W. Domack, Oceanographic controls on benthic foraminifers from the Bellingshausen margin of the Antarctic Peninsula. *Marine Micropaleontology*, *24*, 119–155, 1994.

Ishman, S.E. and M.R. Sperling, Benthic foraminiferal record of Holocene deep-water evolution in the Palmer Deep, western Antarctic Peninsula. *Geology*, *30*, 435-438, 2002.

Johnson, R.A and D.W. Wichern, *Applied Multivariate Statistical Analysis*, 594 pp., Prentice Hall, Englewood Cliffs, New Jersey, 1982.

Kellogg, D.E. and T.B. Kellogg, Microfossil distribution in modern Amundsen Sea sediments. *Marine Micropaleontology, 12*, 203-222, 1987.

Kennett, J.P., Foraminiferal evidence of shallow calcium carbonate solution boundary, Ross Sea, Antarctica, *Science, 153*, 191-193, 1966.

Kennett, J.P., The Fauna of the Ross Sea: Ecology and Distribution of Foraminifera. *New Zealand Department of Scientific and Industrial Research Bulletin, 186*, pp. 48, 1968.

Lena, H., Foraminiferos bentonicos del noroeste de la Peninsula Antarctica. *Physis, 39*, 9-20, 1980.

Lutze, G.H. and W.T. Coulbourn, Recent benthic foraminifera from the continental margin of northwest Africa: community structure and distribution. *Marine Micropaleontology, 8*, 361-401, 1984.

Mackensen, A., H. Grobe, and D.K. Fuetterer, Benthic foraminiferal assemblages from the eastern Weddell Sea between 68 and 73°S: distribution, ecology and fossilization potential, *Marine Micropaleontology, 16*, 241-283, 1990.

Mackensen, A., D.K. Fuetterer, H. Grobe, and G. Schmiedl, Benthic foraminiferal assemblages from the eastern South Atlantic Polar Front region between 35° and 57°S, *Marine Micropaleontology, 22*, 33-69, 1993.

Maiya, S. and Y. Inoue, Abyssal foraminifera from the Bellingshausen Sea. *Report of the Technological Research Center, Japanese National Oil Company, 16*, 1-30, 1982.

Markus, T., C. Kottmeier, and E. Fahrbach, Ice formation in coastal polynyas in the Weddell Sea and their impact on the oceanic salinity, in *Antarctic Sea Ice Physical Processes, Interactions, and Variability, Antarctic Research Series, 74*, edited by M. Jeffries, pp. 273-292, AGU, Washington, D.C., 1998.

McKnight, W.M., Jr., The distribution of foraminifera off parts of the Antarctic coast. *Bulletin of American Paleontology, 44*, 65-158, 1962.

Milam, R.W., and J.B. Anderson, Distribution and ecology of Recent benthonic foraminifera of the Adelie-George V continental shelf and slope, Antarctica. *Marine Micropaleontology, 6*, 297-325, 1981.

Morigi, C., F.J. Jorissen, A. Gervais, S. Guichiard, and A.M. Borsetti, Benthic foraminiferal faunas in surface sediments off NW Africa: Relationship with organic flux to the ocean floor. *Journal of Foraminiferal Research, 31*, 369-384, 2001.

Osterman, L.E., and T.B. Kellogg, Recent benthic foraminifer distributions from the Ross Sea, Antarctica: Relation to ecologic and oceanographic conditions. *Journal of Foraminiferal Research, 9*, 250–269, 1979.

Parkinson, C.L., Length of the sea ice season in the Southern Ocean, 1988-1994, *Antarctic Sea Ice Physical Processes, Interactions, and Variability, Antarctic Research Series, 74*, edited by M. Jeffries, pp. 173-186, AGU, Washington, D.C., 1998.

Pflum, C.E., The distribution of foraminifera in the eastern Ross Sea, Antarctica. *Bulletin of American Paleontology, 50*, 151-209, 1966.

Phleger, F.B. and A. Soutar, Production of benthic foraminifera in three east Pacific oxygen minima. *Micropaleontology, 19*, 100-115, 1973.

Polyak, L., S. Korsun, L.A. Febo, V. Stanovoy, T. Khusid, M. Hald, B.E. Paulsen and D.J. Lubinski, Benthic foraminiferal assemblages from the southern Kara Sea, a river-influenced Arctic marine environment. *Journal of Foraminiferal Research, 32*, 252-273, 2002.

Pudsey, C.J. and J. Evans, First survey of Antarctic sub-ice shelf sediments reveals mid-Holocene ice shelf retreat. *Geology, 29*, 787-790, 2001.

Pudsey, C.J., J. Evans, E.W. Domack, P. Morris, and R.A. del Valle, Bathymetry and acoustic facies beneath the former Larsen-A and Prince Gustav ice shelves, north-west Weddell Sea, *Antarctic Science, 13*, 312-322, 2001.

Rott, H., P. Skvarca, and T. Nagler, Rapid collapse of the northern Larsen Ice Shelf, Antarctica, *Science, 271*, 788-792, 1996.

Rott, H., W. Rack, T. Nagler, and P. Skvarca, Climatically induced retreat and collapse of northern Larsen Ice Shelf, Antarctic Peninsula, *Annals of Glaciology, 27*, 86-92, 1998.

Schwerdtfeger, W., The effect of the Antarctic Peninsula on the temperature regime of the Weddell Sea, *Monthly Weather Review, 103*, 45-51, 1975.

Skvarca, P., Fast recession of the northern Larsen Ice Shelf monitored by space images. *Annals of Glaciology, 17*, 317-321, 1993.

Skvarca, P., W. Rack, H. Rott, and T. Ibarazabal Y Donangelo, Climatic trend and the retreat and disintegration of ice shelves on the Antarctic Peninsula: an overview, *Polar Research, 18*, 151-157, 1999.

Sloan, B.J., L.A. Lawver, and J.B. Anderson, Seismic stratigraphy of the Larsen Basin, eastern Antarctic Peninsula, in *Geology and Seismic Stratigraphy of the Antarctic Margin, Antarctic Research Series, 68*, edited by A.K. Cooper, P.F. Barker, and G. Brancolini, pp. 59-74, AGU, Washington, D.C., 1995.

Vaughan, D.G., and C.S.M. Doake, Recent armospheric warming and retreat of ice shelves on the Antarctic Peninsula. *Nature, 379*, 328-331, 1996.

Ward, B.L., P.J. Barrett, and P. Vella, Distribution and ecology of benthic foraminifera in McMurdo Sound, Antarctica, *Palaeogeography, Palaeoclimatology, Palaeoecology, 58*, 139-153, 1987.

Wollenberg, J.E. and A. Mackensen, Living benthic foraminifers from the central Arctic Ocean: faunal composition, standing stock and diversity. *Marine Micropaleontology, 34*, 153-185, 1998.

S. E. Ishman, Department of Geology, Southern Illinois University, MS 4324, Carbondale, IL 62901-4324.

P. Szymcek, Department of Geology, Southern Illinois University, MS 4324, Carbondale, IL 62901-4324.